Urological Cancer 2022

Urological Cancer 2022

Editors

José I. López
Claudia Manini

MDPI • Basel • Beijing • Wuhan • Barcelona • Belgrade • Manchester • Tokyo • Cluj • Tianjin

Editors

José I. López
Biocruces-Bizkaia Health
Research Institute
Spain

Claudia Manini
San Giovanni Bosco Hospital
Turin, Italy

Editorial Office
MDPI
St. Alban-Anlage 66
4052 Basel, Switzerland

This is a reprint of articles from the Special Issue published online in the open access journal *Cancers* (ISSN 2072-6694) (available at: https://www.mdpi.com/journal/cancers/special_issues/UC_2022).

For citation purposes, cite each article independently as indicated on the article page online and as indicated below:

LastName, A.A.; LastName, B.B.; LastName, C.C. Article Title. *Journal Name* **Year**, *Volume Number*, Page Range.

ISBN 978-3-0365-6968-0 (Hbk)
ISBN 978-3-0365-6969-7 (PDF)

Cover image courtesy of José I. López

Contents

About the Editors

José I. López

José I. López is an Advisor Researcher of the Biomarker in Cancer Unit at the Biocruces-Bizkaia Health Research Institute. He graduated from the Faculty of Medicine, University of the Basque Country, Leioa, Spain, and trained in pathology at the Hospital Universitario 12 de Octubre, Madrid, Spain. He received his PhD degree at the Universidad Complutense of Madrid, Spain. Dr. Lopez has served as a pathologist for more than 30 years in several hospitals in Spain and subspecializes in uropathology, on the topic of which he has published more than 200 peer-reviewed articles and reviews. Dr. Lopez is interested in translational uropathology in general and in renal cancer in particular and collaborates with several international research groups unveiling the genomic landscape of urological cancer. Intratumor heterogeneity, tumor sampling, tumor microenvironment, tumor ecology, immunotherapy, and basic mechanisms of carcinogenesis are his main topics of interest.

Claudia Manini

Claudia Manini is head of the Department of Pathology at the San Giovanni Bosco Hospital in Turin, Italy. She graduated from the Faculty of Medicine and Surgery and post-graduated in surgical pathology from the University of Turin, Turin, Italy. Dr. Manini has served as a pathologist for more than 25 years in several hospitals in Italy, developing expertise in diagnostic uropathology, neuropathology and gynecopathology. She is also affiliated to the Department of Public Health and Pediatric Sciences at the University of Turin. Her main interest is translational pathology.

cancers

MDPI

Editorial

Advances in Urological Cancer in 2022, from Basic Approaches to Clinical Management

Claudia Manini [1,2,*], Estíbaliz López-Fernández [3,4], José I. López [5] and Javier C. Angulo [6,7]

1 Department of Pathology, San Giovanni Bosco Hospital, 10154 Turin, Italy
2 Department of Sciences of Public Health and Pediatrics, University of Turin, 10124 Turin, Italy
3 FISABIO Foundation, 46020 Valencia, Spain
4 Faculty of Health Sciences, European University of Valencia, 46023 Valencia, Spain
5 Biocruces-Bizkaia Health Research Institute, 48903 Barakaldo, Spain
6 Clinical Department, Faculty of Medical Sciences, European University of Madrid, 28005 Madrid, Spain
7 Department of Urology, University Hospital of Getafe, 28907 Madrid, Spain
* Correspondence: claudia.manini@aslcittaditorino.it

Citation: Manini, C.; López-Fernández, E.; López, J.I.; Angulo, J.C. Advances in Urological Cancer in 2022, from Basic Approaches to Clinical Management. *Cancers* **2023**, *15*, 1422. https://doi.org/10.3390/cancers15051422

Received: 13 February 2023
Revised: 14 February 2023
Accepted: 17 February 2023
Published: 23 February 2023

This Special Issue includes 12 articles and 3 reviews dealing with several basic and clinical aspects of prostate, renal, and urinary tract cancer published during 2022 in *Cancers*, and intends to serve as a multidisciplinary chance to share the last advances in urological neoplasms.

This international forum of urological cancer includes different perspectives from 14 different countries: Canada, New Zealand, Italy, the USA, Germany, South Korea, Japan, Spain, Austria, China, the Netherlands, Taiwan, Poland, and the UK. An overview of these contributions shows the great variability of topics currently impacting the urological clinical practice, from the molecular mechanisms underlying prostate cancer development, for example, to the appropriateness of the robotic surgery in radical prostatectomy or partial nephrectomy and the current role of prostate-specific membrane antigen positron-emission tomography (PSMA-PET) imaging in prostate cancer. In addition, this "Urological Cancer 2022" Issue is also an opportunity to highlight some relevant international achievements of the specialty published elsewhere during 2022.

Pellerin et al. [1] analyzed the effects of the chronic exposition to bisphenols in bladder epithelium. Bisphenols A and S are chemical compounds used in the plastic industry to produce polycarbonates necessary for generating epoxy and vinyl ester resins. These worldwide distributed composites are industrially produced by the condensation of phenol and acetone and make up part of several plastics such as PVC. These products are insoluble in water and are present in the urine in normal conditions. Importantly, they are endocrine disruptors that interfere with cellular signaling pathways in urothelial cells [2]. Using normal urothelial cells (pediatric volunteers) and non-invasive (RT4, cell line ATCC HTB-2) and invasive (T24, cell line ATCC HTB-4) bladder cancer cells, the authors evaluate the impact of bisphenols on the energy metabolism, proliferation, migration, and pro-tumorigenic effect in human urothelium. They conclude that a chronic exposure to bisphenols A and S increases the proliferation rate and decreases the migration capacities of normal urothelial cells, which could result in urothelial hyperplasia. By contrast, these chemical products increase the energy metabolism, physiological activity, and cell proliferation, which could eventually promote urothelial cancer progression, especially from non-invasive to invasive variants.

The clinical identification of aggressive variants of prostate carcinoma requires more accurate markers. Reader et al. [3] have analyzed how the variations in the expression of Activins B and C impact the growth of PNT1A and PC3 prostate cancer cell lines. Activins are homo- or hetero-dimers belonging to the transforming growth factor-β family involved in prostate homeostasis, which are dysregulated in prostate cancer [4]. The authors have detected that the expression of Activin B was increased in prostate cancer samples, with a higher Gleason index, and that its overexpression inhibited the growth of PNT1A cells and increased PC3 cells' growth and migration. Interestingly, Activin C showed the opposite expression, with decreased immunostainings in prostate cancer cells with high Gleason grades, an increased overexpression in PNT1A cells, and a decreased growth in PC3 cells. The authors conclude that the combination of Activin B increasing and Activin C decreasing is associated with a higher Gleason grade in prostate adenocarcinoma and suggest its potential usefulness as prognostic biomarkers in this neoplasm.

Tossetta et al. [5] focused on the role of the ciliary neurotrophic factor (CNTF) in prostate cancer. Despite the fact that several crucial pathways such as MAPK/ERK, AKT/PI3K, and Jak/STAT regulate prostate cancer progression are triggered by CNTF, little is known about the effect of this member of the IL-6 family. The authors analyze the immunohistochemical expression of CNTF and its receptor in androgen-responsive (n = 10, radical prostatectomy samples) and castration-resistant (n = 10, transurethral resection samples) prostate cancers. Additionally, CNTF and its receptor expression are analyzed in androgen-dependent (LNCaP) and androgen-independent (11Rv1) prostate cancer cell lines by Western blotting and immunofluorescence. They also show that CNTF treatment down-regulates MAPK/ERK and AKT/PI3K pathways, inhibiting the matrix metalloproteinase-2 (MMP-2), a major component of the extracellular matrix degrader, and what is mainly responsible for tumor invasiveness. The authors conclude that CNTF plays a key role in the remodeling of the prostate cancer environment and suggests that this cytokine may modulate prostate cancer invasion. CNTF could represent a novel therapeutic approach in patients with castration-resistant prostate cancer.

Zapala et al. [6] investigated the usefulness of the preoperative systemic immune-inflammation index (SII) in predicting survival in a retrospective series of 421 patients with non-metastatic prostate cancer treated with radical prostatectomy. They found that a high SII was an independent predictor of overall survival. Furthermore, the combination of high age-adjusted Charlson Comorbidity Index (ACCI), the Cancer of the Prostate Risk Assessment Postsurgical score (CAPRA-S), and the SII identifies patients at the highest risk of death. The authors conclude that SII should be added to the prognosticators of patients with prostate cancer.

Clear cell renal cell carcinoma (CCRCC) is a perfect example of tumor complexity and a permanent target of analysis in recent years. In this collection, Shi et al. [7] analyzed the value of a ferroptotic gene-based signature in the prognosis of several series of CCRCC downloaded from the GEO database. The analyzed genes were obtained from the FerrDb V2 database. They identify a set of nine genes with prognostic implications differentially expressed in CCRCC, and found that the GLS2 enzyme, encoded by the *GLS2* gene and regulated by p53, may be a ferroptotic suppressor in CCRCC. The authors conclude that this nine-gene signature could eventually be an independent prognosticator in this neoplasm and advice for further investigations. Aside from that, several interesting investigations have been performed this year and deserve a short mention. Intratumor heterogeneity (ITH) is a constant, extensively analyzed event in CCRCC and its level has been correlated in 2022 with tumor aggressiveness. A mathematical study based on game theory [8] and a histological analysis [9] confirm that aggressive variants of CCRCC typically display low levels of ITH and agree with a genomic analysis of 101 cases already published in 2018 [10]. Additionally, several investigations have analyzed the influence of tumor growth patterns in the inter-regional genomic variability of these neoplasms [11], supporting the need for a personalized tumor sampling to strengthen tumor analysis [12,13].

In their review, Christenson et al. [14] revisited all the treatment modalities available so far in prostate cancer, focusing especially on the targets to interrupt the biological progression in lethal forms of prostate cancer, that now have a 5-year overall survival of only 30%. They revise current therapies, considering first low-risk and high-risk non-metastatic cancer, and then how to target metastatic cases, including hormone therapy, chemotherapy, PSMA-targeted radiation, genome-targeted precision therapy, and immunotherapy. The authors also review the ETS fusion positive (involving *TMPRSS2*, *SLC45A3*, *ETV1*, *ETV4*, and *FLI1* genes) and negative (involving *SPOP*, *FOXA1*, and *IDH1* genes) molecular subtypes of prostate cancer. The intimate mechanisms regulating metastatic castration-resistant prostate cancer and the androgen receptor ablation for intercepting advanced cancer are also analyzed. The intestinal microbiota as a potential promoter of castration-resistant tumor variants, the innovations in managing neuroendocrine tumors, and the difficulties of applying immune checkpoint inhibition in prostate cancer appear among the future directions in the review. Finally, the authors point also to CRISPR/Cas enzymes-assisted gene editing as a promising arena to develop further in prostate cancer management.

Other clinically oriented topics have also been incorporated in this Special Issue on the advances in urologic cancer regarding patient follow-up, diagnosis, and treatment. A very interesting one is a randomized clinical trial performed in the Netherlands which deals with the transition of care between a specialist and primary care physician [15]. Patients were randomized and allocated to specialist or general practitioner care for a head-to-head comparison. Several advantages of primary care follow-up over specialists' have been identified, including accessibility and more personalized attention, with a similar effectiveness. This study also identifies several challenges that must be addressed before the transition to primary-care follow-up can be a reality, by using quality indicators and improving communication and collaboration. However, another report from the same study shows that from the patient's perspective, hospital-based follow-up is preferred, but efforts should be made to improve physician's knowledge about personal aspects of the patient, improve symptoms management, and promote global health [16].

A series of articles evaluate the diagnostic pitfalls of prostate cancer by using different tools including PSA, multiparametric magnetic resonance imaging (mpMRI), and new generation imaging with the PSMA-PET modality; the latter has many therapeutic implications. In this respect, a study addresses the variables associated with false-positive PSA results using real-world data in a Spanish cohort of 1664 patients followed for two years. The false-positive results were as high as 47%, resulting in a positive predictive value of merely 13% [17]. This rate is much higher than previously reported in trials with screening data [18] Many factors were demonstrated to be associated with the presence of false-positive results, including age, previous PSA evaluation, family history of prostate cancer, and alcohol intake, but these associations were sustained only in asymptomatic patients [17]. These data do not serve to evaluate overdiagnoses and overtreatment but help to sustain that the PSA era in prostate cancer diagnosis should be closing.

A study in this Special Issue addresses the limitations of mpMRI for primary prostate cancer diagnosis in the form of a systematic review and meta-analysis that compare mpMRI with prostate-specific membrane antigen (PSMA) positron emission tomography (PET) imaging [19]. With the limitation of significant heterogeneity observed, PSMA-PET computerized tomography (CT) seems superior to mpMRI in primary cancer diagnosis, but not in the definition of cancer location within the gland. However, as can be expected from a whole-body procedure, PSMA-PET CT has valuable potential for tumor staging. False-positive MRI is another troublesome reality in clinical practice as it is not easy to differentiate false and real positive lesions, thus making evident the difficulty of further advancing in the MRI diagnostic ability [20]. PSMA-PET CT has also been investigated to solve MRI Prostate Imaging Reporting and Data System (PI-RADS) 4–5 lesions and negative biopsy discordance [21] and also the combination of PSMA-PET CT and mpMRI is being currently investigated in the MP4 clinical trial in which PI-RADS 4 or 5 lesions $\geq$10 mm on mpMRI are given the option of a PSMA-PET CT before biopsy. The intention of this approach is to predict better aggressive prostate cancer. That opens a new perspective to evaluate the feasibility of proceeding to prostate cancer surgery directly without a biopsy [22]. Additionally, PROSPET-BX clinical trial, currently undertaken in Italy, might confirm the superiority of PSAM-PET CT/transrectal ultrasound (TRUS) fusion prostate biopsy over mpMRI/TRUS fusion biopsy to further spare unnecessary biopsies [23].

Inspired by all these changes of paradigm, an excellent review on the clinical applications of PSMA-PET CT examination in patients with prostate cancer has been included in this Special Issue [24]. The article addresses the limitations and pitfalls of this new generation diagnostic imaging modality and emphasizes its therapeutic implications also. In fact, prolonged progression free and overall survival has been very recently confirmed in castration-resistant prostate cancer with lutetium Lu 177 vipivotide tetraxetan (^{177}Lu-PSMA-617) radioligand therapy [25]. Additionally, the PSMA-PET CT-guided intensification of radiotherapy is being investigated in a Canadian clinical trial, with special effort on cancer control, long-term toxicity, and health-related quality of life issues [26].

Carbon-ion radiotherapy, another modality to improve the effectiveness of radiation therapy, has been also evaluated in this Special Issue, with a retrospective study focusing on the older population of patients with prostate cancer [27]. This study confirms that carbon-ion radiotherapy is a safe and effective high-dose intensive treatment. This modality of radiation has been popular in different institutions in Japan for the treatment of different urologic cancers [28]. This modern technology provides several unique physical and radiobiologic properties that allow low levels of energy to be deposited in tissues proximal to the target, while the majority of energy is released in the target itself. That may have important advantages, especially in the setting of recurrent disease [29]. Another modality of radiation is extreme hypofractionation with stereotactic body radiation therapy (SBRT) in which treatment is delivered in one to five fractions, an encouraging alternative in the low- and intermediate-risk profile of patients that competes with high-dose brachytherapy [30,31].

Surgery has also seriously evolved to consider robotic prostatectomy the gold standard of surgical care for localized prostate cancer, that improves the functional outcomes of urinary continence and potency. This is also the topic of another article in the Special Issue [32]. Still, the definition of continence "without pads" or "social continence" makes difficult the comparison of the results [33]. New and effective modalities to surgically correct post-prostatectomy incontinence have been developed in recent decades and can be used both for stress urinary incontinence after prostatectomy and after radiation therapy [34–36].

Another interesting application of robotics in surgical urologic oncology is partial nephrectomy. The clinical benefits of indocyanine green florescence in robot-assisted partial nephrectomy are discussed in another element of this Special Issue. Reduced blood loss without a negative impact in the positive surgical margin rate is suggested using green dye [37]. However, future prospective randomized controlled trials are needed to confirm the presumed operative and functional advantages of this approach. Moreover, the issue presents another very interesting collaboration regarding metastatic renal cell carcinoma treatment, a field that has been subject to important paradigm changes in recent years [38]. The German multicenter prospective study PAZOREAL presented by Doehn et al. [39] reveals very interesting data on the effectiveness and safety of pazopanib (first-line), nivolumab (second-line), and everolimus (second- and third-line) in a real-life setting. This sequence is widely used in clinical practice. Targeted treatments for metastatic renal cell carcinoma allow for a more tailored approach, but predictive elements for immune-checkpoint inhibitors or tyrosine kinase inhibitors as a first-line treatment still lack genuine prediction markers [40].

Many studies have faced the optimal management of bladder urothelial malignancy in recent years. Some have searched for new therapeutic alternatives in the scenario of Bacillus Calmette-Guerin (BCG) shortage to prevent urothelial cancer recurrence and progression. Device-assisted intravesical chemotherapy using recirculating hyperthermic mitomycin-C (HIVEC) has been widely used in Spain [41]. Current new evidence favors the use of HIVEC in high-risk non-muscle-invasive bladder cancer [42], but not in the intermediate risk [43]. Many other studies have extended to step beyond classical morphologic parameters and stratify the prognosis of muscle-invasive bladder cancer according to new molecular markers that take into account basal or luminal phenotypes discovered [44,45]. A further step that is currently being undertaken is the evaluation of the intratumor microenvironment landscape, with implications not only in prognosis, but also in the response to systemic immunotherapy [46]. In this sense, immune-checkpoint inhibitors have been recently approved as a second-line treatment for metastatic bladder cancer and are currently being investigated in a neoadjuvant setting in non-metastatic disease [47,48].

Upper urinary tract urothelial carcinoma is another malignancy addressed in the Special Issue. Ha et al. revealed that intravesical recurrence after radical nephroureterectomy is associated with flexible but not with rigid diagnostic ureteroscopy [49]. This specific report opens a new perspective that requires a further evaluation in large-population controlled studies. The issue of intravesical recurrence after upper urinary tract cancer diagnosis and treatment is a big unsolved problem in the comprehensive management of urothelial malignancy. Several meta-analyses have confirmed the higher rate of intravesical recurrence after radical nephroureterectomy in patients who underwent diagnostic ureteroscopy preoperatively [50–52], but with no concurrent impact on long-term survival [52]. Probably the negative impact on intravesical recurrence free survival is due more to endoscopic biopsy than to ureteroscopy itself. As the use of flexible ureteroscopy is routinely recommended by clinical guidelines [53], future studies are needed to assess the role of immediate postoperative intravesical chemotherapy in patients undergoing biopsy during ureteroscopy for suspected upper tract urothelial cancer.

In summary, "Urological Cancer 2022" is a remarkable piece of knowledge that presents new relevant clinical, molecular, imaging, and therapeutic data in the urological field and invites researchers in urologic malignancy to enter a multidisciplinary approach and face some of the most relevant and current topics in urology.

Conflicts of Interest: The authors declare no conflict of interest.

References

1. Pellerin, E.; Pellerin, F.A.; Chabaud, S.; Pouliot, F.; Bolduc, S.; Pelletier, M. Bisphenols A and S alter the bioenergetics and behaviours of normal urothelial and bladder cancer cells. *Cancers* **2022**, *14*, 4011. [CrossRef] [PubMed]
2. Pellerin, E.; Caneparo, C.; Chabaud, S.; Bolduc, S.; Pelletier, M. Endocrine-disrupting effects of bisphenols on urological cancers. *Environ. Res.* **2021**, *195*, 110485. [CrossRef] [PubMed]
3. Reader, K.L.; John-McHaffie, S.; Zellhuber-McMillan, S.; Jowett, T.; Mottershead, D.G.; Cunliffe, H.E.; Gold, E.J. Activin B and activin C have opposing effects on prostate cancer progression and cell growth. *Cancers* **2023**, *15*, 147. [CrossRef] [PubMed]
4. Gold, E.; Risbridger, G. Activins and activin antagonists in the prostate and prostate cancer. *Mol. Cell. Endocrinol.* **2012**, *359*, 107–112. [CrossRef]
5. Tossetta, G.; Fantone, S.; Gesuita, R.; Goteri, G.; Senzacqua, M.; Marcheggiani, F.; Tiano, L.; Marzioni, D.; Mazzucchelli, R. Ciliary neurotrophic factor modulates multiple downstream signaling pathways in prostate cancer inhibiting cell invasiveness. *Cancers* **2022**, *14*, 5917. [CrossRef]
6. Zapala, P.; Garbas, K.; Lewandowski, Z.; Zapala, L.; Slusarczyk, A.; Slusarczyk, C.; Mielczarek, L.; Radziszewski, P. The clinical utility of systemic immune-inflammation index supporting Charlson comorbidity index and CAPRA-S Score in determining survival after radical prostatectomy. A single centre study. *Cancers* **2022**, *14*, 4135. [CrossRef]
7. Shi, Z.; Zheng, J.; Liang, Q.; Liu, Y.; Yang, Y.; Wang, R.; Wang, M.; Zang, Q.; Xuan, Z.; Sun, H.; et al. Identification and validation of a novel ferroptotic prognostic genes-based signature of clear cell renal cell carcinoma. *Cancers* **2022**, *14*, 4690. [CrossRef]
8. Laruelle, A.; Rocha, A.; Manini, C.; Lopez, J.I.; Inarra, E. Effects of heterogeneity on cancer: A game theory perspective. *arXiv* **2022**, arXiv:2202.02211. [CrossRef]
9. Manini, C.; Lopez-Fernandez, E.; Lawrie, C.H.; Laruelle, A.; Angulo, J.C.; Lopez, J.I. Clear cell renal cell carcinomas with aggressive behavior display low intratumor heterogeneity at the histological level. *Curr. Urol. Rep.* **2022**, *23*, 93–97. [CrossRef]
10. Turajlic, S.; Xu, H.; Litchfield, K.; Rowan, A.; Horswell, S.; Chambers, T.; O'Brien, T.; López, J.I.; Watkins, T.B.K.; Nicol, D.; et al. Deterministic evolutionary trajectories influence primary tumor growth. *Cell* **2018**, *173*, 595–610. [CrossRef]
11. Fu, X.; Zhao, Y.; Lopez, J.I.; Rowan, A.; Au, L.; Fendler, A.; Hazell, S.; Xu, H.; Horswell, S.; Shepherd, S.; et al. Spatial patterns of tumour growth impact clonal diversification in a computational model and the TRACERx Renal study. *Nat. Ecol. Evol.* **2022**, *6*, 88–102. [CrossRef] [PubMed]
12. Lopez, J.I.; Cortés, J.M. Multi-site tumor sampling (MSTS): A new tumor selection method to enhance intratumor heterogeneity detection. *Hum. Pathol.* **2017**, *64*, 1–6. [CrossRef] [PubMed]
13. Manini, C.; Lopez-Fernandez, E.; Lopez, J.I. Towards personalized sampling in clear cell renal cell carcinomas. *Cancers* **2022**, *14*, 3381. [CrossRef] [PubMed]
14. Christenson, M.; Song, C.S.; Liu, Y.G.; Chatterjee, B. Precision targets for intercepting the lethal progression of prostate cancer: Potential avenues for personalized therapy. *Cancers* **2022**, *14*, 892. [CrossRef] [PubMed]

15. Wollersheim, B.M.; van Asselt, K.M.; Pos, F.J.; Akdemir, E.; Crouse, S.; van der Poel, H.G.; Aaronson, N.K.; van de Poll-Franse, L.V.; Boekhout, A.H. Specialist versus primary care prostate cancer follow-up: A process evaluation of a randomized controlled trial. *Cancers* **2022**, *14*, 3166. [CrossRef]
16. Wollersheim, B.M.; van der Poel, H.G.; van Asselt, K.M.; Pos, F.J.; Tillier, C.N.; Akdemir, E.; Vis, A.N.; Lampe, M.I.; van den Bergh, R.; Somford, D.M.; et al. Quality of early prostate cancer follow-up care from the patients' perspective. *Support Care Cancer* **2022**, *30*, 10077–10087. [CrossRef]
17. Lumbreras, B.; Parker, L.A.; Caballero-Romeu, J.P.; Gómez-Pérez, L.; Puig-García, M.; López-Garrigós, M.; García, N.; Hernández-Aguado, I. Variables associated with false-positive PSA results: A cohort study with real-world data. *Cancers* **2022**, *14*, 261. [CrossRef]
18. Kilpeläinen, T.P.; Tammela, T.L.; Roobol, M.; Hugosson, J.; Ciatto, S.; Nelen, V.; Moss, S.; Määttänen, L.; Auvinen, A. False-positive screening results in the European randomized study of screening for prostate cancer. *Eur. J. Cancer* **2011**, *47*, 2698–2705. [CrossRef]
19. Zhao, Y.; Simpson, B.S.; Morka, N.; Freeman, A.; Kirkman, A.; Kelly, D.; Whitaker, H.C.; Emberton, M.; Norris, J.M. Comparison of multiparametric magnetic resonance imaging with prostate-specific membrane antigen positron-emission tomography imaging in primary prostate cancer diagnosis: A systematic review and meta-analysis. *Cancers* **2022**, *14*, 3497. [CrossRef]
20. Stavrinides, V.; Syer, T.; Hu, Y.; Giganti, F.; Freeman, A.; Karapanagiotis, S.; Bott, S.R.J.; Brown, L.C.; Burns-Cox, N.; Dudderidge, T.J.; et al. False positive multiparametric magnetic resonance imaging phenotypes in the biopsy-naïve prostate: Are they distinct from significant cancer-associated lesions? Lessons from PROMIS. *Eur. Urol.* **2021**, *79*, 20–29. [CrossRef]
21. Wong, L.-M.; Koschel, S.; Whish-Wilson, T.; Farag, M.; Bolton, D.; Zargar, H.; Corcoran, N.; Lawrentschuk, N.; Christov, A.; Thomas, L. Investigating PSMA-PET/CT to resolve prostate MRI PIRADS4-5 and negative biopsy discordance. *World J. Urol.* **2023**, *41*, 463–469. [CrossRef] [PubMed]
22. Woo, H.H.; Khanani, H.; Thompson, N.J.; Sorensen, B.J.; Baskaranathan, S.; Bergersen, P.; Chalasani, V.; Dean, T.; Dias, M.; Symons, J.; et al. Multiparametric Magnetic Resonance Imaging of the Prostate and Prostate-specific Membrane Positron Emission Tomography Prior to Prostate Biopsy (MP4 Study). *Eur. Urol. Open Sci.* **2022**, *47*, 119–125. [CrossRef] [PubMed]
23. Lopci, E.; Lazzeri, M.; Colombo, P.; Casale, P.; Buffi, N.M.; Saita, A.; Peschechera, R.; Hurle, R.; Marzo, K.; Leonardi, L.; et al. Diagnostic performance and clinical impact of PSMA PET/CT versus mpMRI in patients with a high suspicion of prostate cancer and previously negative biopsy: A prospective trial (PROSPET-BX). *Urol. Int.* **2023**. [CrossRef]
24. Rasul, S.; Haug, A.R. Clinical applications of PSMA PET examination in patients with prostate cancer. *Cancers* **2022**, *14*, 3768. [CrossRef]
25. Mehrens, D.; Kramer, K.K.M.; Unterrainer, L.M.; Beyer, L.; Bartenstein, P.; Froelich, M.F.; Tollens, F.; Ricke, J.; Rübenthaler, J.; Schmidt-Hegemann, N.-S.; et al. Cost-effectiveness analysis of ^{177}Lu-PSMA-617 radioligand therapy in metastatic castration-resistant prostate cancer. *J. Natl. Compr. Cancer Netw.* **2023**, *21*, 43–50.e2. [CrossRef]
26. Petit, C.; Delouya, G.; Taussky, D.; Barkati, M.; Lambert, C.; Beauchemin, M.-C.; Clavel, S.; Mok, G.; Gauthier Paré, A.-S.; Nguyen, T.-V.; et al. PSMA-PET/CT guided intensification of radiotherapy for prostate cancer (PSMAgRT): Findings of detection rate, impact on cancer management, and early toxicity from a phase 2 randomised controlled trial. *Int. J. Radiat. Oncol. Biol. Phys.* **2023**. [CrossRef]
27. Hiroshima, Y.; Ishikawa, H.; Iwai, Y.; Wakatsuki, M.; Utsumi, T.; Suzuki, H.; Akakura, K.; Harada, M.; Sakurai, H.; Ichikawa, T.; et al. Safety and efficacy of carbon-ion radiotherapy for elderly patients with high-risk prostate cancer. *Cancers* **2022**, *14*, 4015. [CrossRef]
28. Ishikawa, H.; Hiroshima, Y.; Kanematsu, N.; Inaniwa, T.; Shirai, T.; Imai, R.; Suzuki, H.; Akakura, K.; Wakatsuki, M.; Ichikawa, T.; et al. Carbon-ion radiotherapy for urological cancers. *Int. J. Urol.* **2022**, *29*, 1109–1119. [CrossRef]
29. Malouff, T.D.; Mahajan, A.; Krishnan, S.; Beltran, C.; Seneviratne, D.S.; Trifiletti, D.M. Carbon ion therapy: A modern review of an emerging technology. *Front. Oncol.* **2020**, *10*, 82. [CrossRef] [PubMed]
30. Gómez-Aparicio, M.A.; Valero, J.; Caballero, B.; García, R.; Hernando-Requejo, O.; Montero, Á.; Gómez-Iturriaga, A.; Zilli, T.; Ost, P.; López-Campos, F.; et al. Extreme hypofractionation with SBRT in localized prostate cancer. *Curr. Oncol.* **2021**, *28*, 2933–2949. [CrossRef] [PubMed]
31. Corkum, M.T.; Achard, V.; Morton, G.; Zilli, T. Ultrahypofractionated radiotherapy for localised prostate cancer: How far can we go? *Clin. Oncol. (R. Coll. Radiol.)* **2022**, *34*, 340–349. [CrossRef] [PubMed]
32. D'Altilia, N.; Mancini, V.; Falagario, U.G.; Martino, L.; Di Nauta, M.; Calò, B.; Del Giudice, F.; Basran, S.; Chung, B.I.; Porreca, A.; et al. A matched-pair analysis after robotic and retropubic radical prostatectomy: A new definition of continence and the impact of different surgical techniques. *Cancers* **2022**, *14*, 4350. [CrossRef] [PubMed]
33. Abrams, P.; Constable, L.D.; Cooper, D.; MacLennan, G.; Drake, M.J.; Harding, C.; Mundy, A.; McCormack, K.; McDonald, A.; Norrie, J.; et al. Outcomes of a noninferiority randomised controlled trial of surgery for men with urodynamic stress incontinence after prostate surgery (MASTER). *Eur. Urol.* **2021**, *79*, 812–823. [CrossRef] [PubMed]
34. Angulo, J.C.; Arance, I.; Esquinas, C.; Dorado, J.F.; Marcelino, J.P.; Martins, F.E. Outcome measures of adjustable transobturator male system with pre-attached scrotal port for male stress urinary incontinence after radical prostatectomy: A prospective study. *Adv. Ther.* **2017**, *34*, 1173–1183. [CrossRef]
35. Angulo, J.C.; Cruz, F.; Esquinas, C.; Arance, I.; Manso, M.; Rodríguez, A.; Pereira, J.; Ojea, A.; Carballo, M.; Rabassa, M.; et al. Treatment of male stress urinary incontinence with the adjustable transobturator male system: Outcomes of a multi-center Iberian study. *Neurourol. Urodyn.* **2018**, *37*, 1458–1466. [CrossRef]

36. Esquinas, C.; Angulo, J.C. Effectiveness of adjustable transobturator male system (ATOMS) to treat male stress incontinence: A systematic review and meta-analysis. *Adv. Ther.* **2019**, *36*, 426–441. [CrossRef]
37. Yang, Y.K.; Hsieh, M.L.; Chen, S.Y.; Liu, C.Y.; Lin, P.H.; Kan, H.C.; Pang, S.T.; Yu, K.J. Clinical benefits of indocyanine green fluorescence in robot-assisted partial nephrectomy. *Cancers* **2022**, *14*, 3032. [CrossRef]
38. Angulo, J.C.; Shapiro, O. The changing therapeutic landscape of metastatic renal cancer. *Cancers* **2019**, *11*, 1227. [CrossRef]
39. Doehn, C.; Bögemann, M.; Grünwald, V.; Welslau, M.; Bedke, J.; Schostak, M.; Wolf, T.; Ehneß, R.; Degenkolbe, E.; Witecy, S.; et al. The non-interventional PAZOREAL Study to assess the effectiveness and safety of pazopanib in a real-life setting: Reflecting a changing mRCC treatment landscape. *Cancers* **2022**, *14*, 5486. [CrossRef]
40. Nunes-Xavier, C.E.; Angulo, J.C.; Pulido, R.; López, J.I. A critical insight into the clinical translation of PD-1/PD-L1 blockade therapy in clear cell renal cell carcinoma. *Curr. Urol. Rep.* **2019**, *20*, 1. [CrossRef]
41. Plata, A.; Guerrero-Ramos, F.; Garcia, C.; González-Díaz, A.; Gonzalez-Valcárcel, I.; De la Morena, J.M.; Díaz-Goizueta, F.J.; Del Álamo, J.F.; Gonzalo, V.; Montero, J.; et al. Long-term experience with hyperthermic chemotherapy (HIVEC) using Mitomycin-C in patients with non-muscle invasive bladder cancer in Spain. *J. Clin. Med.* **2021**, *10*, 5105. [CrossRef] [PubMed]
42. Guerrero-Ramos, F.; González-Padilla, D.A.; González-Díaz, A.; de la Rosa-Kehrmann, F.; Rodríguez-Antolín, A.; Inman, B.A.; Villacampa-Aubá, F. Recirculating hyperthermic intravesical chemotherapy with mitomycin C (HIVEC) versus BCG in high-risk non-muscle-invasive bladder cancer: Results of the HIVEC-HR randomized clinical trial. *World J. Urol.* **2022**, *40*, 999–1004. [CrossRef] [PubMed]
43. Angulo, J.C.; Álvarez-Ossorio, J.L.; Domínguez-Escrig, J.L.; Moyano, J.L.; Sousa, A.; Fernández, J.M.; Gómez-Veiga, F.; Unda, M.; Carballido, J.; Carrero, V.; et al. Hyperthermic Mitomycin C in intermediate-risk non-muscle-invasive bladder cancer: Results of the HIVEC-1 trial. *Eur. Urol. Oncol.* **2023**, *6*, 58–66. [CrossRef] [PubMed]
44. Angulo, J.C.; Lopez, J.I.; Flores, N.; Toledo, J.D. The value of tumour spread, grading and growth pattern as morphological predictive parameters in bladder carcinoma. A critical revision of the 1987 TNM classification. *J. Cancer Res. Clin. Oncol.* **1993**, *119*, 578–593. [CrossRef] [PubMed]
45. Calvete, J.; Larrinaga, G.; Errarte, P.; Martín, A.M.; Dotor, A.; Esquinas, C.; Nunes-Xavier, C.E.; Pulido, R.; López, J.I.; Angulo, J.C. The coexpression of fibroblast activation protein (FAP) and basal-type markers (CK 5/6 and CD44) predicts prognosis in high-grade invasive urothelial carcinoma of the bladder. *Hum. Pathol.* **2019**, *91*, 61–68. [CrossRef] [PubMed]
46. Bian, Z.; Chen, J.; Liu, C.; Ge, Q.; Zhang, M.; Meng, J.; Liang, C. Landscape of the intratumoral microenvironment in bladder cancer: Implications for prognosis and immunotherapy. *Comput. Struct. Biotechnol. J.* **2022**, *21*, 74–85. [CrossRef]
47. Konala, V.M.; Adapa, S.; Aronow, W.S. Immunotherapy in bladder cancer. *Am. J. Ther.* **2022**, *29*, e334–e337. [CrossRef]
48. Rizzo, A.; Mollica, V.; Santoni, M.; Palmiotti, G.; Massari, F. Pathologic complete response in urothelial carcinoma patients receiving neoadjuvant immune checkpoint inhibitors: A Meta-Analysis. *J. Clin. Med.* **2022**, *11*, 1038. [CrossRef]
49. Ha, J.S.; Jeon, J.; Ko, J.C.; Lee, H.S.; Yang, Y.; Kim, D.; Kim, J.S.; Ham, W.S.; Choi, Y.D.; Cho, K.S. Intravesical recurrence after radical nephroureterectomy in patients with upper tract urothelial carcinoma is associated with flexible diagnostic ureteroscopy, but not with rigid diagnostic ureteroscopy. *Cancers* **2022**, *14*, 5629. [CrossRef]
50. Marchioni, M.; Primiceri, G.; Cindolo, L.; Hampton, L.J.; Grob, M.B.; Guruli, G.; Schips, L.; Shariat, S.F.; Autorino, R. Impact of diagnostic ureteroscopy on intravesical recurrence in patients undergoing radical nephroureterectomy for upper tract urothelial cancer: A systematic review and meta-analysis. *BJU Int.* **2017**, *120*, 313–319. [CrossRef]
51. Guo, R.Q.; Hong, P.; Xiong, G.Y.; Zhang, L.; Fang, D.; Li, X.S.; Zhang, K.; Zhou, L.Q. Impact of ureteroscopy before radical nephroureterectomy for upper tract urothelial carcinomas on oncological outcomes: A meta-analysis. *BJU Int.* **2018**, *121*, 184–193. [CrossRef] [PubMed]
52. Nowak, Ł.; Krajewski, W.; Chorbińska, J.; Kiełb, P.; Sut, M.; Moschini, M.; Teoh, J.Y.-C.; Mori, K.; Del Giudice, F.; Laukhtina, E.; et al. The impact of diagnostic ureteroscopy prior to radical nephroureterectomy on oncological outcomes in patients with upper tract urothelial carcinoma: A comprehensive systematic review and meta-analysis. *J. Clin. Med.* **2021**, *10*, 4197. [CrossRef] [PubMed]
53. Rouprêt, M.; Babjuk, M.; Burger, M.; Capoun, O.; Cohen, D.; Compérat, E.M.; Cowan, N.C.; Dominguez-Escrig, J.L.; Gontero, P.; Mostafid, A.H.; et al. European Association of Urology Guidelines on upper urinary tract urothelial carcinoma: 2020 update. *Eur. Urol.* **2021**, *79*, 62–79. [CrossRef] [PubMed]

cancers

MDPI

Article

Bisphenols A and S Alter the Bioenergetics and Behaviours of Normal Urothelial and Bladder Cancer Cells

Ève Pellerin [1], Félix-Antoine Pellerin [1], Stéphane Chabaud [1], Frédéric Pouliot [2,3], Stéphane Bolduc [1,3,*,†] and Martin Pelletier [4,5,*,†]

1 Centre de Recherche en Organogénèse Expérimentale/LOEX, Regenerative Medicine Division, CHU de Québec-Université Laval Research Center, Quebec, QC G1J 1Z4, Canada
2 Oncology Division, CHU de Québec-Université Laval Research Center, Quebec, QC G1R 2J6, Canada
3 Department of Surgery, Faculty of Medicine, Laval University, Quebec, QC G1V 0A6, Canada
4 Infectious and Immune Disease Division, CHU de Québec-Université Laval Research Center, Quebec, QC G1V 4G2, Canada
5 Department of Microbiology-Infectious Diseases and Immunology, Faculty of Medicine, Laval University, Quebec, QC G1V 0A6, Canada
* Correspondence: stephane.bolduc@fmed.ulaval.ca (S.B.); martin.pelletier@crchudequebec.ulaval.ca (M.P.); Tel.: +1-418-525-4444 (ext. 42282) (S.B.); +1-418-525-4444 (ext. 46166) (M.P.)
† These authors contributed equally to this work.

Simple Summary: This research brings new knowledge on the potential roles of bisphenol A and bisphenol S on bladder cancer progression. By assessing the impact of bisphenols A and S on normal urothelial cells and non-invasive and invasive bladder cancer cells, this study aimed to demonstrate that these endocrine-disrupting chemicals could promote bladder cancer progression through the alteration of the bioenergetics and behaviours of healthy and cancerous bladder cells. These results could provide a better understanding of the pathophysiology of bladder cancer and its hormone-sensitive characteristics. Furthermore, this study suggests that bisphenols A and S could affect bladder cancer recurrence, progression and patient prognosis.

Abstract: Bisphenol A (BPA) and bisphenol S (BPS) are used in the production of plastics. These endocrine disruptors can be released into the environment and food, resulting in the continuous exposure of humans to bisphenols (BPs). The bladder urothelium is chronically exposed to BPA and BPS due to their presence in human urine samples. BPA and BPS exposure has been linked to cancer progression, especially for hormone-dependent cancers. However, the bladder is not recognized as a hormone-dependent tissue. Still, the presence of hormone receptors on the urothelium and their role in bladder cancer initiation and progression suggest that BPs could impact bladder cancer development. The effects of chronic exposure to BPA and BPS for 72 h on the bioenergetics (glycolysis and mitochondrial respiration), proliferation and migration of normal urothelial cells and non-invasive and invasive bladder cancer cells were evaluated. The results demonstrate that chronic exposure to BPs decreased urothelial cells' energy metabolism and properties while increasing them for bladder cancer cells. These findings suggest that exposure to BPA and BPS could promote bladder cancer development with a potential clinical impact on bladder cancer progression. Further studies using 3D models would help to understand the clinical consequences of this exposure.

Keywords: bladder cancer; bisphenol A; bisphenol S; energy metabolism; migration; proliferation

Citation: Pellerin, È.; Pellerin, F.-A.; Chabaud, S.; Pouliot, F.; Bolduc, S.; Pelletier, M. Bisphenols A and S Alter the Bioenergetics and Behaviours of Normal Urothelial and Bladder Cancer Cells. *Cancers* **2022**, *14*, 4011. https://doi.org/10.3390/cancers14164011

Academic Editors: José I. López and Claudia Manini

Received: 18 July 2022
Accepted: 15 August 2022
Published: 19 August 2022

1. Introduction

In the last decades, hormone-dependent cancers, such as breast [1] and prostate cancers [2], have increased in industrialized countries. Among several causes, part of this rise could be associated with the growing number of endocrine disruptors found in the environment [3]. More than 350 synthetic molecules in the environment, including bisphenols (BPs), are considered endocrine disruptors because of their capacity to modulate the action or the metabolism of various hormones in the organism [3].

The bisphenol family comprises many molecules, such as bisphenol A (BPA) and bisphenol S (BPS). BPA is used to produce plastics, such as polycarbonate and epoxy resins [4]. It is found in various daily objects (e.g., water bottles and food containers) [5], making this compound ubiquitous in the environment [6]. Exposure to BPA has been associated with cancer development, especially in hormone-dependent cancers such as breast [7] and prostate cancers [8]. Recently, BPS has been used as a safer alternative to BPA in plastic production because of its excellent stability [9]. However, BPS has also been linked with cancer progression [7,10]. BPA is found at measurable concentrations in the urine of >90% of the population [10], whereas BPS is detectable in 78% of urine samples at comparable concentrations to BPA [11]. BPA and BPS are found at similar concentrations in urine, ranging from 1 to 100 nM [11]. The ubiquitous presence of BPA and BPS in the urine results in chronic exposure to the urinary tract, particularly the bladder, where urine can be stored for many hours [12].

The multiple binding capacities of BPA and BPS allow the compounds to alter different signalling pathways associated with cell migration, proliferation and invasion [10,13]. In addition, studies have shown that the bladder's urothelial cells (UCs) express cell receptors targeted by BPA and BPS, such as estrogen receptors (ERs) α and β (ERα and ERβ), the androgen receptor (AR) and the G protein-coupled estrogen receptor (GPER) [14,15]. Although the bladder is not recognized as a hormone-sensitive tissue, studies have shown the role of these sex steroid receptors in bladder cancer initiation and progression. As such, the activation of ERα could promote the proliferation of bladder cancer cells [16], whereas ERβ and AR could promote bladder cancer growth and invasion via the alteration of tumour suppressor gene expression [17,18]. Therefore, the pro-tumorigenic tendencies of BPA and BPS, their binding capacities to cell receptors and the presence of these key receptors in the bladder urothelium could suggest a potential role for BPs in bladder cancer development [19].

It was previously demonstrated that the energy metabolism of healthy bladder fibroblasts decreased after chronic exposure to physiological concentrations of BPA. In contrast, BPA-exposed cancer-associated fibroblasts (CAFs) displayed an increased glycolytic metabolism, leading to extracellular acidification [20]. This enhanced acidification can lead to the inhibition of immune cells, such as monocytes, as well as reorganization of the extracellular matrix [21,22], thus potentially promoting bladder cancer progression [20]. The hypothesis of this study was that BPA and BPS would impact the energy metabolism and properties of normal urothelial cells and bladder cancer cells, i.e., migration, proliferation and expression of cell markers of invasive potential, which could promote bladder cancer initiation and progression.

2. Materials and Methods

2.1. Cell Lines

All procedures involving patients were conducted according to the Helsinki Declaration and were approved by the local Research Ethical Committee. Each specimen was obtained with the donor's consent for tissue harvesting, and all experimental procedures were performed according to the CHU de Québec-Université Laval guidelines. Normal urothelial cells (UCs) were extracted from healthy human urological tissue biopsies and cultured as previously described [23,24]. The UCs were isolated from two healthy paediatric volunteers undergoing surgery for a benign condition (UC1 and UC2) and used as non-transformed primary cell lines.

UCs, RT4 non-invasive bladder cancer cells (ATCC HTB-2) and T24 invasive bladder cancer cells (ATCC HTB-4) were maintained in culture media composed of a 3:1 mix of Dulbecco–Vogt modification of Eagle's medium (DMEM) (Invitrogen, Burlington, ON, Canada) and Ham F12 medium (Invitrogen) supplemented with 5% fetal bovine serum clone II (Hyclone, GE Healthcare Life Science, Wauwatosa, WI, USA), 24.3 µg/mL adenine

(Corning, Tewksbury, MA, USA), 5 μg/mL insulin (Sigma-Aldrich, Oakville, ON, Canada), 0.212 μg/mL isoproterenol (Sandoz, Boucherville, QC, Canada), 0.4 mg/mL hydrocortisone (Calbiochem, San Diego, CA, USA), 10 ng/mL epidermal growth factor (Austral Biologicals, San Ramon, CA, USA), 100 U/mL penicillin (Sigma-Aldrich) and 25 mg/mL gentamicin (Schering-Plough Canada Inc./Merck, Scarborough, ON, Canada), and incubated at 37 °C with 8% CO_2. Media were changed three times per week.

2.2. Seahorse Energy Metabolism Measurements

UCs, RT4 and T24 cells were plated in 96-well Seahorse XF cell culture plates (Agilent/Seahorse Bioscience, Chicopee, MA, USA) and exposed or not to 10^{-8} M BPA (Millipore Sigma, Oakville, ON, Canada) or 10^{-8} M BPS (Millipore Sigma) for 72 h before measurements. Media were changed every day. Seahorse XFe96 sensor cartridge plates (Agilent/Seahorse Bioscience) were hydrated the day before the analysis with the XF Calibrant (Agilent/Seahorse Bioscience) and incubated overnight at 37 °C without CO_2. Before the bioenergetics measurements, cells were washed and incubated for one hour with Glyco Stress media or Mito Stress media. Glyco Stress media contained XF Base Medium (minimal DMEM) (Agilent/Seahorse Bioscience) supplemented with 2 mM L-glutamine (Wisent Bioproducts Inc., Saint-Jean-Baptiste, QC, Canada). Mito Stress media consisted of XF Base Medium supplemented with 2 mM L-glutamine, 1 mM sodium pyruvate (Wisent Bioproducts Inc.) and 10 mM D-(+)-glucose (Millipore Sigma). The extracellular acidification rate (ECAR), representative of glycolytic metabolism, and the oxygen consumption rate (OCR), representative of mitochondrial respiration, were determined using the XFe Extracellular Flux Analyzer (Agilent/Seahorse Bioscience) [20,25].

The glycolytic metabolism was established by the sequential injection of 10 mM D-(+)-glucose (Millipore Sigma), 1.5 μM of the ATP synthase inhibitor oligomycin (Cayman Chemical, Ann Arbor, MI, USA) to inhibit mitochondrial respiration and force the cells to maximize their glycolytic capacity, and 50 mM 2-deoxy-D-glucose (2-DG) (Alfa Aesar, Ward Hill, MA, USA), a competitive inhibitor of the first step of glycolysis.

The mitochondrial respiration was determined by the sequential injection of 1.5 μM of the ATP synthase inhibitor oligomycin (Cayman Chemical), 0.5 μM of the mitochondrial uncoupler trifluoromethoxy carbonylcyanide phenylhydrazone (FCCP) (Cayman Chemical) and a combination of 0.5 μM of the mitochondrial complex I inhibitor rotenone (MP Biomedicals, Santa Ana, CA, USA) and 0.5 μM of the mitochondrial complex III inhibitor antimycin A (Millipore Sigma). The concentrations indicated for each injection represent the final concentrations in the wells. At least three measurement cycles (3 min of mixing + 3 min of measuring) were completed before and after each injection.

The OCR and ECAR were calculated using Wave software v2.6 (Agilent/Seahorse Bioscience). Following the manufacturer's instructions, energy metabolism was normalized according to the number of cells using a CyQuant Cell proliferation assay kit (Invitrogen). The fluorescence of each well was measured at 485 nm/535 nm for 0.1 s using the Victor2 1420 MultiLabel Counter plate reader (Perkin Elmer Life Sciences, Waltham, MA, USA) and Wallac 1420 software (Perkin Elmer). The normalization values were calculated from the fluorescence measurements with Microsoft Excel software (Microsoft, Redmond, WA, USA) and applied to the metabolic values. Metabolic values were presented as percentages with 100% established using the first three measurements. Therefore, the baseline was established before the injection of glucose or oligomycin (see Figures S1–S4). Each experiment included at least three replicates per condition ($n \geq 3$), and each experiment was repeated at least three times ($N \geq 3$).

2.3. Proliferation

On day 0, UCs, RT4 and T24 cells were seeded in 24-well culture plates at 60,000 cells/well density. Cells were incubated for two hours at 37 °C with 8% CO_2 to allow cell adhesion and then treated or not with 10^{-8} M BPA or BPS. The medium, supplemented or not with BPs, was changed daily for three days. On days 1 to 3, cells from three wells were collected with trypsin, centrifuged at 300 g for 10 min, resuspended in 10 mL ISOTON II diluent (Beckman Coulter, Mississauga, ON, Canada) and counted separately using a Z2 Coulter Particle Count and Size Analyzer (Beckman Coulter) [26]. A graph illustrating the numbers of cells per well as a function of time was performed to calculate the proliferation rate. Proliferation values for days 1 to 3 were established as percentages of control (i.e., untreated condition) on day 1. Therefore, the proliferation value of the control on day 1 was established as 100%. Each condition included three replicates (n = 3) for every cell line, and each experiment was repeated independently three times (N = 3).

2.4. Migration

UCs, RT4 and T24 cells were seeded in 12-well culture plates at 150,000 cells/well density. Cells were incubated for two hours at 37 °C with 8% CO_2 to allow cell adhesion. Then, cells were treated or not with 10^{-8} M BPA or BPS and incubated at 37 °C with 8% CO_2. After 72h, a scratch test was performed [27]. Briefly, a vertical scratch was performed in each well using a 200 µL pipet tip. Wells were rinsed twice to remove detached cells with 3:1 DMEM–Ham F12 medium supplemented with 0.5% fetal bovine serum clone II. Two mL of medium supplemented with 0.5% serum with or without 10^{-8} M BPA or BPS was added to each well. This low serum concentration was chosen to avoid cell proliferation and ensure the observed results are due to cell migration. Cell migration was assessed using a Zeiss Axio Imager M2 Time-Lapse microscope equipped with an AxioCam ICc1 camera (Carl Zeiss, Oberkochen, Germany). Images were processed with the AxioVision 40 V4.8.2.0 software (Carl Zeiss). Photographs were taken every hour for a total of 17 h. Analyses of closure area were measured using ImageJ software (NIH, Bethesda, MD, USA). Migration speed was calculated as the slope of the closure area (Y-axis) as a function of time (X-axis) with the formula "ax + b", where "a" represents migration speed. The slopes of each cell line were established as percentages of control (i.e., untreated condition). Each condition included two replicates (n = 2) for every cell line, and each experiment was repeated independently three times (N = 3).

2.5. Flow Cytometry

RT4 and T24 cells were seeded in 12-well culture plates at 125,000 cells/well density. Cancer cells were treated with or without 10^{-8} M BPA for ten days and incubated at 37 °C with 8% CO_2. The medium was changed three times a week. After ten days, cells were collected with trypsin, centrifuged at 300 g for 10 min and resuspended in 100 µL of PBS. Cell suspensions were individually transferred in 100 µL of 3.7% formol to fix cells, and samples were incubated at 4 °C until flow cytometry analyses. Cell samples were washed twice with PBS on the analysis day, followed by blocking using PBS-1% BSA for 45 min at room temperature. Incubation with primary antibody anti-alpha smooth muscle actin (α-SMA) coupled with FITC (1/250 dilution; Abcam) or with isotypic control FITC antibody (1/1000 dilution; Abcam) was performed at room temperature for one hour in PBS-1% BSA. Cells were washed twice with PBS-1% BSA and once with PBS only. Cells were resuspended in 100 µL PBS and samples were analyzed by flow cytometry with a FACSCalibur (Becton Dickson, San Jose, CA, USA) [28]. For each replicate, the value of the isotypic sample was subtracted from the value of the positive sample (cells incubated with antibodies). Each condition included six replicates (n = 6), and each experiment was repeated independently three times (N = 3).

2.6. Statistical Analysis

Graphical representation and statistical analyses were performed using Microsoft Excel (Microsoft) and GraphPad Prism Software v.9.3 (San Diego, CA, USA). The results are expressed as mean ± standard error of the mean (SEM). Statistical analyses were performed using the non-parametric tests, the Mann–Whitney test or the Kruskal–Wallis test. Normality analyses were performed for the data in Figure 1 and Figure 7. The values did not meet the required premises to assume normality, which justifies using non-parametric statistical tests. Statistical significance was established at $p < 0.05$.

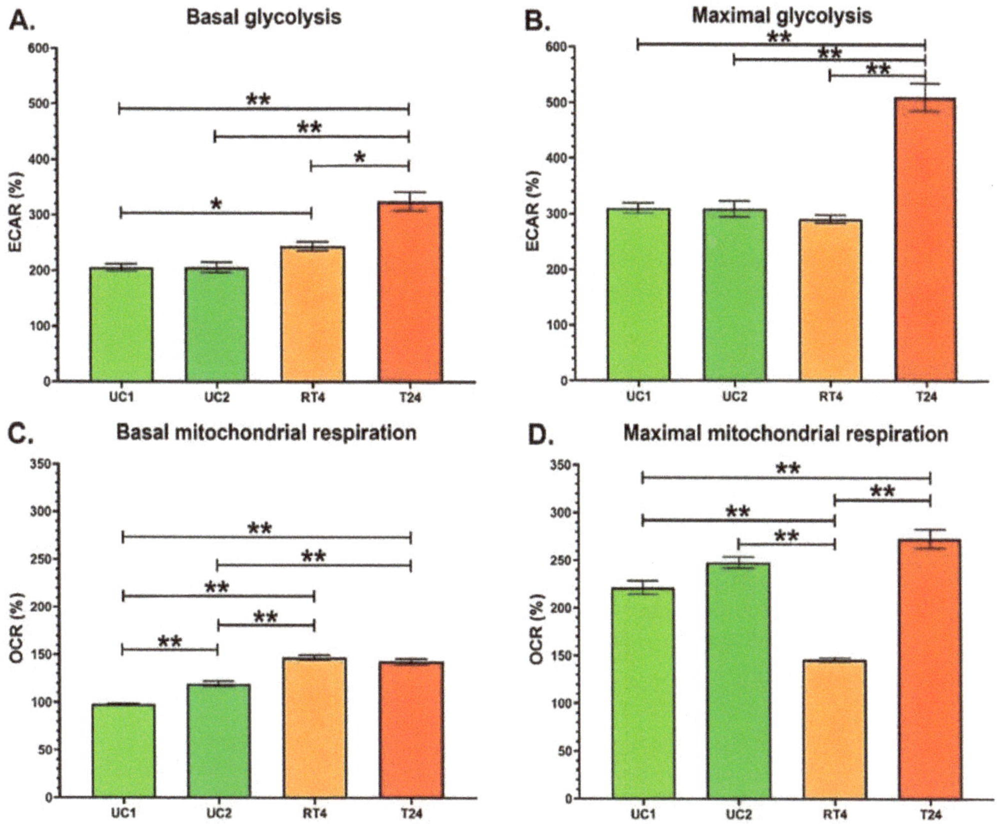

Figure 1. T24 invasive bladder cancer cells exhibit an increased glycolytic capacity compared with RT4 non-invasive bladder cancer cells. (**A**,**B**) ECAR and (**C**,**D**) OCR were determined using the XFe96 Extracellular Flux Analyzer in two populations of normal urothelial cells (UC1 and UC2) and non-invasive (RT4) and invasive (T24) bladder cancer cells to establish (**A**) basal glycolysis, (**B**) maximal glycolytic capacity, (**C**) basal mitochondrial respiration and (**D**) maximal mitochondrial capacity. Data are presented as the mean ± SEM and displayed as percentages of UC1 acting as control (n = 10, N = 4). A baseline (100%) was established before injections (see Figure S1). * $p < 0.05$, ** $p < 0.01$ by Kruskal–Wallis test.

3. Results

3.1. *T24 Invasive Bladder Cancer Cells Exhibit an Increased Glycolytic Capacity Compared with RT4 Non-Invasive Bladder Cancer Cells*

Before evaluating the impact of chronic exposure to BPs, the energy metabolism of two UC populations (UC1 and UC2) and two bladder cancer cells (RT4 non-invasive and T24 invasive) was evaluated to compare their glycolytic and mitochondrial capacities. First, UC1 and UC2 cell populations had a similar glycolytic metabolism (Figure 1A,B). However, UC2 cells exhibited higher OCR levels at basal capacity compared with UC1 cells, whereas the maximal mitochondrial capacity of UC1 and UC2 cells was not significantly different (Figure 1C,D).

Secondly, RT4 non-invasive bladder cancer cells displayed a significantly higher basal glycolysis than UC1 cells (Figure 1A). RT4 cells tended to exhibit slightly higher ECAR levels compared with UC2 cells ($p = 0.07$) (Figure 1A), whereas ECAR levels for maximal glycolytic capacity were comparable with UCs (Figure 1B). RT4 cells had higher OCR levels for basal mitochondrial respiration than both UCs (Figure 1C) but exhibited the lowest maximal mitochondrial capacity compared with the other cell lines (Figure 1D).

Thirdly, T24 invasive bladder cancer cells exhibited the highest ECAR levels for basal and maximal glycolytic capacity compared with the other cell lines (Figure 1A,B). A general representation of the energy metabolism of T24 cells shows that these cancer cells are highly glycolytic (Figure S1). T24 cells exhibit similar OCR levels for basal mitochondrial respiration compared with RT4 cells but significantly higher levels for maximal mitochondrial respiration (Figure 1C,D). As observed for the glycolytic capacity, T24 cells exhibited significantly higher levels of OCR for maximal mitochondrial respiration compared with RT4 non-invasive bladder cancer cells and UCs. Therefore, initial analyses showed that T24 cells displayed a greater glycolytic capacity than RT4 cells, whereas the two populations of UCs exhibited roughly similar energy metabolism.

3.2. Chronic Exposure to Physiological Concentrations of BPA or BPS Does Not Modulate the Glycolysis and Mitochondrial Respiration of Normal Urothelial Cells

Two UC populations were exposed to physiological concentrations of BPA or BPS to evaluate the impact of these endocrine disruptors on their bioenergetics. In vivo, UCs can be in contact with BPA and BPS through urine [11,29]. Chronic exposure to 10^{-8} M BPA or BPS did not seem to impact the energy metabolism of UCs. A slight but not significant decrease in ECAR levels was observed for the glycolytic capacity of both compounds (Figure 2A,B), as well as a subtly decreased OCR level in basal mitochondrial respiration when exposed to BPS ($p = 0.12$) (Figure 2C). Exposure to BPs did not affect OCR levels associated with maximal mitochondrial respiration (Figure 2D). Overall, 72 h chronic exposure to physiological concentrations of BPA or BPS did not significantly affect the bioenergetics of normal UCs (Figure S2).

3.3. RT4 Non-Invasive Bladder Cancer Cells Chronically Exposed to Physiological Concentrations of BPs Exhibit Increased Bioenergetics

RT4 cells were exposed to physiological concentrations of BPA or BPS to evaluate the impact of these endocrine disruptors on non-invasive bladder cancer cells. Like UCs, bladder cancer cells can be exposed to BPs through urine [11,29]. BPA exposure did not influence the ECAR levels of the basal glycolysis of RT4 cells, whereas chronic exposure of RT4 cells to 10^{-8} M BPS significantly increased the ECAR levels of basal glycolysis (Figure 3A). Chronic exposure to BPA had a non-significant effect on the ECAR levels associated with the maximal glycolytic capacity. In contrast, exposure to physiological concentrations of BPS tended to increase ECAR levels ($p = 0.07$) (Figure 3B). RT4 chronically exposed to 10^{-8} M BPA exhibited significantly increased OCR levels associated with basal and maximal mitochondrial respiration (Figure 3C,D). Although not significantly different from the control condition for mitochondrial metabolism, BPS exposure seemed to have similar biological effects to BPA, suggesting this alternative compound could have comparable effects. Therefore, chronic exposure to physiological concentrations of BPs increased the bioenergetics of RT4 non-invasive bladder cancer cells (Figure S3).

3.4. T24 Invasive Bladder Cancer Cells Chronically Exposed to Physiological Concentrations of BPA or BPS Exhibit an Increased Glycolytic Metabolism

T24 cells were exposed to physiological concentrations of BPA or BPS to evaluate the impact of these compounds on invasive bladder cancer cell metabolism. Chronic exposure to 10^{-8} M BPA or BPS increased the ECAR levels of basal glycolysis of invasive bladder cancer cells (Figure 4A), whereas only 10^{-8} M BPS significantly increased the ECAR levels of the maximal glycolytic capacity (Figure 4B). Although not significant, BPA seemed to slightly increase the ECAR levels of the maximal glycolysis of T24 cells ($p = 0.096$) (Figure 4B). Chronic exposure to BPA and BPS did not affect the OCR levels associated with the basal and maximal mitochondrial respiration of T24 cells (Figure 4C,D).

As previously observed with RT4 cells, BPA and BPS seemed to have similar effects on the energy metabolism of invasive bladder cancer cells (Figure S4). Overall, T24 invasive bladder cancer cells exhibited an increased glycolytic metabolism when chronically exposed to physiological concentrations of BPs.

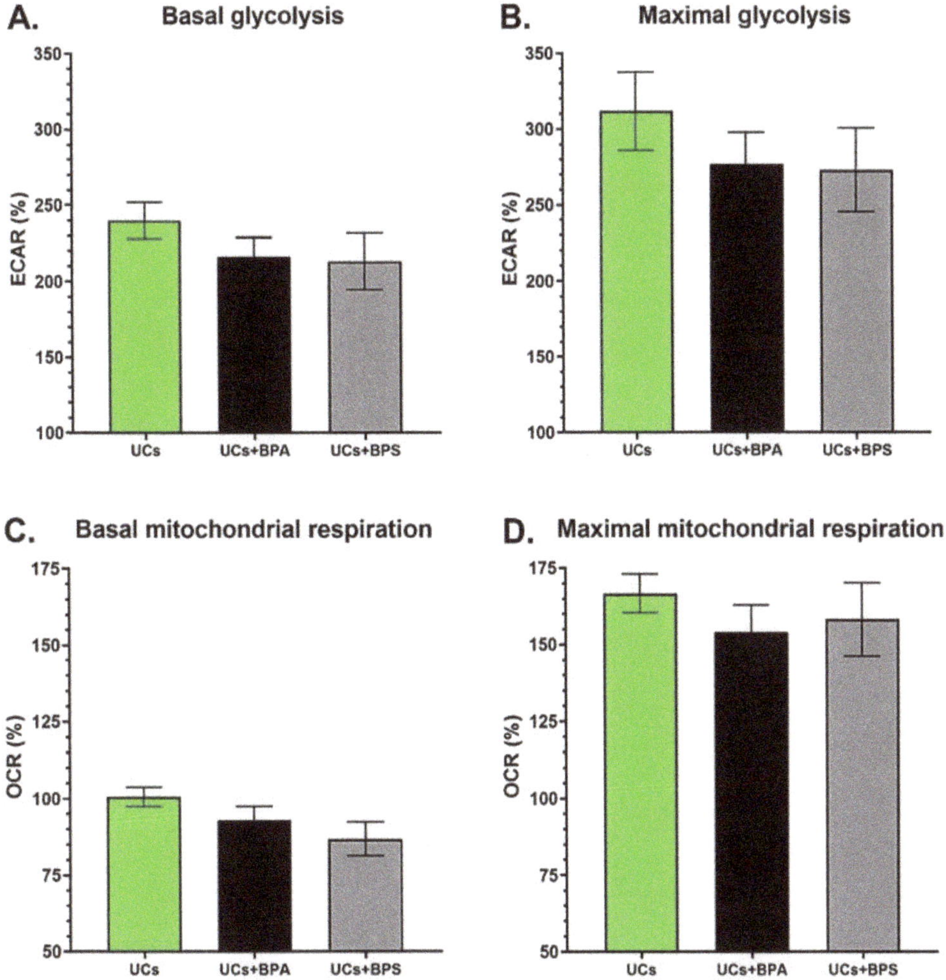

Figure 2. Chronic exposure to physiological concentrations of BPA or BPS has no significant effect on the glycolysis and mitochondrial respiration of normal urothelial cells. (**A**,**B**) ECAR and (**C**,**D**) OCR were determined using the XFe96 Extracellular Flux Analyzer for normal urothelial cells (UCs) with or without chronic exposure to physiological concentrations of BPA or BPS to establish (**A**) basal glycolysis, (**B**) maximal glycolytic capacity, (**C**) basal mitochondrial respiration and (**D**) maximal mitochondrial capacity. Analyses represent the results for two populations of normal urothelial cells (UC1 and UC2). Data are presented as the mean ± SEM and displayed as percentages of controls (i.e., untreated condition) ($n \geq 3$, $N = 4$). A baseline (100%) was established before injections (see Figure S2). $p < 0.05$ by Mann–Whitney test.

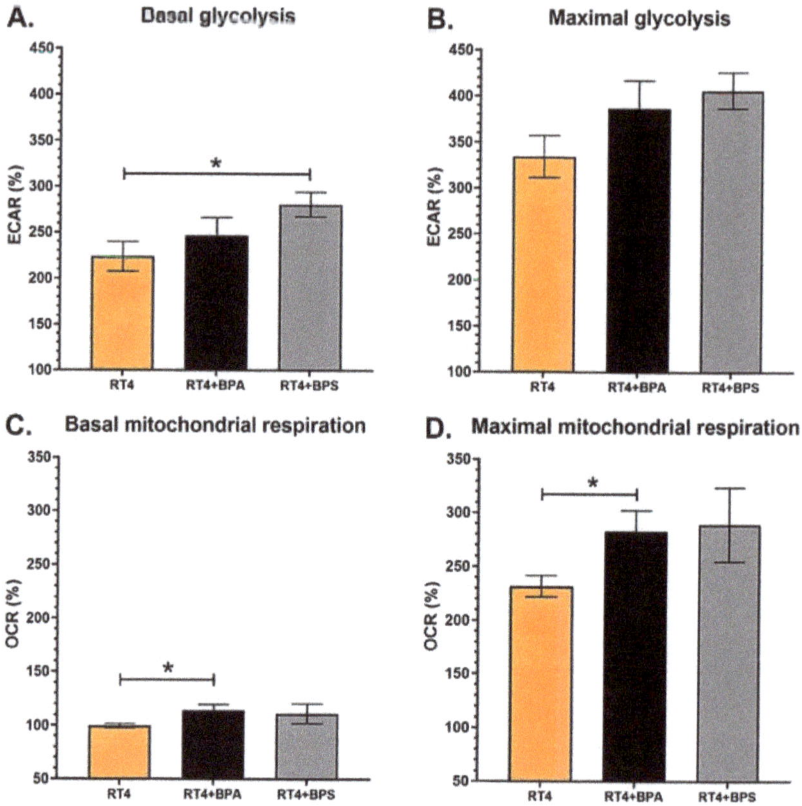

Figure 3. RT4 non-invasive bladder cancer cells chronically exposed to physiological concentrations of BPs exhibit increased bioenergetics. (**A**,**B**) ECAR and (**C**,**D**) OCR were determined using the XFe96 Extracellular Flux Analyzer for RT4 non-invasive bladder cancer cells with or without chronic exposure to physiological concentrations of BPA or BPS to establish (**A**) basal glycolysis, (**B**) maximal glycolytic capacity, (**C**) basal mitochondrial respiration and (**D**) maximal mitochondrial capacity. Data are presented as the mean ± SEM and displayed as percentages of controls (i.e., untreated condition) ($n \geq 3$, $N = 3$). A baseline (100%) was established before injections (see Figure S3). * $p < 0.05$ by Mann–Whitney test.

3.5. Chronic Exposure to Physiological Concentrations of BPA or BPS Increases the Proliferation Rate of RT4 Non-Invasive Bladder Cancer Cells and Induces an Initial Boost of Proliferation for UCs and T24 Cells

The proliferation rate of urothelial, RT4 and T24 cells was evaluated for three days under chronic exposure to 10^{-8} M BPA or BPS. First, UCs exposed to BPA or BPS exhibited a significantly higher cell number on days 1 to 3 compared with the control (Figure 5A). However, it is possible to observe that, following this initial increase, the proliferation rate seemed to stabilize, suggesting that the proliferation rate could be enhanced in the first 24 h of BPA or BPS exposure, resulting in an initial boost. Secondly, RT4 non-invasive bladder cancer cells chronically exposed to 10^{-8} M BPA or BPS exhibited a significantly higher proliferation rate on day 3 than the control (Figure 5B). However, exposure to BPs did not impact RT4 cell proliferation on days 1 and 2. Thirdly, chronic exposure of T24 cells to BPA or BPS did not affect the proliferation rate on days 2 and 3. However, exposure to BPA significantly increased the proliferation rate of T24 cells on day 1 (Figure 5C), but its clinical impact is probably not significant. This observation could be associated with an initial increase in proliferation due to BPA exposure since the initial effect observed following BPA or BPS exposure was not maintained on days 2 and 3. Therefore, chronic exposure to physiological concentrations of BPA or BPS increased the proliferation rate of RT4 cells and induced an initial boost of proliferation in the first 24 h for normal UCs and T24 cells.

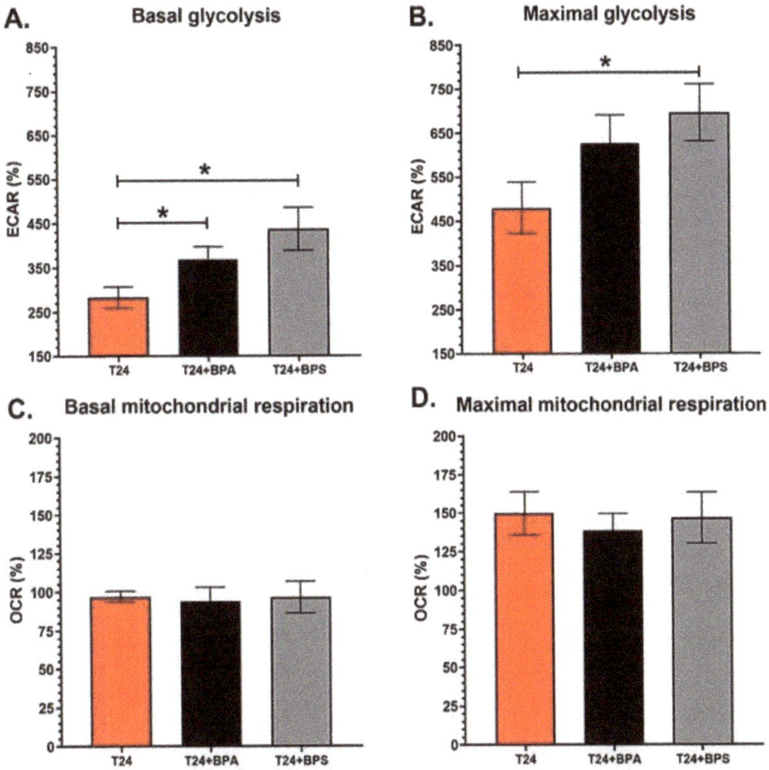

Figure 4. T24 invasive bladder cancer cells chronically exposed to BPA or BPS physiological concentrations exhibit an increased glycolytic metabolism. (**A**,**B**) ECAR and (**C**,**D**) OCR were determined using the XFe96 Extracellular Flux Analyzer for T24 invasive bladder cancer cells with or without chronic exposure to physiological concentrations of BPA or BPS to establish (**A**) basal glycolysis, (**B**) maximal glycolytic capacity, (**C**) basal mitochondrial respiration and (**D**) maximal mitochondrial capacity. Data are presented as the mean ± SEM and displayed as percentages of controls (i.e., untreated condition) (n ≥ 3, N = 3). A baseline (100%) was established before injections (see Figure S4). * $p < 0.05$ by Mann–Whitney test.

3.6. Chronic Exposure to Physiological Concentrations of BPA or BPS Decreases the Migration of Normal Urothelial Cells While Increasing the Migration Speed of Bladder Cancer Cells

The migration of urothelial, RT4 and T24 cells was evaluated under chronic exposure to 10^{-8} M BPA or BPS. On the one hand, chronic exposure to physiological concentrations of BPA significantly decreased the migration speed of UCs. In contrast, BPS exposure only resulted in a slight reduction in migration (p = 0.056) (Figure 6A). On the other hand, 10^{-8} M BPA tended to increase the migration speed of RT4 non-invasive bladder cancer cells (p = 0.06), but BPS did not affect the migration of RT4 cells (Figure 6B). BPA and BPS significantly increased the migration speed of T24 invasive bladder cancer cells (Figure 6C). Therefore, chronic exposure to physiological concentrations of BPA or BPS decreased normal UCs while increasing the migration of T24 invasive bladder cancer cells.

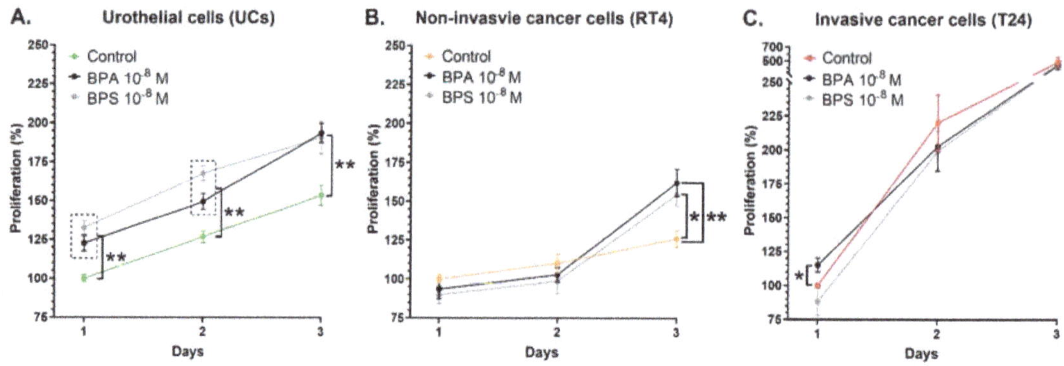

Figure 5. Chronic exposure to physiological concentrations of BPA or BPS increases the proliferation rate of RT4 non-invasive bladder cancer cells and induces an initial boost of proliferation for UCs and T24 cells. The proliferation rate of (**A**) normal urothelial cells (UCs), (**B**) RT4 non-invasive bladder cancer cells and (**C**) T24 invasive bladder cancer cells was established over three days with or without chronic exposure to physiological concentrations of BPA or BPS. The proliferation rate is illustrated by the number of cells as a function of time. Data are presented as the mean ± SEM and displayed as percentages of controls at day 1 (i.e., untreated condition) (n = 3, N = 3). * $p < 0.05$, ** $p < 0.01$ by Mann–Whitney test.

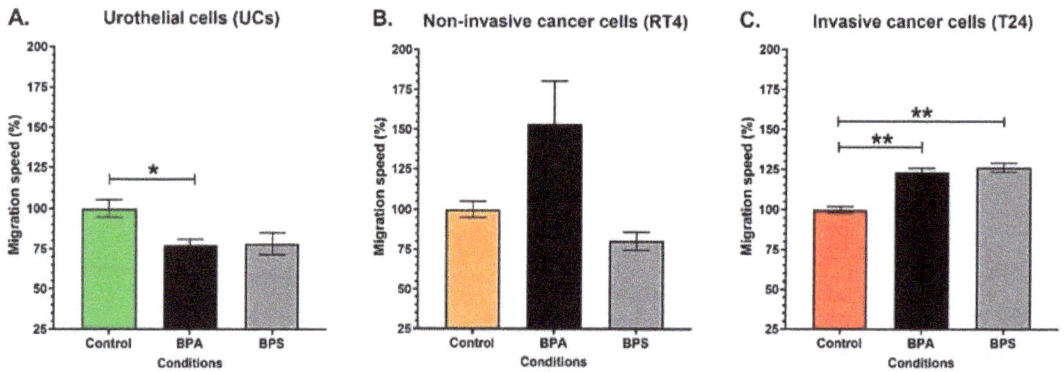

Figure 6. Chronic exposure to physiological concentrations of BPA or BPS decreases the migration speed of normal urothelial cells while increasing the migration speed of bladder cancer cells. The migration speed of (**A**) normal urothelial cells (UCs), (**B**) RT4 non-invasive bladder cancer cells and (**C**) T24 invasive bladder cancer cells was evaluated by time-lapse microscopy with or without chronic exposure to physiological concentrations of BPA or BPS. Data are presented as the mean ± SEM and displayed as percentages of controls (i.e., untreated condition) (n = 2, N = 3). The 100% migration value of UCs represents a mean of 12,06 cm^2/h, for RT4 1,68 cm^2/h and for T24 4,77 cm^2/h. * $p < 0.05$, ** $p < 0.01$ by Mann–Whitney test.

3.7. RT4 Non-Invasive Bladder Cancer Cells Chronically Exposed to Physiological Concentrations of BPA Exhibit an Increased Expression of α-SMA Expression

RT4 and T24 cancer cells were chronically exposed to 10^{-8} M BPA and analyzed by flow cytometry to evaluate the impact of this endocrine disruptor on the expression of alpha-smooth muscle actin (α-SMA), which can be used as a cell marker for the invasive potential of cancer cells [30]. Compared with T24 invasive bladder cancer cells, RT4 non-invasive cancer cells expressed significantly lower levels of α-SMA (Figure 7). However, when chronically exposed to 10^{-8} M BPA, RT4 cells exhibited an increased expression of α-SMA. Furthermore, α-SMA levels of RT4 cells exposed to BPA were not significantly different from the α-SMA levels of T24 cells. The chronic exposure of T24 cells to 10^{-8} M BPA did not substantially affect the α-SMA expression levels. See Figure S5 for an example of gating analysis. Overall, chronic exposure of RT4 non-invasive bladder cancer cells to physiological concentrations of BPA increased the expression of α-SMA.

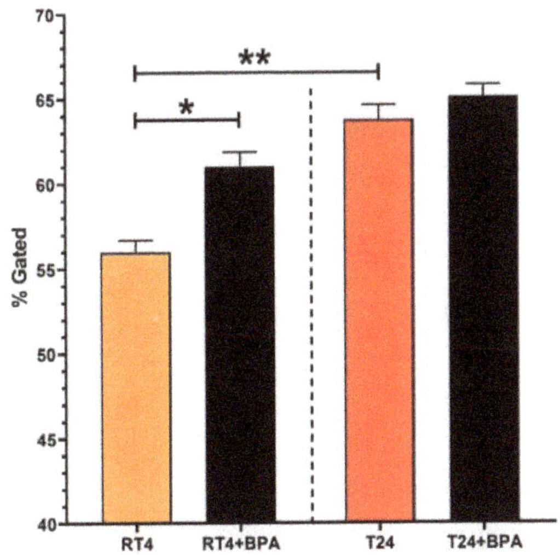

Figure 7. RT4 non-invasive bladder cancer cells chronically exposed to BPA's physiological concentrations exhibit an increased α-SMA expression. The expression of α-smooth muscle actin (α-SMA) by RT4 non-invasive and T24 invasive bladder cancer cells was measured by flow cytometry with or without chronic exposure to physiological concentrations of BPA. Data are presented as the mean ± SEM (n = 6, N = 3). * $p < 0.05$, ** $p < 0.01$ by Kruskal–Wallis test.

4. Discussion

Since the effects of BPs on bladder cancer have not yet been established, the impact of chronic exposure to physiological concentrations of BPA or BPS on normal urothelial cells and non-invasive and invasive bladder cancer cells was studied. Therefore, the effects of BPs on energy metabolism, proliferation, migration and α-SMA expression were examined.

Before evaluating the impact of chronic exposure to physiological concentrations of BPs, the energy metabolism of UCs and RT4 non-invasive and T24 invasive bladder cancer cells was evaluated to compare their glycolytic and mitochondrial capacities. The two different populations of normal UCs (UC1 and UC2) showed similar levels of energy metabolism, and RT4 non-invasive bladder cancer cells displayed higher basal glycolysis and mitochondrial respiration than UCs. In contrast, T24 invasive bladder cancer cells exhibited a greater glycolytic capacity than normal UCs and RT4 non-invasive bladder cancer cells.

Cancer cells typically exhibit an increased metabolic rate due to enhanced physiological activity, characterized by an uncontrolled proliferation rate [31]. Cancer cells are characterized by increased glycolytic metabolism, at the expense of mitochondrial respiration, even in the presence of oxygen and functioning mitochondria [31]. This metabolic switch is called the Warburg effect, a phenomenon that allows cancer cells to have an adapted energy metabolism to support cell growth and proliferation and promote cell invasion [31]. The increased intake of glucose results in a high synthesis of pyruvate. Since the glycolytic rate is superior to the mitochondrial capacity, the excess pyruvate is converted to lactate through the enzyme lactate dehydrogenase [32].. The increased use of the glycolytic pathway results in enhanced lactate production, acidifying the extracellular microenvironment. Lactate can inhibit immune cells, such as monocytes, that are responsible for eliminating unhealthy cells, including cancer [33]. The accumulation of lactate also reorganizes the extracellular matrix, which could promote tumour invasion [22]. However, this enhanced glycolysis was only observed in the RT4 cells for basal glycolysis, not maximal glycolysis. RT4 cells had a significantly higher basal glycolytic metabolism than UC1, but not UC2, and had a similar maximal glycolytic metabolism compared with both UCs. The basal mitochondrial capacity of RT4 cells was increased compared with UCs. The absence of an enhanced maximal glycolytic capacity when compared with T24 cells could be because RT4 cells are non-invasive, therefore, are less prone to produce an excess of lactate to invade the extracellular matrix.

Furthermore, RT4 cells had a slower doubling rate than T24 cells. The comparison of the proliferation rate of untreated RT4 and T24 cells demonstrates an important difference between both cell lines. RT4 cells have a doubling time of 37 to 66 h [34,35], whereas T24 cells are reported to have a doubling time of 19 to 24 h [34–36]. Therefore, RT4 non-invasive bladder cancer cells have a lower level of physiological activity, which could explain the absence of metabolic switch observed. Finally, UCs exhibited a similar energy metabolism even though UC2 cells had higher levels of basal mitochondrial respiration. This difference could be due to individual variations associated with genetics or environmental factors, such as drugs, chemical exposure and health [37].

In vivo, the basal layer's UCs are protected from exposure to urine and its contaminants, such as BPs. However, cancer cell growth can alter the impermeability of the urothelium by disrupting cell–cell adhesion [38], thus resulting in the exposure of UCs surrounding the tumour and underlying UCs to urine and potentially BPs. UCs can also be exposed to BPs through the bladder vascular system perfusing the bladder stroma [39]. BPA and BPS had a weak effect on the bioenergetics of UCs. Although not significant, this tendency could be associated with inhibiting certain enzymes related to cell metabolism by BPs. BPA has been shown to inhibit metabolizing enzymes, such as cytochromes P450, glucose transporters and enzymes from the electron transport chain, that can impact some main metabolic pathways, for example, mitochondrial respiration [40,41]. It is, therefore, essential to confirm the metabolic analyses with physiological parameters to evaluate the functional impact of BPA or BPS on UCs.

The alteration in the energy metabolism of RT4 cells after chronic exposure to BPs could be associated with the multiple binding capacities of BPA and BPS. BPA was shown to bind to ERs with a median inhibitory concentration (IC_{50}) value of 3.3–73 nM [42], which corresponds to the range of concentrations of BPs found in urine (1–100 nM) [11]. Therefore, the increased energy metabolism observed could be related to the enhanced physiological activity of RT4 cells following exposure to BPs. Although the impact of BPS on mitochondrial capacity was not significantly different from the control condition, chronic exposure to this compound seems to induce similar effects to BPA. These results need to be correlated through the impact of BPA or BPS on the physiological activity of RT4 cells. The relevance of these results depends on the associated biological consequences, for example, proliferation and migration.

Thirdly, T24 invasive bladder cancer cells chronically exposed to physiological concentrations of BPA or BPS did not seem to impact the mitochondrial capacities but exhibited an increased glycolytic metabolism. The increased glycolytic capacity observed in T24 cells following chronic exposure to BPA or BPS could lead to higher lactate production and enhance the acidification of the tumour microenvironment. Although T24 cells are invasive cancer cells, chronic exposure to BPA or BPS could accentuate the consequences of matrix acidification [31], further inhibiting the local immune cells [33], thus facilitating invasion and metastases formation through a major matrix remodelling.

The increased proliferation rate of RT4 non-invasive cancer cells chronically exposed to physiological concentrations of BPA or BPS could suggest a more aggressive phenotype. Indeed, studies have shown that BPA could promote cell proliferation by binding to the GPER in breast cancer cells, which activates the phosphoinositide 3-kinase (PI3K)/protein kinase B (Akt)/mammalian target of rapamycin (mTOR) signalling pathway [43]. Furthermore, BPA can bind to the ERRγ of endometrial cancer cells, which can activate the epidermal growth factor receptor (EGFR)/extracellular signal-regulated kinase (ERK) pathway associated with proliferation [44]. On the other hand, BPS has also been shown to promote cell proliferation by altering the PI3K/Akt signalling pathway in breast cancer cells [45]. BPA or BPS exposure to T24 invasive bladder cancer cells did not impact the proliferation rate in the long term. Still, BPA induced an increased proliferation in the first 24 h. BPA can alter multiple signalling pathways associated with proliferation, which could explain the increased proliferation rate. However, this difference was not observed on days 2 and 3, which could be explained by the already high proliferation rate of T24 cells, therefore obscuring the impact of BPs in the long term.

Next, the impact of chronic exposure to BPA or BPS on cell migration was evaluated. UCs chronically exposed to physiological concentrations of BPA or BPS exhibited decreased migration. This observation could impact the urothelium tissue repair in the case of disease or bladder injury. Cell migration is essential in wound healing to allow epithelialization and wound closure [46]. The decreased migration capacity of UCs could result in a slower wound closure of the bladder wall, increasing the exposure of underlying tissues to the toxic substances found in urine, such as urea and carcinogens [47]. Chronic exposure to BPA or BPS did not impact RT4 cell migration. RT4 cells have a low migration capacity as these cells form and stay in clusters, resulting in minimal migration. Therefore, the impact of BPs could potentially be obscured by the fact that RT4 cells simply do not tend to migrate. However, chronic exposure to BPA or BPS increased the migration of T24 invasive bladder cancer cells. On the one hand, BPA has been shown to increase the migration of lung cancer cells through the GPER/EGFR/ERK1/2 signalling pathway [48] and increase the migration of triple-negative breast cancer cells through ERRγ [7]. Studies by Derouiche et al. demonstrated that exposure to BPA can promote the cell migration of prostate cancer cells through the modulation of ion channel protein expression associated with calcium entry [8]. On the other hand, exposure to BPS has been shown to promote the migration of human non-small cell lung cancer cells through ERK1/2, mediated by the transforming growth factor β (TGF-β)/Smad-2/3 signalling pathway [11]. Deng et al. have reported that BPS can also promote the migration of triple-negative breast cancer cells in vitro through the GPER/Hippo signalling pathway, resulting in the activation and nuclear accumulation of yes-associated protein (YAP), thus upregulating downstream genes [49]. Therefore, these results concur with the literature and the metabolic analyses, demonstrating an enhanced energy metabolism in T24 cells when exposed to BPA or BPS.

In brief, chronic exposure to physiological concentrations of BPs tended to decrease the energy metabolism of UCs, increase their proliferation in the first 24 h and decrease their migration. In addition, exposure to BPs increased the energy metabolism and the proliferation of non-invasive RT4 bladder cancer cells but had little to no effect on their migration. Finally, exposure to BPs increased the glycolytic capacity and the migration of invasive T24 bladder cancer cells but did not impact their mitochondrial respiration or proliferation (Table 1). These results suggest that chronic exposure to BPA and BPS alters the energy metabolism and behaviours of UCs and non-invasive and invasive bladder cancer cells, which could potentially promote bladder cancer progression.

Table 1. Impact of chronic exposure to physiological concentrations of BPA and BPS on the energy metabolism, proliferation and migration of UCs and RT4 non-invasive and T24 invasive bladder cancer cells. UCs are generally negatively affected by exposure to BPs, illustrated by the decreased bioenergetics and migration. Conversely, exposure to BPs generally enhances the bioenergetics, proliferation and migration of cancer cells.

Cell Types	Parameters		BPA	BPS
UCs	Glycolysis	Basal	↓	↓
		Maximal	↓	↓
	Mitochondrial respiration	Basal	Ø	↓
		Maximal	Ø	Ø
	Proliferation		↑	↑
	Migration		↓↓	↓
RT4 cells	Glycolysis	Basal	Ø	↑↑
		Maximal	Ø	↑
	Mitochondrial respiration	Basal	↑↑	↑
		Maximal	↑↑	↑
	Proliferation		↑↑	↑↑
	Migration		↑	Ø
T24 cells	Glycolysis	Basal	↑↑	↑↑
		Maximal	↑	↑↑
	Mitochondrial respiration	Basal	Ø	Ø
		Maximal	Ø	Ø
	Proliferation		Ø	Ø
	Migration		↑↑	↑↑

Legend: Ø = No impact; ↑/↓ = Tendency; ↑↑/↓↓ = Significant.

The impact of chronic exposure to BPA on the α-SMA expression of RT4 and T24 cells was established. α-SMA can be used as an epithelial–mesenchymal transition (EMT) marker, during which α-SMA expression is increased [30]. When cancer cells transit from a non-invasive to an invasive and metastatic phenotype, they undergo various steps, including EMT [50]. Increased α-SMA expression is associated with an invasive phenotype and a greater capacity to produce metastases [30], therefore resulting in a poorer prognosis in multiple cancers such as lung [51] and breast cancer [52]. Consequently, invasive cancer cells tend to express higher levels of α-SMA than non-invasive cancer cells [30], allowing α-SMA expression to be used as a cell marker for cancer cell aggressiveness. The results show that RT4 non-invasive bladder cancer cells expressed significantly lower levels of α-SMA when compared with T24 invasive bladder cancer cells. Chronic exposure of RT4 cells to a physiological concentration of BPA increased the expression of α-SMA. Chronic exposure to BPA could, therefore, enhance the aggressiveness of non-invasive cancer cells, thus promoting the transition from a non-invasive to an invasive phenotype. Clinically, these results suggest that chronic exposure to BPs could promote bladder cancer recurrence and progression.

This study is not without limitations. First, the inability to ensure a bisphenol-free control represents a limitation [53]. Since bisphenol-based plastics are abundantly used in laboratory equipment and scientific instruments, it is challenging to avoid external bisphenol contamination. However, there were still significant differences between controls and the BPA or BPS conditions. Second, using an in vitro 2D cell culture is a notable limitation. Using a 3D bladder cancer model [54] would better represent the physiological impact of chronic exposure to BPA or BPS on UCs and bladder cancer tumours. RT4 and

T24 cells were selected in this study for their ability to form spheroids, as we are planning to further investigate the effects of BPs using our 3D bladder cancer model [54].

Measuring the presence of BPs in different stages of bladder cancer samples would be an interesting perspective for this study to determine if advanced bladder cancers exhibit higher levels of BPs than low-grade bladder cancer patients. Furthermore, evaluating the impact of BPs' metabolites on bladder cancer progression would be valuable. A significant proportion of BPs is metabolized and excreted through urine [55]. However, studies have shown that these metabolites can remain active in the body [56], suggesting they could also potentially affect cancer development. Unfortunately, the impact of BPs' metabolites has never been investigated in bladder cancer.

5. Conclusions

Few studies have evaluated the potential impact of chronic exposure to BPs on bladder cancer, despite the presence of these endocrine disruptors in human urine and the accumulating literature demonstrating their pro-tumorigenic capacities. This study has brought valuable insights into the effects of chronic exposure to a physiological concentration of BPA or BPS on bladder cancer progression. The impact of these BPs on the energy metabolism, proliferation, migration and α-SMA expression of normal UCs and RT4 non-invasive and T24 invasive bladder cancer cells was established. The results show that chronic exposure to BPA or BPS increases the proliferation rate of UCs while decreasing their migration, which could result in hyperplasia and altered wound healing capacities. Exposure to BPA or BPS also increases cancer cells' energy metabolism and physiological activity, particularly the metabolism, proliferation and α-SMA expression of RT4 non-invasive bladder cancer cells, which could promote bladder cancer progression from a non-invasive to an invasive phenotype. The ubiquitous and continuous exposure to these endocrine disruptors, through food and the environment, could have a meaningful clinical impact on bladder cancer recurrence, progression and patient prognosis.

Supplementary Materials: The following supporting information can be downloaded at https://www.mdpi.com/article/10.3390/cancers14164011/s1, Figure S1:Glycolytic and mitochondrial metabolism of normal urothelial cells (UC1 and UC2) and non-invasive (RT4) and invasive (T24) bladder cancer cells; Figure S2: Impact of BPA and BPS on the glycolytic and mitochondrial metabolism of two populations of normal urothelial cells (UC1 and UC2); Figure S3: Impact of BPA and BPS on the glycolytic and mitochondrial metabolism of RT4 non-invasive bladder cancer cells; Figure S4: Impact of BPA and BPS on the glycolytic and mitochondrial metabolism of T24 invasive bladder cancer cells; Figure S5: Example of gating analysis for the impact of BPA on the expression of α-SMA of RT4 cells.

Author Contributions: Experimental design: È.P., S.C., S.B. and M.P.; data acquisition: È.P. and F.-A.P.; data analysis: È.P.; data interpretation: È.P., S.C., S.B. and M.P.; writing—original manuscript preparation: È.P.; writing—revision and editing: È.P., S.C., F.P., S.B. and M.P.; final approval: È.P., S.B. and M.P. All authors have read and agreed to the published version of the manuscript.

Funding: This work was supported by the Natural Sciences and Engineering Research Council (NSERC) of Canada (grant number RGPIN-2015-05413), Bladder Cancer Canada, Fondation du CHU de Québec/Concert pour la vie, and the Canadian Urological Association Scholarship Fund (CUASF). The acquisition of the extracellular flux analyzer was supported by John R. Evans Leaders Equipment and Infrastructure Grant 33805 from the Canada Foundation for Innovation awarded to M.P.

Institutional Review Board Statement: This study was conducted according to the guidelines of the Declaration of Helsinki and approved by the Ethics Committee of the CHU de Québec (protocol code 2012-1341, DR-002-1190, approved and renewed annually since December 2013).

Informed Consent Statement: Informed consent was obtained from all subjects who provided tissue samples for this study.

Data Availability Statement: The data presented in this study are available on request from the corresponding author.

Acknowledgments: The authors acknowledge the Fonds de recherche du Québec-Santé (FRQS) for the Senior research scholar awarded to S.B. and the Junior 2 research scholar awarded to M.P. We also acknowledge the Canadian Institutes of Health Research (CIHR) and Fonds de recherche du Québec-Santé (FRQS) for the scholarships awarded to E.P. as well as the support from the Québec Cell, Tissue and Gene Therapy Network—ThéCell, supported by the FRQS.

Conflicts of Interest: The authors declare no conflict of interest.

References

1. Bakken, K.; Alsaker, E.; Eggen, A.E.; Lund, E. Hormone replacement therapy and incidence of hormone-dependent cancers in the Norwegian Women and Cancer study. *Int. J. Cancer* **2004**, *112*, 130–134. [CrossRef] [PubMed]
2. Kelly, S.P.; Anderson, W.F.; Rosenberg, P.S.; Cook, M.B. Past, Current, and Future Incidence Rates and Burden of Metastatic Prostate Cancer in the United States. *Eur. Urol. Focus* **2018**, *4*, 121–127. [CrossRef] [PubMed]
3. Rochefort, H. Endocrine disruptors (EDs) and hormone-dependent cancers: Correlation or causal relationship? *Comptes Rendus Biol.* **2017**, *340*, 439–445. [CrossRef]
4. Murata, M.; Kang, J.H. Bisphenol A (BPA) and cell signaling pathways. *Biotechnol. Adv.* **2018**, *36*, 311–327. [CrossRef] [PubMed]
5. Shafei, A.; Ramzy, M.M.; Hegazy, A.I.; Husseny, A.K.; El-Hadary, U.G.; Taha, M.M.; Mosa, A.A. The molecular mechanisms of action of the endocrine disrupting chemical bisphenol A in the development of cancer. *Gene* **2018**, *647*, 235–243. [CrossRef]
6. Di Donato, M.; Cernera, G.; Giovannelli, P.; Galasso, G.; Bilancio, A.; Migliaccio, A.; Castoria, G. Recent advances on bisphenol-A and endocrine disruptor effects on human prostate cancer. *Mol. Cell. Endocrinol.* **2017**, *457*, 35–42. [CrossRef]
7. Zhang, X.L.; Liu, N.; Weng, S.F.; Wang, H.S. Bisphenol A Increases the Migration and Invasion of Triple-Negative Breast Cancer Cells via Oestrogen-related Receptor Gamma. *Basic Clin. Pharmacol. Toxicol.* **2016**, *119*, 389–395. [CrossRef]
8. Derouiche, S.; Warnier, M.; Mariot, P.; Gosset, P.; Mauroy, B.; Bonnal, J.L.; Slomianny, C.; Delcourt, P.; Prevarskaya, N.; Roudbaraki, M. Bisphenol A stimulates human prostate cancer cell migration via remodelling of calcium signalling. *Springerplus* **2013**, *2*, 54. [CrossRef]
9. Jia, Y.; Sun, R.; Ding, X.; Cao, C.; Yang, X. Bisphenol S Triggers the Migration and Invasion of Pheochromocytoma PC12 Cells via Estrogen-Related Receptor alpha. *J. Mol. Neurosci.* **2018**, *66*, 188–196. [CrossRef]
10. Li, S.; Wang, B.; Tang, Q.; Liu, J.; Yang, X. Bisphenol A triggers proliferation and migration of laryngeal squamous cell carcinoma via GPER mediated upregulation of IL-6. *Cell Biochem. Funct.* **2017**, *35*, 209–216. [CrossRef]
11. Song, P.; Fan, K.; Tian, X.; Wen, J. Bisphenol S (BPS) triggers the migration of human non-small cell lung cancer cells via upregulation of TGF-beta. *Toxicol. Vitr.* **2019**, *54*, 224–231. [CrossRef]
12. Lewis, S.A. Everything you wanted to know about the bladder epithelium but were afraid to ask. *Am. J. Physiol. Ren. Physiol.* **2000**, *278*, F867–F874. [CrossRef]
13. Wang, Z.; Liu, H.; Liu, S. Low-Dose Bisphenol A Exposure: A Seemingly Instigating Carcinogenic Effect on Breast Cancer. *Adv. Sci.* **2017**, *4*, 1600248. [CrossRef]
14. Teng, J.; Wang, Z.Y.; Prossnitz, E.R.; Bjorling, D.E. The G protein-coupled receptor GPR30 inhibits human urothelial cell proliferation. *Endocrinology* **2008**, *149*, 4024–4034. [CrossRef]
15. Li, P.; Chen, J.; Miyamoto, H. Androgen Receptor Signaling in Bladder Cancer. *Cancers* **2017**, *9*, 20. [CrossRef]
16. Bernardo, C.; Santos, J.; Costa, C.; Tavares, A.; Amaro, T.; Marques, I.; Gouveia, M.J.; Félix, V.; Afreixo, V.; Brindley, P.J.; et al. Estrogen receptors in urogenital schistosomiasis and bladder cancer: Estrogen receptor alpha-mediated cell proliferation. *Urol. Oncol.* **2020**, *38*, 738.e723–738.e735. [CrossRef]
17. Ou, Z.; Wang, Y.; Chen, J.; Tao, L.; Zuo, L.; Sahasrabudhe, D.; Joseph, J.; Wang, L.; Yeh, S. Estrogen receptor β promotes bladder cancer growth and invasion via alteration of miR-92a/DAB2IP signals. *Exp. Mol. Med.* **2018**, *50*, 1–11. [CrossRef]
18. Ide, H.; Mizushima, T.; Jiang, G.; Goto, T.; Nagata, Y.; Teramoto, Y.; Inoue, S.; Li, Y.; Kashiwagi, E.; Baras, A.S.; et al. FOXO1 as a tumor suppressor inactivated via AR/ERβ signals in urothelial cells. *Endocr. Relat. Cancer* **2020**, *27*, 231–244. [CrossRef]
19. Pellerin, E.; Caneparo, C.; Chabaud, S.; Bolduc, S.; Pelletier, M. Endocrine-disrupting effects of bisphenols on urological cancers. *Environ. Res.* **2021**, *195*, 110485. [CrossRef]
20. Pellerin, È.; Chabaud, S.; Pouliot, F.; Pelletier, M.; Bolduc, S. Bisphenol A Alters the Energy Metabolism of Stromal Cells and Could Promote Bladder Cancer Progression. *Cancers* **2021**, *13*, 5461. [CrossRef]
21. Young, V.J.; Brown, J.K.; Maybin, J.; Saunders, P.T.; Duncan, W.C.; Horne, A.W. Transforming growth factor-β induced Warburg-like metabolic reprogramming may underpin the development of peritoneal endometriosis. *J. Clin. Endocrinol. Metab.* **2014**, *99*, 3450–3459. [CrossRef] [PubMed]
22. Niu, D.; Luo, T.; Wang, H.; Xia, Y.; Xie, Z. Lactic acid in tumor invasion. *Clin. Chim. Acta* **2021**, *522*, 61–69. [CrossRef] [PubMed]
23. Magnan, M.; Berthod, F.; Champigny, M.F.; Soucy, F.; Bolduc, S. In vitro reconstruction of a tissue-engineered endothelialized bladder from a single porcine biopsy. *J. Pediatr. Urol.* **2006**, *2*, 261–270. [CrossRef] [PubMed]
24. Cattan, V.; Bernard, G.; Rousseau, A.; Bouhout, S.; Chabaud, S.; Auger, F.A.; Bolduc, S. Mechanical stimuli-induced urothelial differentiation in a human tissue-engineered tubular genitourinary graft. *Eur. Urol.* **2011**, *60*, 1291–1298. [CrossRef]
25. Pelletier, M.; Billingham, L.K.; Ramaswamy, M.; Siegel, R.M. Extracellular flux analysis to monitor glycolytic rates and mitochondrial oxygen consumption. *Methods Enzymol.* **2014**, *542*, 125–149. [CrossRef]

26. Pellerin, F.A.; Caneparo, C.; Pellerin, È.; Chabaud, S.; Pelletier, M.; Bolduc, S. Heat-Inactivation of Fetal and Newborn Sera Did Not Impair the Expansion and Scaffold Engineering Potentials of Fibroblasts. *Bioengineering* **2021**, *8*, 184. [CrossRef] [PubMed]
27. Goulet, C.R.; Champagne, A.; Bernard, G.; Vandal, D.; Chabaud, S.; Pouliot, F.; Bolduc, S. Cancer-associated fibroblasts induce epithelial-mesenchymal transition of bladder cancer cells through paracrine IL-6 signalling. *BMC Cancer* **2019**, *19*, 137. [CrossRef]
28. Ringuette Goulet, C.; Bernard, G.; Tremblay, S.; Chabaud, S.; Bolduc, S.; Pouliot, F. Exosomes Induce Fibroblast Differentiation into Cancer-Associated Fibroblasts through TGFβ Signaling. *Mol. Cancer Res.* **2018**, *16*, 1196–1204. [CrossRef]
29. Bushnik, T.; Haines, D.; Levallois, P.; Levesque, J.; Van Oostdam, J.; Viau, C. Lead and bisphenol A concentrations in the Canadian population. *Health Rep.* **2010**, *21*, 7–18.
30. Anggorowati, N.; Ratna Kurniasari, C.; Damayanti, K.; Cahyanti, T.; Widodo, I.; Ghozali, A.; Romi, M.M.; Sari, D.C.; Arfian, N. Histochemical and Immunohistochemical Study of α-SMA, Collagen, and PCNA in Epithelial Ovarian Neoplasm. *Asian Pac. J. Cancer Prev.* **2017**, *18*, 667–671. [CrossRef]
31. Liberti, M.V.; Locasale, J.W. The Warburg Effect: How Does it Benefit Cancer Cells? *Trends Biochem. Sci.* **2016**, *41*, 211–218. [PubMed]
32. Farhana, A.; Lappin, S.L. Biochemistry, Lactate Dehydrogenase. In *StatPearls*; StatPearls Publishing LLC.: Treasure Island, FL, USA, 2022.
33. Goetze, K.; Walenta, S.; Ksiazkiewicz, M.; Kunz-Schughart, L.A.; Mueller-Klieser, W. Lactate enhances motility of tumor cells and inhibits monocyte migration and cytokine release. *Int. J. Oncol.* **2011**, *39*, 453–463. [CrossRef] [PubMed]
34. Vallo, S.; Michaelis, M.; Rothweiler, F.; Bartsch, G.; Gust, K.M.; Limbart, D.M.; Rödel, F.; Wezel, F.; Haferkamp, A.; Cinatl, J., Jr. Drug-Resistant Urothelial Cancer Cell Lines Display Diverse Sensitivity Profiles to Potential Second-Line Therapeutics. *Transl. Oncol.* **2015**, *8*, 210–216. [CrossRef] [PubMed]
35. Masters, J.R.; Hepburn, P.J.; Walker, L.; Highman, W.J.; Trejdosiewicz, L.K.; Povey, S.; Parkar, M.; Hill, B.T.; Riddle, P.R.; Franks, L.M. Tissue culture model of transitional cell carcinoma: Characterization of twenty-two human urothelial cell lines. *Cancer Res.* **1986**, *46*, 3630–3636. [PubMed]
36. Bubeník, J.; Baresová, M.; Viklický, V.; Jakoubková, J.; Sainerová, H.; Donner, J. Established cell line of urinary bladder carcinoma (T24) containing tumour-specific antigen. *Int. J. Cancer* **1973**, *11*, 765–773. [CrossRef]
37. Lobo, I. Environmental influences on gene expression. *Nat. Educ.* **2008**, *1*, 39.
38. Coradini, D.; Casarsa, C.; Oriana, S. Epithelial cell polarity and tumorigenesis: New perspectives for cancer detection and treatment. *Acta Pharmacol. Sin.* **2011**, *32*, 552–564. [CrossRef]
39. Werb, Z.; Lu, P. The Role of Stroma in Tumor Development. *Cancer J.* **2015**, *21*, 250–253. [CrossRef]
40. Quesnot, N.; Bucher, S.; Fromenty, B.; Robin, M.A. Modulation of metabolizing enzymes by bisphenol a in human and animal models. *Chem. Res. Toxicol.* **2014**, *27*, 1463–1473. [CrossRef]
41. Khan, S.; Beigh, S.; Chaudhari, B.P.; Sharma, S.; Aliul Hasan Abdi, S.; Ahmad, S.; Ahmad, F.; Parvez, S.; Raisuddin, S. Mitochondrial dysfunction induced by Bisphenol A is a factor of its hepatotoxicity in rats. *Environ. Toxicol.* **2016**, *31*, 1922–1934. [CrossRef]
42. Liu, X.; Sakai, H.; Nishigori, M.; Suyama, K.; Nawaji, T.; Ikeda, S.; Nishigouchi, M.; Okada, H.; Matsushima, A.; Nose, T.; et al. Receptor-binding affinities of bisphenol A and its next-generation analogs for human nuclear receptors. *Toxicol. Appl. Pharmacol.* **2019**, *377*, 114610. [CrossRef] [PubMed]
43. Dairkee, S.H.; Luciani-Torres, M.G.; Moore, D.H.; Goodson, W.H., 3rd. Bisphenol-A-induced inactivation of the p53 axis underlying deregulation of proliferation kinetics, and cell death in non-malignant human breast epithelial cells. *Carcinogenesis* **2013**, *34*, 703–712. [CrossRef] [PubMed]
44. Yaguchi, T. The endocrine disruptor bisphenol A promotes nuclear ERRγ translocation, facilitating cell proliferation of Grade I endometrial cancer cells via EGF-dependent and EGF-independent pathways. *Mol. Cell. Biochem.* **2019**, *452*, 41–50. [CrossRef] [PubMed]
45. Huang, W.; Zhao, C.; Zhong, H.; Zhang, S.; Xia, Y.; Cai, Z. Bisphenol S induced epigenetic and transcriptional changes in human breast cancer cell line MCF-7. *Environ. Pollut.* **2019**, *246*, 697–703. [CrossRef] [PubMed]
46. Schultz, G.S.; Chin, G.A.; Moldawer, L.; Diegelmann, R.F. Principles of Wound Healing. In *Mechanisms of Vascular Disease: A Reference Book for Vascular Specialists*; Fitridge, R., Thompson, M., Eds.; University of Adelaide Press© The Contributors 2011: Adelaide, Australia, 2011.
47. Hecht, S.S. Human urinary carcinogen metabolites: Biomarkers for investigating tobacco and cancer. *Carcinogenesis* **2002**, *23*, 907–922. [CrossRef] [PubMed]
48. Zhang, K.S.; Chen, H.Q.; Chen, Y.S.; Qiu, K.F.; Zheng, X.B.; Li, G.C.; Yang, H.D.; Wen, C.J. Bisphenol A stimulates human lung cancer cell migration via upregulation of matrix metalloproteinases by GPER/EGFR/ERK1/2 signal pathway. *Biomed. Pharmacother.* **2014**, *68*, 1037–1043. [CrossRef] [PubMed]
49. Deng, Q.; Jiang, G.; Wu, Y.; Li, J.; Liang, W.; Chen, L.; Su, Q.; Li, W.; Du, J.; Wong, C.K.C.; et al. GPER/Hippo-YAP signal is involved in Bisphenol S induced migration of triple negative breast cancer (TNBC) cells. *J. Hazard. Mater.* **2018**, *355*, 1–9. [CrossRef]
50. Kalluri, R.; Weinberg, R.A. The basics of epithelial-mesenchymal transition. *J. Clin. Investig.* **2009**, *119*, 1420–1428. [CrossRef]

51. Lee, H.W.; Park, Y.M.; Lee, S.J.; Cho, H.J.; Kim, D.H.; Lee, J.I.; Kang, M.S.; Seol, H.J.; Shim, Y.M.; Nam, D.H.; et al. Alpha-smooth muscle actin (ACTA2) is required for metastatic potential of human lung adenocarcinoma. *Clin. Cancer Res.* **2013**, *19*, 5879–5889. [CrossRef]
52. Jeon, M.; You, D.; Bae, S.Y.; Kim, S.W.; Nam, S.J.; Kim, H.H.; Kim, S.; Lee, J.E. Dimerization of EGFR and HER2 induces breast cancer cell motility through STAT1-dependent ACTA2 induction. *Oncotarget* **2017**, *8*, 50570–50581. [CrossRef]
53. Vom Saal, F.S.; Welshons, W.V. Evidence that bisphenol A (BPA) can be accurately measured without contamination in human serum and urine, and that BPA causes numerous hazards from multiple routes of exposure. *Mol. Cell. Endocrinol.* **2014**, *398*, 101–113. [CrossRef] [PubMed]
54. Ringuette Goulet, C.; Bernard, G.; Chabaud, S.; Couture, A.; Langlois, A.; Neveu, B.; Pouliot, F.; Bolduc, S. Tissue-engineered human 3D model of bladder cancer for invasion study and drug discovery. *Biomaterials* **2017**, *145*, 233–241. [CrossRef] [PubMed]
55. Genuis, S.J.; Beesoon, S.; Birkholz, D.; Lobo, R.A. Human excretion of bisphenol A: Blood, urine, and sweat (BUS) study. *J. Environ. Public Health* **2012**, *2012*, 185731. [CrossRef] [PubMed]
56. Thoene, M.; Dzika, E.; Gonkowski, S.; Wojtkiewicz, J. Bisphenol S in Food Causes Hormonal and Obesogenic Effects Comparable to or Worse than Bisphenol A: A Literature Review. *Nutrients* **2020**, *12*, 532. [CrossRef] [PubMed]

cancers

MDPI

Article

Activin B and Activin C Have Opposing Effects on Prostate Cancer Progression and Cell Growth

Karen L. Reader [1,*], Simon John-McHaffie [1], Sylvia Zellhuber-McMillan [1], Tim Jowett [2], David G. Mottershead [3], Heather E. Cunliffe [1] and Elspeth J. Gold [4]

1 Department of Pathology, University of Otago, Dunedin 9054, New Zealand
2 Department of Mathematics and Statistics, University of Otago, Dunedin 9054, New Zealand
3 School of Pharmacy and Bioengineering, Keele University, Newcastle-under-Lyme ST5 5BG, UK
4 Department of Anatomy, University of Otago, Dunedin 9054, New Zealand
* Correspondence: karen.reader@otago.ac.nz

Simple Summary: Prostate cancer is one of the leading causes of death in men. Current methods for grading and determining treatment options are not as accurate as they need to be with some men either undergoing unnecessary, life-changing treatment or not being treated and developing late-stage disease. We assessed the effect of proteins called activins on the growth of prostate cell lines and examined their expression in prostate cancer biopsy samples from patients with different grades of tumor. Activin B and activin C were shown to have increased and decreased expression, respectively, in higher grade prostate cancer and to have opposing effects on prostate cell growth and migration. Therefore, these proteins are potential markers for distinguishing aggressive from indolent prostate cancer and may also provide novel targets for prostate cancer treatment.

Abstract: Current prognostic and diagnostic tests for prostate cancer are not able to accurately distinguish between aggressive and latent cancer. Members of the transforming growth factor-β (TGFB) family are known to be important in regulating prostate cell growth and some have been shown to be dysregulated in prostate cancer. Therefore, the aims of this study were to examine expression of TGFB family members in primary prostate tumour tissue and the phenotypic effect of activins on prostate cell growth. Tissue cores of prostate adenocarcinoma and normal prostate were immuno-stained and protein expression was compared between samples with different Gleason grades. The effect of exogenous treatment with, or overexpression of, activins on prostate cell line growth and migration was examined. Activin B expression was increased in cores containing higher Gleason patterns and overexpression of activin B inhibited growth of PNT1A cells but increased growth and migration of the metastatic PC3 cells compared to empty vector controls. In contrast, activin C expression decreased in higher Gleason grades and overexpression increased growth of PNT1A cells and decreased growth of PC3 cells. In conclusion, increased activin B and decreased activin C expression is associated with increasing prostate tumor grade and therefore have potential as prognostic markers of aggressive prostate cancer.

Keywords: prostate cancer; biomarker; Gleason grade; transforming growth factor-β family; activin; INHBA; INHBB; INHBC

Citation: Reader, K.L.; John-McHaffie, S.; Zellhuber-McMillan, S.; Jowett, T.; Mottershead, D.G.; Cunliffe, H.E.; Gold, E.J. Activin B and Activin C Have Opposing Effects on Prostate Cancer Progression and Cell Growth. *Cancers* **2023**, *15*, 147. https://doi.org/10.3390/cancers15010147

Academic Editors: José I. López and Claudia Manini

Received: 1 November 2022
Revised: 21 December 2022
Accepted: 21 December 2022
Published: 27 December 2022

1. Introduction

Prostate cancer (PCa) is the second most commonly diagnosed male cancer and the fifth leading cause of cancer deaths in men [1]. Gleason scoring of prostate needle biopsies is currently the best method for determining tumor aggressiveness and metastatic potential, however it is unable to provide a definitive signal about an individual patient's prognosis. Two studies have reported that a significant proportion of patients meeting the criteria for active surveillance who chose to have a radical prostatectomy, had more advanced cancer than predicted by the Gleason score [2,3]. Prostate tumors are heterogeneous, and biopsies may omit to sample regions containing higher-grade tumour, leading to their misclassification as low-risk cancer. The morphology of the biopsy provides a snapshot of the tumor in time but cannot confirm if a low-grade tumor will remain indolent or grow rapidly. For example, Gleason pattern 3 tumors have been shown to be of the same clonal origin to adjacent Gleason pattern 4 tumors and are likely to develop into higher grade cancers [4]. Changes to protein expression can occur throughout the cancer, irrespective of the morphology or grade of the tumor, and can potentially provide information about the overall tumor biology and thus better predict the rate of tumor growth. Therefore, the identification of biomarkers that can be easily quantified in biopsy cores without the need to differentiate between different cell types and tumor grades, that can predict clinical outcome, would aid clinical assessment of progression risk. However, we have a limited understanding of the molecular changes occurring in PCa and which genes are important for driving cancer progression [5]. Recent studies have attempted to identify biomarkers that can predict PCa outcome and while some clinical tests have been developed these are yet to be proven to change clinical decision making and patient outcome [6,7]. Despite numerous large-scale genomic studies, only a small number of single nucleotide polymorphisms have been associated with clinical outcome in PCa [5].

Members of the transforming growth factor-β (TGFB) family, including ligands, receptors and signaling molecules, are known to play an important role in prostate homeostasis and are dysregulated in a number of different cancers, including PCa [8–10]. Activins are members of the TGFB family and consist of homo- or hetero-dimers of the inhibin β subunits (βA, βB, βC and βE) to form, for example, activin A (βA:βA), activin B (βB:βB) or activin AB (βA:βB). The subunit gene names are inhibin-βA (INHBA; activin A), INHBB (activin B), INHBC (activin C) and INHBE (activin E). Activins A and B initiate canonical signaling by binding to type 2 serine/threonine kinase receptors which then recruit type 1 receptors to the receptor complex at the cell surface. This initiates binding and phosphorylation of the transcription factors, mothers against decapentaplegic homolog 2 and 3 (SMAD2 and SMAD3) which, along with the co-factor SMAD4, translocate to the nucleus to initiate gene transcription [11]. Other non-canonical signaling pathways have also been shown to be involved in activin A signaling in different cell types, such as the p38 MAP kinase (MAPK) and c-Jun amino-terminal kinase (JNK) pathways. For a review of the function and roles of activins and inhibins see Namwanje, Brown [12].

There is some evidence that activins are involved in the regulation of PCa growth. Activin A (INHBA) has been shown to inhibit proliferation of LNCaP cells and both follistatin (FST) and activin C (INHBC) can block INHBA growth inhibition of these cells [13,14]. Chen et al. [15] have reported that high expression of INHBA is associated with reduced survival and in metastasis compared to corresponding primary sites in prostate cancer. However, little is known about the effect of INHBB or INHBC on the growth of normal and malignant prostate cells. We have recently reported that INHBC protein expression was increased in human PCa tissue from cases with extra-capsular spread when compared to normal prostate tissue or organ-confined cases [16]. These results need to be validated in a greater number of cases, and the expression of other TGFB family ligands, receptors and signaling molecules should be examined.

The first aim of this study was to determine if the level of protein expression of activins, their receptors and signaling molecules differ between human PCa biopsy samples with different Gleason patterns using immunohistochemistry with commercially available antibodies. A second aim was to examine the effect of recombinant INHBA, INHBB and INHBC on proliferation of normal, immortalized prostate cells (PNT1A) and PCa cell lines (DU145, LNCaP and PC3), and to compare levels of activin protein expression between these different cell lines. The third aim was to examine the effect of ectopic INHBB or INHBC overexpression on growth and migration of prostate cell lines.

2. Materials and Methods

All reagents were supplied by Thermo Fisher Scientific, Auckland, NZ unless otherwise stated.

2.1. Immunohistochemistry

Tissue microarray (TMA) slides (PR483b, PR752, PR8011a; US Biomax, Derwood, MD, USA) containing cores from human cases of primary prostate adenocarcinoma (134 cores, 88 cases) and normal prostate tissue (15 cores, 13 cases) were immuno-stained following methods described by Marino et al. [17]. The antibodies used for immunohistochemistry are listed in Table 1. Secondary antibody detection was performed using the DAKO EnVision + Dual Link System-HRP Rabbit/Mouse (DAKO, Glostrup, Denmark) with a 15 min diaminobenzidine (DAB) incubation followed by hematoxylin counterstaining. Primary antibody concentrations were optimized using sample TMA slides (US Biomax; T191) and sections of normal human prostate tissue. Negative control mouse and normal goat antibodies were used on either a TMA slide (T191), or sections of human prostate, at the highest concentration used for the corresponding species and antibody type, with either the secondary antibody used above, or polyclonal rabbit anti-goat IgG/HRP (DAKO). These slides were negative for DAB staining. Immunohistochemistry was performed for each individual antibody on all three TMA slides in a single run to allow immuno-reactive scores to be compared across TMAs. Immuno-reactive scores were semi-quantified digitally using Fiji software [18] according to the methods previously described by Helps et al. [19] and Fuhrich et al. [20]. Briefly, TMAs were imaged using an Aperio Digital Slide Scanner (Leica Biosystems, Melbourne, Australia) and individual core images thresholded (minimum) using Fiji to select only the area of the tissue. Hematoxylin and DAB staining were digitally separated using the color deconvolution plugin HDAB [21] and the DAB image thresholded (default) to select only the area stained by DAB. This was overlaid on the selected area of the tissue and the percent area of DAB stained tissue calculated by the software. Histogram analysis was carried out on the deconvoluted DAB image and the intensity of DAB staining calculated by the formula:

$$\text{DABwt\%} = \left(\frac{\text{DABwt}}{\Sigma(\text{Histogram Value Counts}) \times 255}\right) \times 100$$

Table 1. Antibodies used for immunohistochemistry.

Antibody	Supplier	Cat. No	µg/mL or Dilution	Species & Clonality
INHBA	Abcam	ab56057	20	Rabbit polyclonal
INHBB	R & D Systems	MAB659	2.5	Mouse monoclonal
INHBC	Abcam	ab73904	1	Mouse monoclonal
INHA	Abcam	ab81234	0.2	Rabbit polyclonal
ACVR2A	Abcam	ab10595	5	Goat polyclonal
ACVR2B	Abcam	ab76940	12.5	Mouse monoclonal
SMAD2	Abcam	ab47083	5	Rabbit polyclonal
SMAD3	Abcam	ab40854	14	Rabbit monoclonal
FST	Abcam	ab157471	8	Rabbit monoclonal
BMP4	Abcam	ab39973	1	Rabbit polyclonal
AR	Abcam	ab3509	1:300	Rabbit polyclonal
MKI67	Abcam	ab66155	2.5	Rabbit polyclonal
MYC	Abcam	ab32	2.5	Mouse monoclonal
TP53	Abcam	ab26	5	Mouse monoclonal
BCL2	Abcam	ab7973	0.6	Rabbit polyclonal
Negative mouse	DAKO	X0931	12.5	Mouse monoclonal
Normal goat	Santa Cruz	Sc-2028	5	Goat IgG

Abcam (Cambridge, UK); R & D Systems (Minneapolis, MN, USA); DAKO (Santa Clara, CA, USA); Santa Cruz (Dallas, TX, USA).

Immuno-reactive scores (IRS) were calculated for each TMA core by multiplying the % area of DAB staining by the DABwt%. A TMA slide stained for INHBC was also manually scored by two experienced, blinded, independent observers as described by

Marino et al. [22] and the IRS compared between the two methods. Results from the manual and digital scoring methods were significantly correlated with a Pearson R^2 of 0.7227 and $p < 0.001$.

2.2. Online Analysis of Activin mRNA Expression

An online analysis of publicly available PCa patient datasets was performed with cumulated data from Oncomine (Compendia Bioscience https://www.oncomine.org/resource/main.html accessed on 19 May 2021). A FireBrowse analysis of PCa patient datasets from The Cancer Genome Atlas (TCGA; data version 2016-01-28) that compared gene expression between normal and PCa samples was also performed. Presented data is modified from Oncomine or FireBrowse. The FireBrowse analysis did not provide statistical significance values.

2.3. Recombinant INHBC Expression and Purification

Commercially available recombinant INHBC, consisting of the mature region only, did not alter PCa cell proliferation in our assays, whereas full-length, unpurified INHBC had been previously shown to alter the effect of INHBA on LNCaP cell proliferation [13]. Therefore, we produced full length, purified, recombinant INHBC to examine the effect of INHBC on PCa cell proliferation. An expression cassette was synthesized encoding the wild-type sequence of human *INHBC* modified to incorporate the rat serum albumin signal sequence at the N-terminal end, followed by a His8 affinity tag for purification and a Strep II tag to allow detection of the proregion (GenScript, Piscataway, NJ, USA). The cDNA fragment was cloned into the pEFIRESp expression vector [23] and the plasmid transfected into CHO-K1 cells using Lipofectamine 3000 following the manufacturer's instructions. Stable cell lines (polyclonal) were established using puromycin selection (Sigma, Auckland, New Zealand). Cells were grown to near confluence and the growth media replaced with serum-free production media consisting of DMEM with 0.1 mg/mL BSA (Sigma), 2 mM GlutaMAX, 100 IU/mL penicillin and 100 µg/mL streptomycin, and cultured for 48 h. Production media was harvested, and INHBC protein purified using HisPur NiNTA resin according to the methods described by Mottershead et al. [24]. The presence of the imidazole elution buffer was found to increase growth of LNCaP cells and inhibit growth of the other prostate cell lines and was therefore removed by dialyzing the INHBC preparation in three changes of Dulbecco's PBS using a Slide-A-Lyzer MINI Dialysis Device (3.5K MWCO). SDS-PAGE and Western blotting was performed on purified INHBC under reducing conditions, according to the methods described by Marino et al. [17]. Intercept blocking buffer (Millenium Science, Auckland, New Zealand) was used for the blocking steps and dilution of primary and secondary antibodies. Nitrocellulose membranes were incubated overnight with a 1:1000 dilution of INHBC antibody (Abcam ab73904; 0.5 µg/mL). Secondary antibody detection was performed using IRDye 800CW goat anti-mouse IgG1 (Li-Cor, Millennium Science, Auckland, New Zealand) and imaged with an Odyssey Classic scanner and Image Studio Lite software. The concentration of the purified INHBC mature region monomer was estimated by Western blot analysis relative to known amounts of recombinant human INHBC (R&D Systems, Minneapolis, MN, USA).

2.4. Prostate Cell Lines

Four human prostate cell lines were used in these studies: PNT1A (Sigma; 95012614), LNCaP (ATCC; CRL-1740), DU145 (ATCC; HTB-81) and PC3 (ATCC; CRL-1435). PNT1A cells are immortalized prostatic epithelial cells that represent normal prostate epithelium [25]. LNCaP, DU145 and PC3 cells originate from prostate carcinoma metastases from lymph node, brain and bone, respectively. All PCa cell lines in our laboratory have been recently validated using STR profiling. All cells were maintained in DMEM with 10% fetal bovine serum (FBS) (Moregate Biotech, Hamilton, New Zealand), 2 mM GlutaMAX-1,

100 U/mL penicillin and 100 µg/mL streptomycin and incubated at 37 °C in a humidified atmosphere of 5% CO_2 in air. FBS was heat-inactivated for 30 min at 56 °C.

2.5. Stable Transfection of Prostate Cell Lines with INHBB and INHBC

The human INHBB pCMV3-SP-N-FLAG expression plasmid (Cat# HG10814-NF; Sino Biological, Beijing, China), the pCMV3-SP-N-FLAG negative control vector (Cat# CV020; Sino Biological), the modified human INHBC pEFIRES expression plasmid (as above) or the pEFIRES negative control vector (Cat#1849; GenScript) were transfected into PNT1A and PC3 cells. Prior to transfection, plasmid stocks were prepared by transforming the DNAs into chemically competent DH5alpha *E. coli* prepared using the Inoue method [26], and plasmid minipreps were validated by restriction endonuclease digestion. Bulk DNA was then isolated for transfection using the Qiagen endotoxin-free Plasmid Maxi Kit (Cat#12362; Qiagen, Auckland, New Zealand). Prostate cells were seeded into 6-well plates with 300,000 PC3 cells or 600,000 PNT1A cells per well, and allowed to attach overnight. During transfection, growth media was replaced with Opti-MEM 30 min before transfection. Plasmid DNA (5 µg) mixed with Lipofectamine 3000 (3.75 µL, Invitrogen) was added and cells were incubated for 4 h at 37 °C in 5% CO_2 in air. After incubation, the Opti-MEM and DNA/Lipofectamine mix was replaced with growth media containing 20% FBS and incubated at 37 °C, 5% CO_2 for 48 h. After the 48 h growth period, stably transfected cells were selected for using hygromycin (pCMV3) or puromycin (pEFIRES) following optimization of the concentration for each antibiotic and cell line to give 100% cell death within 10 days.

2.6. Western Blot Quantitation of Activins in Prostate Cell Lines

Protein isolation was performed as described by Marino et al. [17] from up to three separate passages of each cell line (PNT1A, LNCaP, PC3, cell lines overexpressing INHBB or INHBC, and empty vector controls) and quantified using the Pierce BCA Protein Assay Kit (Life Technologies, Auckland, New Zealand). For each cell line, 60 µg of protein from each cell passage was separated on a 12% SDS-PAGE acrylamide gel, under reducing conditions, and Western blotting analysis performed as described above. This was repeated in three separate Western blots for each cell lysate sample. Proteins were detected with the primary antibodies and Li-Cor secondary antibodies listed in Table 2. Relative signal intensities were compared between cell lines for each protein, normalized to GAPDH, using the Odyssey infra-red imaging system and Image Studio Lite software (Li-Cor).

Table 2. Antibodies used for Western blots.

Antibody	Supplier	Catalogue No	Dilution	Species & Clonality
INHBA (pro-region)	Abcam	ab128958	1:1000	Rabbit Monoclonal IgG1
INHBB (mature)	Abcam	ab128944	1:2000	Rabbit polyclonal IgG
INHBC (mature)	Abcam	ab73904	1:1000	Mouse monoclonal IgG1
GAPDH	Abcam	ab9484	1:6000	Mouse monoclonal IgG2b
GAPDH	Abcam	ab181602	1:6000	Rabbit monoclonal
Anti-rabbit	LI-COR	C20322-01	1:10,000	800 CW
Anti-rabbit	LI-COR	C10628-01	1:25,000	680 LT
Anti-mouse	LI-COR	926-32350	1:10,000	IgG1 800 CW
Anti-mouse	LI-COR	926-68052	1:25,000	IgG2b 680 LT

2.7. Cell Growth Assays

The effect of recombinant activins, signaling pathway inhibitors and overexpression of activins on the growth of human prostate cell lines was tested using an MTS colourimetric cell proliferation assay (CellTiter 96). Cells were cultured for 24 h in 96-well plates in DMEM with 5% FBS with 5000 cells per well. Media was then replaced with either DMEM with 2% FBS (control) or the same medium containing recombinant human INHBC (either our full-length INHBC or mature region from R&D Systems; 1629-AC), human INHBA (mature region, R&D systems; 338-AC), human INHBB (mature region, R&D systems; 659-AB),

SB431542 ALK4/5/7 inhibitor (Tocris, In Vitro Technologies, Auckland, New Zealand), LY294002 PI3K/AKT inhibitor (Tocris), SB202190 MAPK inhibitor (Tocris), LDN193181 ALK2/3 inhibitor (Selleckchem #S2618) or DMSO vehicle control. Cells were incubated for either 72 h (LNCaP, PC3) or 144 h (PNT1A, DU145) to allow for the different growth rates of the cell lines. Treatments were replicated in three or six wells using a minimum of three different cell passages for each cell line. At the end of the culture period, 20 μL Promega CellTiter 96 AQueous One solution (In Vitro Technologies, Auckland, New Zealand) was added to each well and incubated for 1 h at 37 °C and the absorbance read at a wavelength of 490 nm in a VICTOR X4 multimode plate reader (PerkinElmer, Buckinghamshire, UK).

2.8. Migration Assays

To determine if overexpression of INHBB or INHBC altered cell migration, cells were seeded in 8 μm pore transwell permeable inserts (Corning, #3428) in 24-well companion plates (Falcon, 353504). Cells were seeded at either 25,000 PNT1A or 50,000 PC3 cells per insert and were cultured for 24 h in DMEM with 2% FBS. After incubation, inserts were washed with PBS and cells remaining on the top of the insert were removed. Migrated cells were fixed using 100% methanol for 25 min, stained overnight with 0.5% (*w*/*v*) crystal violet, washed and air-dried. Five separate fields of view per insert were imaged at 20× magnification using a DMi1 microscope with Leica Application Suite, version 3.2.0 (Leica Microsystems, Balgach, Switzerland). Cell counts were taken as the total of all stained cells in five fields of view, and were performed manually. Three independent experiments were performed using separate cell passages with triplicate inserts.

2.9. Statistical Analysis

All statistical analysis was carried out using R software [27]. Three different general methods were used. A General linear model (GLM) assuming a negative binomial response distribution was used to fit the Western blot normalised fluorescent intensity data for the different PCa cell lines. This model was able to accommodate the tendency for the variance of the fluorescence values to increase with increasing mean. Mean fluorescence was modelled using a linear function of the main effects, cell line and activin. The interaction effect between cell line and activin was excluded from the final model because it did not significantly improve the main effects only model. The model was fitted using the *glm.nb* function from the *MASS* package [28].

Generalised least squares (GLS) models were fitted to the immunohistochemistry IRS data and the MTS absorbance data for the modified PC3 cells treated with ALK2/3 inhibitor. The GLS method was used to account for differences in measurement variance across the different treatments. For the IRS data, both the mean and variance was modelled as a function of the "Highest Gleason" treatment effect. The models were fitted using the *gls* function from the *nlme* package [29].

The normal linear model (NLM) was used to fit models to absorbance data from the cell proliferation and migration assays. In all of these models, in addition to the treatment effect of interest, the date of testing was included as a predictor variable to control for any variance in experiment conditions or cell pools across the different dates. The models were fitted using the *lm* function [27]

A NLM with fluorescence fold-change response was found to be the most efficient way to model the Western blot data from the transfected cell lines. Individual fold-change values were calculated by dividing the INHBB or INHBC transfected cell line fluorescent intensities by the empty vector control values for each Western blot date. This approach helped to control for the variance in Western blot intensities across the different dates. The model was fitted using the *lm* function [27].

The validity of the underlying assumptions of all final models were tested using residual diagnostics and were found to be acceptable. All post hoc statistical testing (including adjustment for multiple comparisons) and means estimation was preformed using the R package *emmeans* [30].

3. Results

3.1. Immuno-Reactive Scores in Relation to Gleason Score

Gleason scores were provided for each adenocarcinoma patient core by US Biomax and the IRS were analysed relative to the highest Gleason pattern present in each core. Gleason patterns 1 and 2 were combined into one group due to the low number of cases. Graphs of each protein showing the IRS for each core and Gleason pattern are presented in Figure 1. Representative images of immunohistochemical staining for each protein in normal and high-grade PCa tissue cores are presented in Supplementary Figure S1.

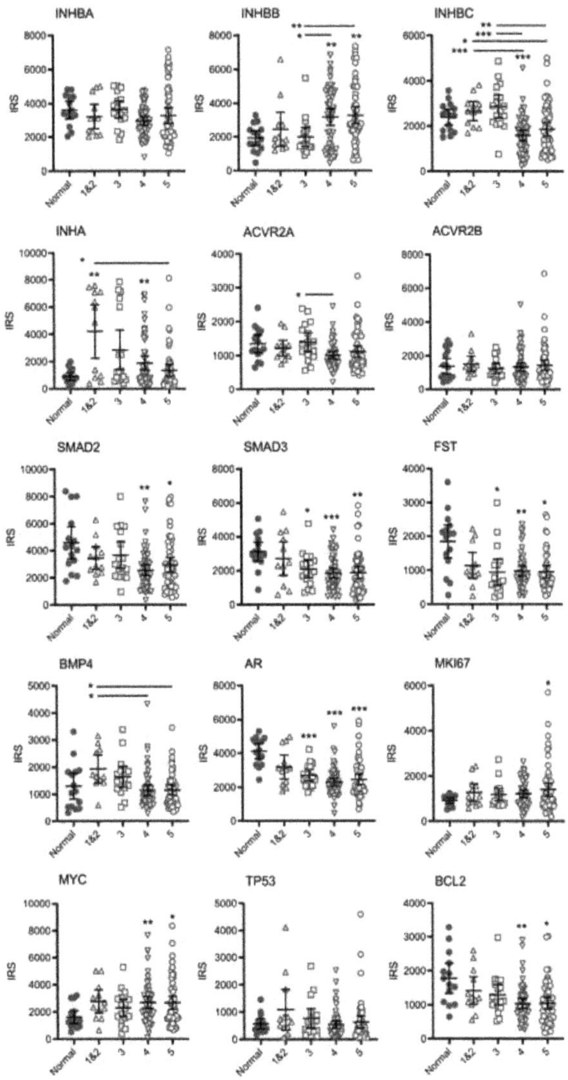

Figure 1. Graphs of IRS for individual TMA cores immunostained for TGFB family and other proteins relative to the highest Gleason pattern present. Bars represent mean +/− 95% confidence limits. Asterisks alone indicate significant differences compared to normal tissue and asterisks beside horizontal lines above bars indicate differences between Gleason patterns. * $p < 0.05$, ** $p < 0.01$, *** $p < 0.001$.

There were no differences in mean IRS for INHBA, however, expression varied greatly in the Gleason pattern 5 cases with a cluster of cores with higher IRS. For INHBB the mean IRS for pattern 4 and pattern 5 cases was greater than the normal and pattern 3 cases. The IRS in the higher-grade cores was variable and appeared to separate into clusters of either high or low IRS. In contrast, the mean IRS for INHBC Gleason pattern 4 was significantly lower than normal tissue, while both patterns 4 and 5 were lower than patterns 1&2 and pattern 3. The mean IRS for INHA (inhibin α) in Gleason pattern 1&2, 3 ($p = 0.0504$) and 4 tumors was higher than normal tissue. However, the IRS were variable and again, the cases segregated into clusters of high or low scores. There was only one significantly different group in the activin receptor 2A (ACVR2A) stained cases (pattern 4 less than pattern 3) and no differences between groups for ACVR2B indicating that receptor expression does not change with increasing Gleason pattern. Expression of SMAD2, SMAD3, follistatin (FST), androgen receptor (AR) and BCL2 decreased while MYC and marker of proliferation Ki-67 (MKI67) expression increased in the higher Gleason patterns compared to normal tissue with no differences between Gleason groups. Bone morphogenetic protein 4 (BMP4) expression appeared to increase in pattern 1&2 (not significant) compared to normal tissue and then decrease again in patterns 4 and 5. Tumour protein p53 (TP53) expression did not change with increasing Gleason pattern.

Spearman correlation coefficient analysis showed significant positive correlation in IRS between most of the proteins examined (data not shown). There was a general pattern of an increase in IRS for INHBB and a decrease in INHBC with increasing Gleason pattern and these were the only two proteins to have a negative Spearman correlation coefficient r-value (-0.253, $p = 0.014$).

3.2. INHBB and INHBC RNA Expression in Prostate Tissue

To determine if the differences in INHBB and INHBC protein expression observed between tumours with different Gleason patterns corresponded to differences in RNA expression in PCa, the Oncomine cancer microarray database was explored. Three studies were identified that compared INHBB gene expression in normal and pathological prostate tissue cases [31–33]. The Yu dataset showed a 1.59-fold increase ($p < 0.0001$) in INHBB expression in prostate carcinoma compared to healthy prostate (Figure 2A). The Luo dataset (Figure 2B) also showed increased INHBB expression (fold change = 1.630) in prostate carcinoma when compared to benign prostatic hyperplasia (BPH; $p = 0.024$). The Tomlins dataset showed a 1.60-fold increase ($p < 0.001$) in INHBB expression in prostatic intraepithelial neoplasia (PIN) compared to the healthy prostate ($n = 25$) (Figure 2C).

Oncomine data for the expression of INHBC in the Liu Prostate dataset [34] showed elevated INHBC (fold change of 1.16; $p = 0.003$) expression in PCa when compared to the healthy prostate (Figure 2D). Oncomine analysis of the TCGA prostate dataset version 2012_10_12 also showed a significant increase ($p < 0.0001$) in INHBC expression compared to the prostate gland (Figure 2E); however, the expression change is minimal with a fold change of 1.02.

Analysis of RNA-Seq data for INHBB using FireBrowse showed a 1.33-fold increase in INHBB expression in tumour samples compared to normal (Figure 2F). In contrast, the FireBrowse INHBC analysis showed a 0.388-fold decrease in INHBC expression in tumour samples compared to normal prostate samples (Figure 2G).

3.3. Prostate Cell Lines Express Different Levels of INHBA, INHBB and INHBC

Western blot analysis for INHBA in the prostate cell lines showed a single band at approximately 45 kDa (Figure 3A). This antibody recognises the pro-region of the protein. LNCaP and PC3 cells had significantly higher expression of INHBA than PNT1A cells (Figure 3D).

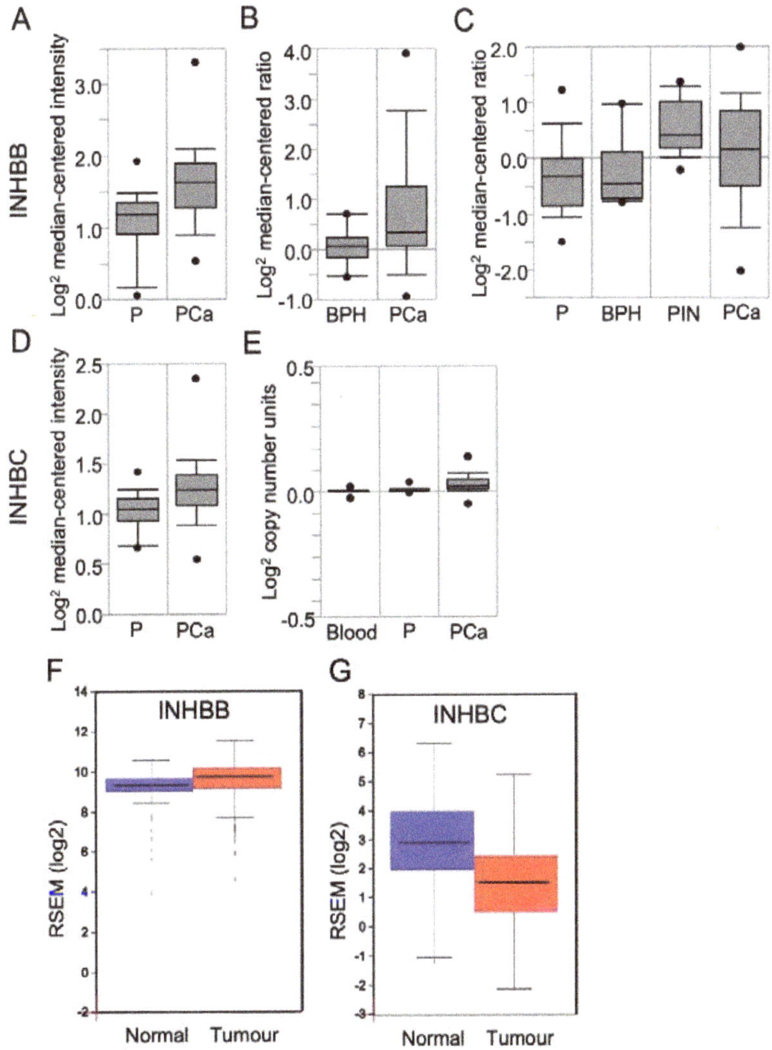

Figure 2. Box plots derived from gene expression data in Oncomine or FireBrowse, comparing results for human prostate expression of INHBB and INHBC. INHBB expression was shown in data sets from (**A**) Yu Prostate, comparing the prostate gland (P; n = 23), and prostate carcinoma (PCa; n = 65); (**B**) Luo Prostate, comparing benign prostatic hyperplasia (BPH; n = 9) and PCa (n = 16); (**C**) Tomlins Prostate data, comparing the normal prostate (P; n = 25), BPH (n = 10), prostatic intraepithelial neoplasia (PIN; n = 13), and PCa (n = 46). INHBC expression was shown in data sets from (**D**) Liu Prostate data, comparing prostate gland samples (P; n = 13) and PCa (n = 44); (**E**) The Cancer Genome Atlas (TCGA) comparing blood (n = 148), the prostate gland (P; n = 61) and PCa (n = 45). (**F**) FireBrowse RNASeq data comparing INHBB expression in normal prostate samples (blue; n = 52) and tumour samples (red; n = 498). (**G**) FireBrowse RNASeq data comparing INHBC expression in normal prostate samples (blue; n = 45) and tumour samples (red; n = 457). Figures are presented as box plots. The box midline is the median, top and bottom of the box represent 75th and 25th percentiles, and top and bottom whiskers represent 90th and 10th percentiles.

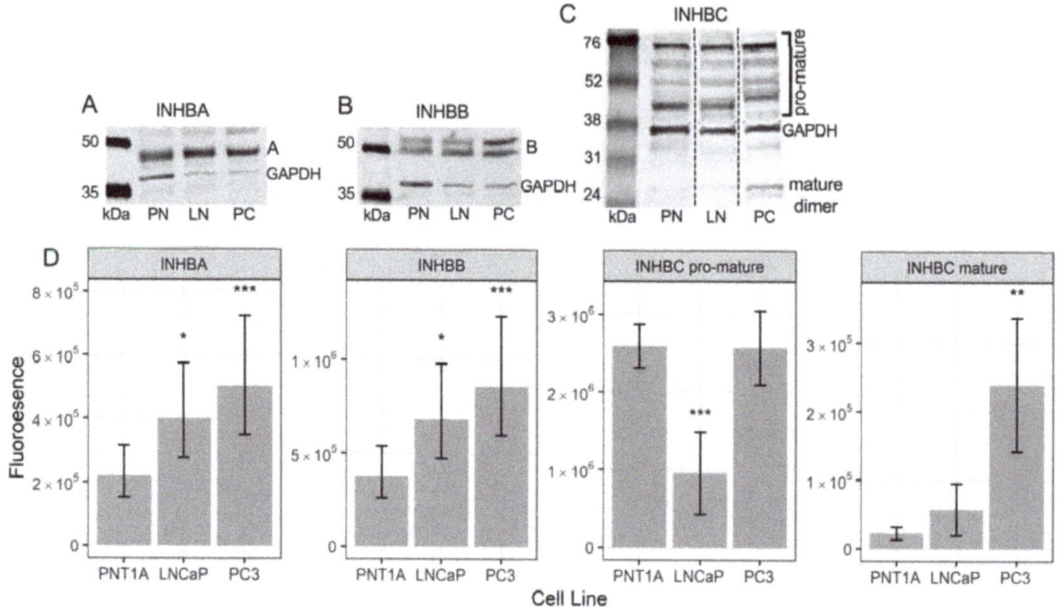

Figure 3. Representative Western blots of (**A**) INHBA, (**B**) INHBB, and (**C**) INHBC protein in PNT1A (PN), LNCaP (LN) and PC3 (PC) cells with GAPDH loading control. (**D**) Graphs showing relative fluorescent intensity of INHBA, INHBB, INHBC pro-mature multimers and INHBC mature dimer normalised to GAPDH in PNT1A, LNCaP and PC3 cells. Data is presented as the estimated marginal mean (emmean) ± 95% confidence limits. * $p < 0.05$, ** $p < 0.01$ and *** $p < 0.001$ relative to PNT1A cells from a General Linear Model analysis. Protein was isolated from one passage of each cell line for INHBA and INHBB and three separate passages for INHBC and quantified in three Western blots for each protein.

The INHBB antibody used for Western blotting was made to a c-terminal (mature region) sequence. Western blotting for INHBB in protein extractions from the three prostate cell lines showed bands at approximately 50 and 55 kDa (Figure 3B) with weak bands at 12 kDa (not shown). Both LNCaP and PC3 cells exhibited significantly higher expression of INHBB when compared to PNT1A cells (Figure 3D).

The INHBC antibody recognises the mature region of the protein which was present in the Western blot as multimers of the pro- and mature regions of INHBC, between 45 and 75 kDa, and dimers of the mature region at approximately 25 kDa (Figure 3C). Bands of INHBC mature region monomer were visible in some blots at around 12 kDa but were very weak relative to the mature region dimer and higher molecular weight bands. PNT1A and PC3 cells expressed similar levels of INHBC pro-mature multimers and significantly more than the LNCaP cells while PC3 cells expressed more INHBC mature dimer than both PNT1A and LNCaP cells (Figure 3D).

3.4. Recombinant INHBC

Western blot analysis of our purified full-length recombinant INHBC, separated under reducing conditions, showed a band at approximately 12 kDa, the expected size of the mature chain monomer of INHBC (Supplementary Figure S2). A larger 55 kDa band was also present which is likely to consist of the unprocessed full-length protein, i.e., pro-region and mature region. The R&D Systems INHBC exhibited mature region monomer (12 kDa) and dimer (24 kDa) only and no pro-region as expected.

3.5. Effect of Recombinant Activins on Prostate Cell Growth

Recombinant INHBA inhibited growth of PNT1A, LNCaP and PC3 cells but not DU145 cells (Figure 4). INHBB inhibited growth of PNT1A and LNCaP cells but not PC3 or DU145 cells (Figure 4). In contrast, INHBC had no effect on growth of any of the cell lines either alone, or in combination with INHBA or INHBB, except for a trend towards an increase in absorbance when the combination of INHBA and INHBC was compared to INHBA alone in PC3 cells ($p = 0.0697$).

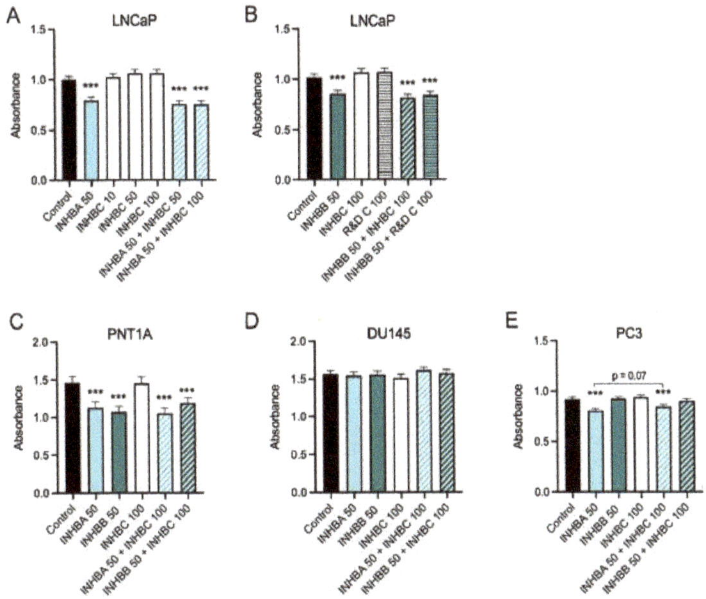

Figure 4. (**A–E**) MTS cell growth assays for prostate cell lines treated with exogenous recombinant INHBA, INHBB, INHBC, or INHBA or INHBB combined with INHBC (ng/mL). Data are presented as the emmean absorbance ± 95% confidence limits from a minimum of three separate biological replicates with three to six technical replicates per treatment per assay. Asterisks indicate significant differences between treated and untreated (control) cells *** $p < 0.001$ analysed by a NLM.

3.6. Overexpression of INHBB or INHBC Alters Growth and Migration of PCa Cell Lines

Western blot analysis showed expression of INHBB was higher in PC3 cell line engineered to stably overexpress INHBB when compared to the empty vector control cells (Figure 5A,C). However, the mean fold-change was not statistically different ($p = 0.1882$). INHBC was also overexpressed in PNT1A ($p = 0.0018$) and PC3 cells ($p = 0.0001$) when compared to empty vector controls (Figure 5B,D). PNT1A cells transfected with INHBB grew very slowly compared to those transfected with the negative control plasmid and were unable to be grown in sufficient numbers for further experiments, suggesting INHBB overexpression inhibits growth of or is genotoxic to these cells. The overexpression of INHBB in the PC3 cell line resulted in a robust increase in cell growth ($p < 0.0001$) and migration ($p < 0.0001$) when compared to the empty vector controls (Figure 6A,D). PNT1A cells overexpressing INHBC showed increased cell growth ($p < 0.001$) and no change in cell migration when compared to the empty vector controls (Figure 6B,E). In contrast, PC3 cells overexpressing INHBC showed decreased cell growth ($p < 0.05$) and increased cell migration ($p < 0.0001$) when compared to the empty vector controls (Figure 6C,F).

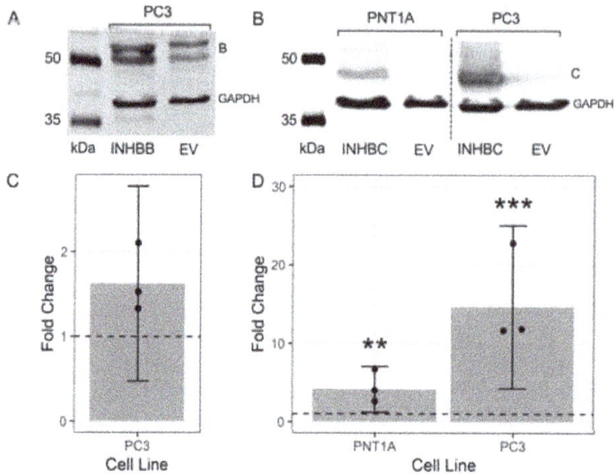

Figure 5. Representative Western blots showing the overexpression of (**A**) INHBB in transfected PC3 cells and (**B**) INHBC in transfected PNT1A and PC3 cell lines. Graphs show fold-change of the fluorescent intensity relative to the EV control cells (dashed line at 1.0) for (**C**) INHBB and (**D**) INHBC. Data presented as individual fold-change values (dots) and the emmean fold-change ± 95% confidence limits (grey bar and lines) from 3 Western blots. ** $p < 0.01$ and *** $p < 0.001$ analysed by a NLM.

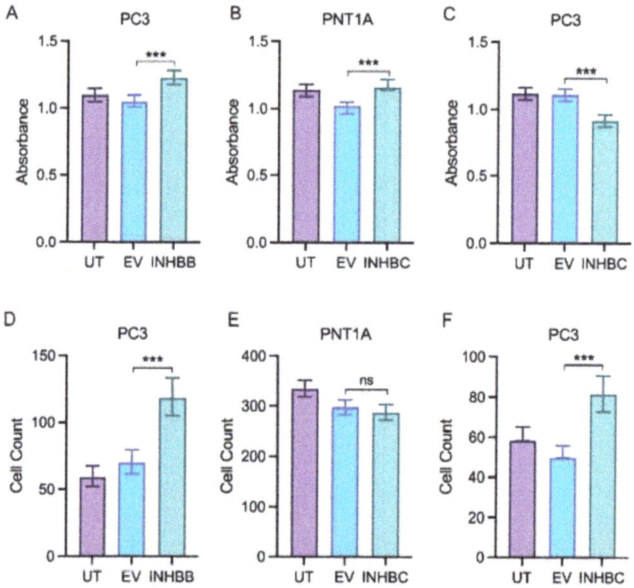

Figure 6. In vitro proliferation and migration assays for prostate cell lines either untransfected (UT), transfected with empty vector control (EV) or overexpressing INHBB or INHBC. MTS growth assay for (**A**) PC3 cells overexpressing INHBB, (**B**) PNT1A cells overexpressing INHBC and (**C**) PC3 cells overexpressing INHBC. Migration assays for (**D**) PC3 cells overexpressing INHBB, (**E**) PNT1A cells overexpressing INHBC and (**F**) PC3 cells overexpressing INHBC. Data are presented as estimated mean absorbance or cell count ± 95% confidence limits from a minimum of three separate passages with triplicate wells per assay. *** $p < 0.001$ analysed by NLM. ns: Insignificance is denoted by ns.

3.7. Response to Pathway Inhibitors Varies between Prostate Cell Lines

While the ALK4/5/7 (Smad2/3) pathway is reported to be the canonical signaling pathway for INHBA and INHBB, other pathways have also been shown to be activated by these ligands. Therefore, dose response growth assays were performed to determine the effect of the ALK4/5/7, p38 MAPK, PI3K/AKT and ALK2/3 pathway inhibitors on PNT1A, LNCaP and PC3 cell lines (Figure 7). Inhibition of ALK4/5/7 had no effect on PNT1A cell growth, but decreased LNCaP and increased PC3 cell growth at 5 μM and 10 μM. The p38 MAPK inhibitor increased growth in both PNT1A and PC3 cells with a greater effect on the PC3 cells. In contrast, inhibition of this pathway decreased growth of LNCaP cells. All three cell lines showed decreased growth in response to the PI3K/AKT inhibitor but the PC3 cells were less sensitive and only responded to the 10 μM dose ($p < 0.05$). The ALK2/3 pathway inhibitor was the only one that consistently decreased growth at the lowest concentrations across all three of the cell lines (Figure 7). There was no significant effect of the DMSO vehicle control at any of the concentrations tested in any cell line.

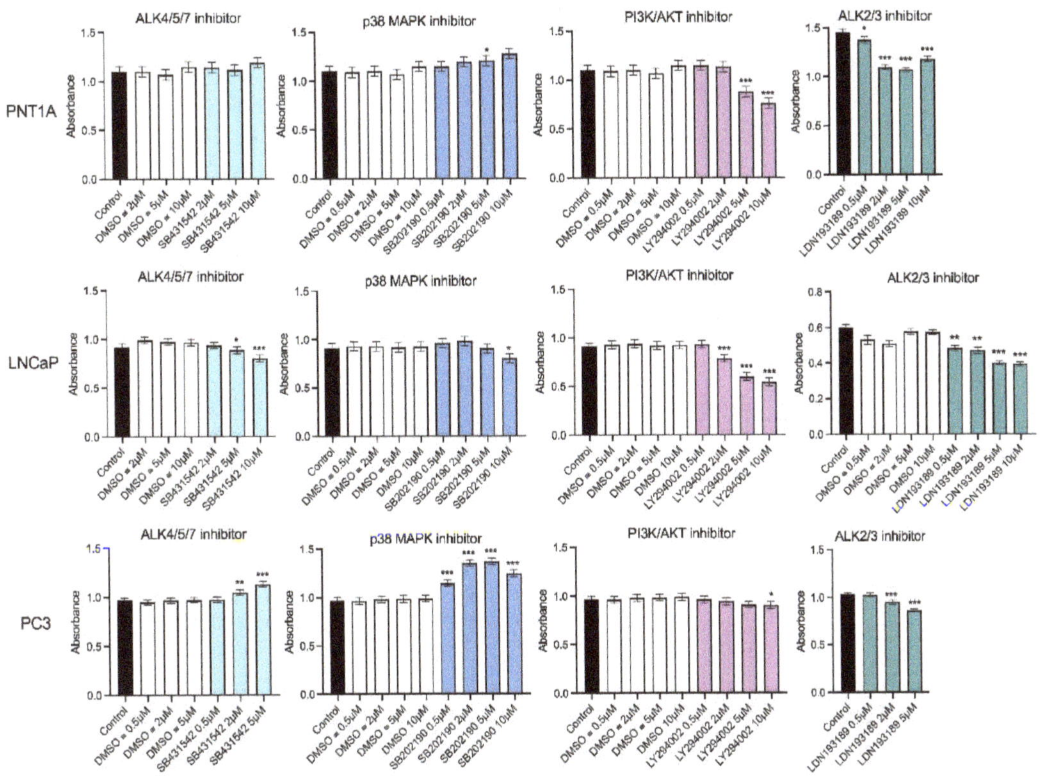

Figure 7. MTS growth assays for prostate cell lines treated with DMSO vehicle control or small molecule pathway inhibitors of ALK4/5/7 (SB431542), p38 MAPK (SB202190), PI3K/AKT (LY294002) and ALK2/3 (LDN193189). The DMSO was used at the equivalent dilution for each inhibitor dose. Data are presented as estimated mean absorbance ± 95% confidence limits from a minimum of three separate passages with three to six wells per assay. * $p < 0.05$; ** $p < 0.01$; *** $p < 0.001$ relative to the DMSO control, analysed by a NLM.

3.8. Summary of Results

Table 3 presents a summary of the different expression levels and effects of activins on prostate cell lines and tissue.

Table 3. Summary of INHBA, INHBB, INHBC expression and function in prostate cells.

Prostate Tissue or Cell Line	Activin Expression			Exogenous Treatment			Overexpression			
				Growth			Growth		Migration	
	A	B	C-Mat	A	B	C	B	C	B	C
Normal	mod	mod	mod							
Gleason 4 & 5	mod	↑	↓							
PNT1A	low	low	low	↓	↓	=	↓	↑	-	=
LNCaP	↑	↑	low	↓	↓	=	-	-	-	-
PC3	↑	↑↑	↑	↓	=	=	↑	↓	↑	↑

A (INHBA); B (INHBB); C-mat (INHBC mature region); C (INHBC); ↑ (increased); ↓ (decreased); mod (moderate); = (no change); - (not done).

4. Discussion

This study has demonstrated for the first time that INHBB and INHBC appear to have opposing roles in regulating PCa cell growth and migration. INHBB protein and RNA expression was increased in tumors with higher Gleason patterns or in PCa compared to normal prostate tissue. The PCa cell lines LNCaP and PC3 also had higher expressin of INHBB protein relative to immortalized prostate epithelial cells (PNT1A). In contrast INHBC protein expression decreased in higher-grade tumors and RNA expression appeared lower in tumor samples compared to normal prostate tissue in a relatively large data set of RNAseq data from the TCGA.

In the absence other clinical data, such as patient survival, we analyzed the IRS of the TMA cores in relation to the highest Gleason pattern present. The morphological changes associated with increasing Gleason pattern represent cell and tissue level changes associated with more advanced cancer. While Gleason scores and the new ISUP grading system are good predictors of survival they still cannot accurately predict which tumors are faster growing or more invasive [35,36]. Tumor heterogeneity is high in PCa and this can be seen in the varied IRS scores across the different proteins and cases presented here. Therefore, any variation in IRS in individual cores relative to normal prostate tissue could indicate a protein is involved in tumor progression irrespective of the Gleason score assigned to the tumor biopsy, and these proteins warrant further study.

There were many cases with significantly increased IRS for INHBB in cores containing Gleason patterns 4 and 5 (Figure 1) and these tumors likely contain more cells undergoing rapid proliferation or epithelial-mesenchymal transition. Our current results indicate that INHBB could be used as a positive biomarker for distinguishing between latent and aggressive PCa tumors. Particularly if combined with decreasing INHBC expression as the IRS for these two proteins were negatively correlated and both changed when Gleason pattern 4 was present, compared to Gleason pattern 3 (Figure 1).

These results appear to differ to our previously published data that showed INHBC expression was higher, and INHBB expression was lower in tumors that had undergone extracapsular spread compared to normal tissue [16,37]. A digital scoring method was employed in the current study which removed the inter/intra-observer error associated with manual scoring methods used in earlier studies and an increased number of cases were assessed. In the previous studies, extracapsular spread was determined from the TNM grading for the patients which do not necessarily correspond to the Gleason scores of the individual biopsy cores on the TMA slide, due to heterogeneity within the tumor. Staining intensity generally differs between stromal and epithelial compartments and thus scores can vary depending on the composition of the core. Our current study has not differentiated between cell types within each biopsy core in order to identify biomarkers that could form the basis of a simple, robust test that does not require expert pathological examination to distinguish and score the different cell compartments.

INHBA has been more extensively studied than the other activins and Kang et al. [38] reported increased INHBA protein and mRNA in primary PCa in patients who developed bone metastasis compared to those who did not develop bone metastasis. Chen et al. [15] have also shown expression of INHBA is elevated in PCa metastases and correlates with poorer survival. We did not observe any differences in INHBA expression, nor the activin receptors, ACVR2A (except for a slight decrease in IRS between patterns 3 and 4) and ACVR2B, between Gleason patterns (Figure 1) but PC3 cells expressed higher levels of INHBA than the immortalized epithelial PNT1A cells (Figure 3D). The mean IRS for inhibin-α (INHA) was higher in Gleason patterns 1 and 2 relative to normal tissue and then decreased again in higher Gleason patterns (Figure 1). INHA may be an early marker of PCa, although each pattern had some cases with high INHA IRS. In a study by Balanathan et al. [39], INHA staining intensity was higher in the stroma and benign regions of PCa tumors with extracapsular spread, but not the epithelial cancer regions, compared to organ confined cancer. There were no obvious changes to stromal INHA staining in our study, however, as mentioned previously, we did not score the different cell types separately. Similarly, BMP4 appears to have a transitional increase in expression in tumors exhibiting only Gleason patterns 1 or 2, followed by a decline in expression with patterns 4 and 5 (Figure 1). Whether the later stage tumors that continue to express higher levels of BMP4 are growing faster or slower is unknown but an increase in expression of this protein may be an early indicator of carcinoma. The remainder of the TGFB family proteins we have examined, SMAD2, SMAD3 and FST all decreased with increasing Gleason patterns compared to normal tissue but did not differ between patterns (Figure 1). Therefore, they would not be useful markers for differentiating between indolent and aggressive PCa. Shipitsin et al. [6] reported a negative correlation between SMAD2 expression and PCa lethality while Lu et al. [40] showed no correlation between SMAD2 and Gleason score but an increased expression of SMAD3 with increasing Gleason score. The loss of FST expression, an inhibitor of INHBA and INHBB may allow increased signaling by these proteins [41]. However, the concurrent loss of the SMAD signaling molecules that INHBA and INHBB have been shown to act through would negate this. Overall, our data highlight the important role of the TGFB family in maintaining normal prostate function as expression of many of these proteins was decreased in prostate tumors compared to normal prostate tissue.

Androgen receptor expression also declined with increasing Gleason grade (Figure 1). Published data for AR protein expression in PCa is contradictory and studies vary in their methodology. For example, some count nuclear staining or measure overall staining, others compare stromal vs. epithelial or primary cancer vs. metastasis. Several studies have, however, reported an inverse relationship between AR protein expression and Gleason grade, particularly when comparing stromal tissue to epithelial cells [42–44]. The current study did not differentiate between the stromal and epithelial compartments and the IRS for AR is significantly lower in all pathologies compared to normal prostate. Unfortunately, markers that decrease expression with increasing tumor grade are not as compatible with a robust prognostic test as those that increase expression. Expression of MKI67 and MYC increased in the higher Gleason pattern cores while BCL2 decreased (Figure 1). This is as expected as the first two are well known markers of proliferation associated with cancer growth while a decrease in BCL2 expression is associated with loss of apoptosis and increased proliferation. Mutations in *TP53* are commonly found in PCa but only some result in increased protein expression as observed in a small subgroup of biologically aggressive PCa [45]. The TP53 isoform Δ133TP53β has also been shown to be increased in higher grade PCa and is associated with immune cell infiltration [46]. These may explain why only a few of the TMA cases across different Gleason patterns in our study exhibited increased staining for TP53 (Figure 1).

The findings from Oncomine and FireBrowse analysis of RNA expression are consistent with our protein expression results showing INHBB expression is higher, and INHBC is lower, in prostate tumours when compared to healthy prostate (Figure 2). This implies that increased expression of INHBB and decreased expression of INHBC may be associated with a poorer prognosis in PCa patients.

Exogenous treatment with recombinant mature-region INHBA inhibited growth in all of the cell lines except for DU145 while both DU145 and PC3 cells were unaffected by treatment with exogenous mature-region INHBB (Figure 4). Addition of full-length recombinant INHBC had no effect on the growth of any of the prostate cell lines tested. The different responses of the cell lines may be due to differing endogenous expression levels of activins, their receptors or activin inhibitors such as FST [47,48]. McPherson et al. [47] showed INHBA inhibited LNCaP cells but in their study DU145 cells were also inhibited and PC3 cells did not respond to either INHBA or INHBB. In contrast, Simon et al. [49] have reported INHBA treatment of PC3 cells increased proliferation. However, this was in the presence of 1% FBS that had not been heat-inactivated. They also showed that INHBA antibody decreased non heat-inactivated serum induced proliferation of PC3 cells indicating this effect was due to an interaction between INHBA and another active component in serum. The proliferation assays in our current study were all performed in the presence of 2% heat inactivated FBS with the same media, conditions and endpoint which has enabled us to directly compare the responses of the different prostate cell lines to the different activins.

The modification of PNT1A or PC3 cells to overexpress INHBC increased growth of PNT1A cells but decreased growth and increased migration of PC3 cells (Figure 6). This indicates that either our recombinant, purified INHBC protein is not bioactive or endogenous expression of INHBC is required for function. Overexpression of INHBB in PNT1A cells decreased the growth and survival of these cells which is in concordance with the effect of exogenous recombinant INHBB on these cells. In contrast, INHBB overexpression increased growth and migration of PC3 cells (Figure 6) suggesting these cells either respond to the endogenous presence of INHBB, or the full-length protein but not the mature region alone. The opposing effects of overexpression of INHBB and INHBC on the immortalized epithelial PNT1A cells and metastatic PCa PC3 cells indicates a switch in the function of these activins from inhibiting or regulating growth in normal epithelial cells to being pro-oncogenic in tumor cells. A similar switch in response to TGFB and INHBA in PCa has been previously described [50,51].

The in vitro assay data corresponds to the protein expression results for both the prostate biopsy cores and the cell lines which suggest an increase in INHBB expression in higher grade tumors and metastatic cell lines may be driving the development of more aggressive cancer, while a decrease in INHBC expression may be reducing growth inhibition in cases of prostate carcinoma. However, expression of INHBC pro- and mature-region was low in LNCaP cells but not PC3 cells, while PNT1A cells had low mature-region and high pro-region protein (Figure 3).

The commercially available INHBC, consisting of the mature chain only, had no effect on LNCaP cell proliferation with or without INHBA or INHBB (Figure 4A,B). The pro-region of TGFB family proteins is known to be important for mediating folding and dimerization of the C-terminal mature domains, binding to extracellular matrix and regulation of signaling [41,52,53]. We hypothesized that the activity of INHBC may require the presence of the N-terminal pro-region, as is the case for cumulin, another TGFB family member [24]. However, our purified, full length, recombinant INHBC containing both the pro-region and mature region proteins was unable to inhibit INHBA or INHBB suppression of growth in LNCaP cells but did trend towards an inhibition of INHBA function in PC3 cells. This differs to results previously reported in LNCaP cells using an unpurified INHBC preparation produced in CHO cells which was able to block INHBA inhibition of LNCaP cell growth [13,14]. CHO cells secrete many factors that can alter cell growth and therefore the effect reported from this INHBC preparation could potentially have been due to the presence of contaminating proteins. Goebel et al. [54] have also demonstrated that mature-region INHBC does not directly antagonize INHBA signaling. However, inhibin-α knock-out mice that have increased INHBA and develop testicular and ovarian tumors, were shown to have reduced tumor formation when modified to overexpress INHBC [17]. Other studies have reported that INHBC can reduce the formation of homodimeric INHBA and increase heterodimeric INHBAC which has a lower affinity for the ACVR2B receptor [14,55]. Marino et al. [22] have also shown that INHBC can directly bind the ACVR2A and 2B receptors but does not alter SMAD2/3 or ERK signaling, indicating INHBC may block INHBA receptor binding.

Inhibition of the ALK4/5/7 signaling pathway had no effect on PNT1A cell growth indicating this pathway is not very active in normal prostate epithelial cells (Figure 7). This may be because INHBA that is known to activate this pathway had low expression in PNT1A cells. Blocking the ALK4/5/7 pathway in PC3 cells however, increased growth, likely due to inhibition of INHBA signaling which would normally decrease growth of these cells. Unexpectedly, LNCaP cells had decreased growth when exposed to the ALK4/5/7 inhibitor. MAPK inhibition decreased growth of both PNT1A and PC3 cells but again had the opposite effect on LNCaP cells. The ALK2/3 inhibitor was the only one to significantly inhibit growth of all three prostate cell lines and therefore the use of the LDN193189 small molecule inhibitor for the treatment of PCa should be further explored.

5. Conclusions

To summarize, our results suggest that combining quantification of INHBB and INHBC in prostate biopsies may enable the identification of tumors containing cells that will be particularly malignant. These, and other TGFB family proteins, could provide more accurate predictions of clinical outcome and guide decision making in relation to surgery, active surveillance, or watchful waiting, and thus improve life expectancy and quality for PCa patients. Inhibition of INHBB expression and/or targeting the ALK2/3 pathway may also be novel therapeutic targets for the treatment of PCa. Further work is needed to elucidate the mechanisms of INHBC cell signaling and to corroborate the TGFB family expression differences observed in a larger cohort of PCa patients with more comprehensive clinical data.

Supplementary Materials: The following supporting information can be downloaded at: https://www.mdpi.com/article/10.3390/cancers15010147/s1, Figure S1: Representative images of immunohistochemical staining for each TGF-β family protein in normal prostate tissue (left) and ISUP Gleason grade 5 prostate cancer tissue cores (right). INHBA, INHBB, INHBC, inhibin-α (INHA), activin receptor type 2A (ACVR2A), activin receptor type 2B (ACVR2B), mothers against decapentaplegic homolog 2 or 3 (SMAD2, SMAD3), follistatin (FST), bone morphogenetic protein 4 (BMP4), androgen receptor (AR), marker of proliferation Ki-67 (MKI67), MYC proto-oncogene (MYC), tumor protein p53 (TP53), B-cell CLL/lymphoma 2 (BCL2); Figure S2: Western blot of recombinant INHBC from R&D Systems and our laboratory (rActC) showing presence of pro-mature and mature-region bands in our full-length recombinant protein.

Author Contributions: Conceptualization K.L.R., D.G.M., H.E.C. and E.J.G.; funding acquisition K.L.R. and E.J.G.; methodology, K.L.R., S.J.-M. and S.Z.-M. with supervision by H.E.C.; data analysis, T.J.; writing—original draft preparation, K.L.R.; writing—review and editing, all other authors (except E.J.G.). All authors have read and agreed to the published version of the manuscript except E.J.G. (deceased).

Funding: This research was funded by a Lottery Health Translational Research Project Grant AP353014 to the late Elspeth Gold and a Lottery Health Postdoctoral Fellowship LHR-2017-49017 to Karen Reader.

Institutional Review Board Statement: Not applicable for studies not involving humans or animals.

Informed Consent Statement: Not applicable.

Data Availability Statement: The data presented in this study are available on request from the corresponding author.

Acknowledgments: We are grateful to Helen Nicholson for her mentorship of KLR and Edward Ottley and Andrew McNaughton for assistance with developing the IRS protocol.

Conflicts of Interest: The authors declare no conflict of interest.

References

1. Wang, L.; Lu, B.; He, M.; Wang, Y.; Wang, Z.; Du, L. Prostate Cancer Incidence and Mortality: Global Status and Temporal Trends in 89 Countries From 2000 to 2019. *Front. Public Health* **2022**, *10*, 811044. [CrossRef] [PubMed]
2. Leyh-Bannurah, S.R.; Abou-Haidar, H.; Dell'Oglio, P.; Schiffmann, J.; Tian, Z.; Heinzer, H.; Karakiewicz, P.I. Primary Gleason pattern upgrading in contemporary patients with D'Amico low-risk prostate cancer: Implications for future biomarkers and imaging modalities. *BJU Int.* **2017**, *119*, 692–699. [CrossRef] [PubMed]
3. Thaxton, C.S.; Loeb, S.; Roehl, K.A.; Kan, D.; Catalona, W.J. Treatment outcomes of radical prostatectomy in potential candidates for 3 published active surveillance protocols. *Urology* **2010**, *75*, 414–418. [CrossRef] [PubMed]
4. Sowalsky, A.G.; Ye, H.; Bubley, G.J.; Balk, S.P. Clonal progression of prostate cancers from Gleason grade 3 to grade 4. *Cancer Res.* **2013**, *73*, 1050–1055. [CrossRef]
5. Boyd, L.K.; Mao, X.; Lu, Y.J. The complexity of prostate cancer: Genomic alterations and heterogeneity. *Nat. Rev. Urol.* **2012**, *9*, 652–664. [CrossRef]
6. Shipitsin, M.; Small, C.; Choudhury, S.; Giladi, E.; Friedlander, S.; Nardone, J.; Blume-Jensen, P. Identification of proteomic biomarkers predicting prostate cancer aggressiveness and lethality despite biopsy-sampling error. *Br. J. Cancer* **2014**, *111*, 1201–1212. [CrossRef]
7. Ross, A.E.; D'Amico, A.V.; Freedland, S.J. Which, when and why? Rational use of tissue-based molecular testing in localized prostate cancer. *Prostate Cancer Prostatic Dis.* **2016**, *19*, 1–6. [CrossRef]
8. Loomans, H.A.; Andl, C.D. Intertwining of Activin A and TGFbeta Signaling: Dual Roles in Cancer Progression and Cancer Cell Invasion. *Cancers* **2014**, *7*, 70–91. [CrossRef]
9. Croxford, K.P.; Reader, K.L.; Nicholson, H.D. The potential role of transforming growth factor beta family ligand interactions in prostate cancer. *AIMS Mol. Sci.* **2017**, *4*, 41–61. [CrossRef]
10. Jung, B.; Staudacher, J.J.; Beauchamp, D. Transforming Growth Factor Beta Super Family Signaling in Development of Colorectal Cancer. *Gastroenterology* **2016**, *152*, 36–52. [CrossRef]
11. Lodberg, A. Principles of the activin receptor signaling pathway and its inhibition. *Cytokine Growth Factor Rev.* **2021**, *60*, 1–17. [CrossRef] [PubMed]
12. Namwanje, M.; Brown, C.W. Activins and Inhibins: Roles in Development, Physiology, and Disease. *Cold Spring Harb. Perspect Biol.* **2016**, *8*, a021881. [CrossRef] [PubMed]
13. Gold, E.; Jetly, N.; O'Bryan, M.K.; Meachem, S.; Srinivasan, D.; Behuria, S.; Risbridger, G. Activin C antagonizes activin A in vitro and overexpression leads to pathologies in vivo. *Am. J. Pathol.* **2009**, *174*, 184–195. [CrossRef] [PubMed]
14. Gold, E.; Marino, F.E.; Harrison, C.; Makanji, Y.; Risbridger, G. Activin-beta(c) reduces reproductive tumour progression and abolishes cancer-associated cachexia in inhibin-deficient mice. *J. Pathol.* **2013**, *229*, 599–607. [CrossRef]
15. Chen, L.; De Menna, M.; Groenewoud, A.; Thalmann, G.N.; Kruithof-de Julio, M.; Snaar-Jagalska, B.E. A NF-kB-Activin A signaling axis enhances prostate cancer metastasis. *Oncogene* **2020**, *39*, 1634–1651. [CrossRef]
16. Ottley, E.C.; Reader, K.L.; Lee, K.; Marino, F.E.; Nicholson, H.D.; Risbridger, G.P.; Gold, E. Over-Expression of Activin-betaC Is Associated with Murine and Human Prostate Disease. *Horm Cancer* **2017**, *8*, 100–107. [CrossRef]
17. Marino, F.E.; Risbridger, G.; Gold, E. Activin-beta C modulates gonadal, but not adrenal tumorigenesis in the inhibin deficient mice. *Mol. Cell Endocrinol.* **2015**, *409*, 41–50. [CrossRef]
18. Schindelin, J.; Arganda-Carreras, I.; Frise, E.; Kaynig, V.; Longair, M.; Pietzsch, T.; Cardona, A. Fiji: An open-source platform for biological-image analysis. *Nat. Methods* **2012**, *9*, 676–682. [CrossRef]
19. Helps, S.C.; Thornton, E.; Kleinig, T.J.; Manavis, J.; Vink, R. Automatic nonsubjective estimation of antigen content visualized by immunohistochemistry using color deconvolution. *Appl. Immunohistochem. Mol. Morphol.* **2012**, *20*, 82–90. [CrossRef]
20. Fuhrich, D.G.; Lessey, B.A.; Savaris, R.F. Comparison of HSCORE assessment of endometrial beta3 integrin subunit expression with digital HSCORE using computerized image analysis (ImageJ). *Anal. Quant Cytopathol. Histpathol.* **2013**, *35*, 210–216.
21. Ruifrok, A.C.; Johnston, D.A. Quantification of histochemical staining by color deconvolution. *Anal. Quant Cytol. Histol.* **2001**, *23*, 291–299. [PubMed]
22. Marino, F.E.; Risbridger, G.; Gold, E. The inhibin/activin signalling pathway in human gonadal and adrenal cancers. *Mol. Hum. Hum. Reprod.* **2014**, *20*, 1223–1237. [CrossRef] [PubMed]
23. Hobbs, S.; Jitrapakdee, S.; Wallace, J.C. Development of a bicistronic vector driven by the human polypeptide chain elongation factor 1alpha promoter for creation of stable mammalian cell lines that express very high levels of recombinant proteins. *Biochem. Biophys. Res. Commun.* **1998**, *252*, 368–372. [CrossRef] [PubMed]
24. Mottershead, D.G.; Sugimura, S.; Al-Musawi, S.L.; Li, J.J.; Richani, D.; White, M.A.; Gilchrist, R.B. Cumulin, an Oocyte-secreted Heterodimer of the Transforming Growth Factor-beta Family, Is a Potent Activator of Granulosa Cells and Improves Oocyte Quality. *J. Biol. Chem.* **2015**, *290*, 24007–24020. [CrossRef] [PubMed]
25. Cussenot, O.; Berthon, P.; Berger, R.; Mowszowicz, I.; Faille, A.; Hojman, F.; Calvo, F. Immortalization of human adult normal prostatic epithelial cells by liposomes containing large T-SV40 gene. *J. Urol.* **1991**, *146*, 881–886. [CrossRef]
26. Sambrook, J.; Russell, D.W. The inoue method for preparation and transformation of competent e. Coli: "Ultra-competent" cells. *CSH Protoc.* **2006**, *1*, 10–1101. [CrossRef]
27. R Core Team. R: A Language and Environment for Statistical Computing. 2020. Available online: https://www.R-project.org/ (accessed on 17 August 2022).

28. Venables, W.N.; Ripley, B.D. *Modern Applied Statistics with S*, 4th ed.; Springer: New York, NY, USA, 2002.
29. Pinheiro, J.; Bates, D.; Debroy, S.; Sarkar, D. Nlme: Linear and Nonlinear Mixed Effects Models. 2022. Available online: https://CRAN.R-project.org/package=nlme (accessed on 17 August 2022).
30. Length, R.V. Emmeans: Estimated Marginal Means, aka Least-Squares Means. 2022. Available online: https://CRAN.R-project.org/package=emmeans (accessed on 17 August 2022).
31. Yu, Y.P.; Landsittel, D.; Jing, L.; Nelson, J.; Ren, B.; Liu, L.; Luo, J.H. Gene expression alterations in prostate cancer predicting tumor aggression and preceding development of malignancy. *J. Clin. Oncol. Off. J. Am. Soc. Clin. Oncol.* **2004**, *22*, 2790–2799. [CrossRef]
32. Luo, J.; Duggan, D.J.; Chen, Y.; Sauvageot, J.; Ewing, C.M.; Bittner, M.L.; Isaacs, W.B. Human prostate cancer and benign prostatic hyperplasia: Molecular dissection by gene expression profiling. *Cancer Res.* **2001**, *61*, 4683–4688.
33. Tomlins, S.A.; Mehra, R.; Rhodes, D.R.; Cao, X.; Wang, L.; Dhanasekaran, S.M.; Chinnaiyan, A.M. Integrative molecular concept modeling of prostate cancer progression. *Nat. Genet.* **2007**, *39*, 41–51. [CrossRef]
34. Liu, P.; Ramachandran, S.; Ali Seyed, M.; Scharer, C.D.; Laycock, N.; Dalton, W.B.; Moreno, C.S. Sex-determining region Y box 4 is a transforming oncogene in human prostate cancer cells. *Cancer Res.* **2006**, *66*, 4011–4019. [CrossRef]
35. Epstein, J.I.; Zelefsky, M.J.; Sjoberg, D.D.; Nelson, J.B.; Egevad, L.; Magi-Galluzzi, C.; Klein, E.A. A Contemporary Prostate Cancer Grading System: A Validated Alternative to the Gleason Score. *Eur. Urol.* **2016**, *69*, 428–435. [CrossRef] [PubMed]
36. Mason, M. Getting better at treating prostate cancer: What clinicians should want from scientists. *J. Cancer Metastasis Treat.* **2017**, *3*, 271–277. [CrossRef]
37. Sangkop, F.; Singh, G.; Rodrigues, E.; Gold, E.; Bahn, A. Uric acid: A modulator of prostate cells and activin sensitivity. *Mol. Cell Biochem.* **2016**, *414*, 187–199. [CrossRef] [PubMed]
38. Kang, H.Y.; Huang, H.Y.; Hsieh, C.Y.; Li, C.F.; Shyr, C.R.; Tsai, M.Y.; Huang, K.E. Activin A enhances prostate cancer cell migration through activation of androgen receptor and is overexpressed in metastatic prostate cancer. *J. Bone Miner Res.* **2009**, *24*, 1180–1193. [CrossRef] [PubMed]
39. Balanathan, P.; Williams, E.D.; Wang, H.; Pedersen, J.S.; Horvath, L.G.; Achen, M.G.; Risbridger, G.P. Elevated level of inhibin-alpha subunit is pro-tumourigenic and pro-metastatic and associated with extracapsular spread in advanced prostate cancer. *Br. J. Cancer* **2009**, *100*, 1784–1793. [CrossRef]
40. Lu, S.; Lee, J.; Revelo, M.; Wang, X.; Lu, S.; Dong, Z. Smad3 is overexpressed in advanced human prostate cancer and necessary for progressive growth of prostate cancer cells in nude mice. *Clin Cancer Res.* **2007**, *13*, 5692–5702. [CrossRef] [PubMed]
41. Walton, K.L.; Makanji, Y.; Harrison, C.A. New insights into the mechanisms of activin action and inhibition. *Mol. Cell Endocrinol.* **2012**, *359*, 2–12. [CrossRef]
42. Gyftopoulos, K.; Sotiropoulou, G.; Varakis, I.; Barbalias, G.A. Cellular distribution of retinoic acid receptor-alpha in benign hyperplastic and malignant human prostates: Comparison with androgen, estrogen and progesterone receptor status. *Eur. Urol.* **2000**, *38*, 323–330. [CrossRef]
43. Leach, D.A.; Need, E.F.; Toivanen, R.; Trotta, A.P.; Palenthorpe, H.M.; Tamblyn, D.J.; Buchanan, G. Stromal androgen receptor regulates the composition of the microenvironment to influence prostate cancer outcome. *Oncotarget* **2015**, *6*, 16135–16150. [CrossRef]
44. Filipovski, V.; Kubelka-Sabit, K.; Jasar, D.; Janevska, V. Androgen Receptor Expression in Epithelial and Stromal Cells of Prostatic Carcinoma and Benign Prostatic Hyperplasia. *Open Access Maced. J. Med. Sci.* **2017**, *5*, 608–612. [CrossRef]
45. Schlomm, T.; Iwers, L.; Kirstein, P.; Jessen, B.; Köllermann, J.; Minner, S.; Erbersdobler, A. Clinical significance of p53 alterations in surgically treated prostate cancers. *Mod. Pathol.* **2008**, *21*, 1371–1378. [CrossRef] [PubMed]
46. Kazantseva, M.; Mehta, S.; Eiholzer, R.A.; Gimenez, G.; Bowie, S.; Campbell, H.; Braithwaite, A.W. The Δ133p53β isoform promotes an immunosuppressive environment leading to aggressive prostate cancer. *Cell Death Dis.* **2019**, *10*, 631. [CrossRef] [PubMed]
47. McPherson, S.J.; Thomas, T.Z.; Wang, H.; Gurusinghe, C.J.; Risbridger, G.P. Growth inhibitory response to activin A and B by human prostate tumour cell lines, LNCaP and DU145. *J. Endocrinol.* **1997**, *154*, 535–545. [CrossRef] [PubMed]
48. McPherson, S.J.; Mellor, S.L.; Wang, H.; Evans, L.W.; Groome, N.P.; Risbridger, G.P. Expression of activin A and follistatin core proteins by human prostate tumor cell lines. *Endocrinology* **1999**, *140*, 5303–5309. [CrossRef] [PubMed]
49. Simon, D.P.; Vadakkadath Meethal, S.; Wilson, A.C.; Gallego, M.J.; Weinecke, S.L.; Bruce, E.; Lyons, P.F.; Haasl, R.J.; Bowen, R.L.; Atwood, C.S.; et al. Activin receptor signaling regulates prostatic epithelial cell adhesion and viability. *Neoplasia* **2009**, *11*, 365–376. [CrossRef] [PubMed]
50. Ao, M.; Williams, K.; Bhowmick, N.A.; Hayward, S.W. Transforming growth factor-beta promotes invasion in tumorigenic but not in nontumorigenic human prostatic epithelial cells. *Cancer Res.* **2006**, *66*, 8007–8016. [CrossRef] [PubMed]
51. Dalkin, A.C.; Gilrain, J.T.; Bradshaw, D.; Myers, C.E. Activin inhibition of prostate cancer cell growth: Selective actions on androgen-responsive LNCaP cells. *Endocrinology* **1996**, *137*, 5230–5235. [CrossRef] [PubMed]
52. McIntosh, C.J.; Lun, S.; Lawrence, S.; Western, A.H.; McNatty, K.P.; Juengel, J.L. The proregion of mouse BMP15 regulates the cooperative interactions of BMP15 and GDF9. *Biol. Reprod.* **2008**, *79*, 889–896. [CrossRef]

53. McIntosh, C.J.; Lawrence, S.; Smith, P.; Juengel, J.L.; McNatty, K.P. Active immunization against the proregions of GDF9 or BMP15 alters ovulation rate and litter size in mice. *Reproduction* **2012**, *143*, 195–201. [CrossRef]
54. Goebel, E.J.; Ongaro, L.; Kappes, E.C.; Vestal, K.; Belcheva, E.; Castonguay, R.; Thompson, T.B. The orphan ligand, activin C, signals through activin receptor-like kinase 7. *Elife* **2022**, *11*, e78197. [CrossRef]
55. Mellor, S.L.; Ball, E.M.; O'Connor, A.E.; Ethier, J.F.; Cranfield, M.; Schmitt, J.F.; Risbridger, G.P. Activin betaC-subunit heterodimers provide a new mechanism of regulating activin levels in the prostate. *Endocrinology* **2003**, *144*, 4410–4419. [CrossRef] [PubMed]

cancers

MDPI

Article

Ciliary Neurotrophic Factor Modulates Multiple Downstream Signaling Pathways in Prostate Cancer Inhibiting Cell Invasiveness

Giovanni Tossetta [1,†], Sonia Fantone [1,†], Rosaria Gesuita [2,3], Gaia Goteri [4], Martina Senzacqua [1], Fabio Marcheggiani [5], Luca Tiano [5], Daniela Marzioni [1,*] and Roberta Mazzucchelli [4]

1 Department of Experimental and Clinical Medicine, Università Politecnica delle Marche, 60126 Ancona, Italy
2 Centre of Epidemiology, Biostatistics and Medical Information Technology, Università Politecnica delle Marche, 60126 Ancona, Italy
3 Department of Biomedical Sciences and Public Health, Università Politecnica delle Marche, 60126 Ancona, Italy
4 Department of Biomedical Sciences and Public Health, Section of Pathological Anatomy, Università Politecnica delle Marche, 60126 Ancona, Italy
5 Department of Life and Environmental Sciences, Università Politecnica delle Marche, 60131 Ancona, Italy
* Correspondence: d.marzioni@univpm.it
† These authors equally contributed to this study.

Simple Summary: Prostate cancer (PCa) is one of the major cancers affecting men. Localized and loco-regional PCa is usually treated by radical prostatectomy, radiation therapy, cryosurgery or with HIFU (High-Intensity Focused Ultrasound). Advanced or metastatic cancers are most often treated with androgen deprivation therapy, but too often such patients progress to lethal castration-resistant PCa. IL-6 plays a key role in prostate cancer but no data on ciliary neurotrophic factor (CNTF), a member of interleukin-6 cytokine family, are known. We detected CNTF and its receptor CNTFRα in human androgen-responsive and in castration-resistant prostate cancer (CRPC) tissues. In addition, we showed that CNTF downregulated MMP-2 and GLUT-1 expression by MAPK/ERK, AKT/PI3K and Jak/STAT pathways in a CRPC in vitro model. This suggests a pivotal role of CNTF as negative modulator of invasion processes in this PCa model.

Abstract: Background: Prostate cancer (PCa) remains the most common diagnosed tumor and is the second-leading cause of cancer-related death in men. If the cancer is organ-confined it can be treated by various ablative therapies such as RP (radical prostatectomy), RT (radiation therapy), brachytherapy, cryosurgery or HIFU (High-Intensity Focused Ultrasound). However, advanced or metastatic PCa treatment requires systemic therapy involving androgen deprivation, but such patients typically progress to refractory disease designated as castration-resistant prostate cancer (CRPC). Interleukin-6 (IL-6) has been established as a driver of prostate carcinogenesis and tumor progression while less is known about the role of ciliary neurotrophic factor (CNTF), a member of the IL-6 cytokine family in prostate cancer. Moreover, MAPK/ERK, AKT/PI3K and Jak/STAT pathways that regulate proliferative, invasive and glucose-uptake processes in cancer progression are triggered by CNTF. Methods: We investigate CNTF and its receptor CNTFRα expressions in human androgen-responsive and castration-resistant prostate cancer (CRPC) by immunohistochemistry. Moreover, we investigated the role of CNTF in proliferative, invasive processes as well as glucose uptake using two cell models mimicking the PCa (LNCaP cell line) and CRPC (22Rv1 cell line). Conclusions: Our results showed that CNTF and CNTFRa were expressed in PCa and CRPC tissues and that CNTF has a pivotal role in prostate cancer environment remodeling and as a negative modulator of invasion processes of CRPC cell models.

Keywords: CNTF; CNTFRα; prostate cancer; LNCaP; 22Rv1; castration resistant; castration sensitive; STAT3; ERK; MMP

Citation: Tossetta, G.; Fantone, S.; Gesuita, R.; Goteri, G.; Senzacqua, M.; Marcheggiani, F.; Tiano, L.; Marzioni, D.; Mazzucchelli, R. Ciliary Neurotrophic Factor Modulates Multiple Downstream Signaling Pathways in Prostate Cancer Inhibiting Cell Invasiveness. *Cancers* **2022**, *14*, 5917. https://doi.org/10.3390/cancers14235917

Academic Editors: José I. López and Claudia Manini

Received: 18 October 2022
Accepted: 26 November 2022
Published: 30 November 2022

1. Introduction

Prostate cancer is one of the major cancers affecting men and remains the second leading cause of cancer-related death [1]. It is a biologically heterogeneous disease; the majority of instances are localized or loco-regional and can be treated by surgery or other ablative procedures, while more advanced or metastatic patients who remain androgen-dependent at the time of diagnosis are treated with androgen deprivation therapy (ADT). The response to ADT is variable and many such patients become refractory to treatment and will progress to lethal castration-resistant prostate cancer (CRPC) [2]. IL-6 has been established as a driver of carcinogenesis and tumor progression in PCa [3–5], while less is known about the role of ciliary neurotrophic factor (CNTF), a member of IL-6 type cytokine family [6]. CNTF, and other cytokines of this group (e.g., IL-6, LIF, OMS and IL-11), require additional non-signaling receptors (long signal-transducing α-receptor glycoprotein chains) that are bound to the membrane [7]. CNTFRα-chain non-signaling receptor subunit (CNTFRα) is the high affinity non-signaling receptor for CNTF. CNTFRα is associated with a heterodimeric β-receptor complex LIFR and gp130, and this association allows signal transduction due to phosphorylation of tyrosine residue of CNTFRα after its interaction with CNTF [7]. The CNTF receptor complex triggers simultaneous activation of different intracellular signaling pathways such as MAPK/ERK, AKT/PI3K and Jak/STAT pathways which mediate survival and/or differentiation in different cell types [8,9]. Moreover, it has been reported that PCa progression and metastasis are correlated both to PTEN/AKT/PI3K axis alteration and RAS/MAPK/ERK signaling activation, while the RAS/MAPK/ERK pathway alone is significantly activated in both primary and metastatic lesions [10,11]. In addition, similar to many other tumors, glucose metabolism in PCa is different between the early and advanced stages of the disease. Moreover, the glycolytic metabolic profile differs in androgen-sensitive and insensitive PCa cells [12,13]. There are fourteen different members of GLUT transporters (GLUT 1–12, GLUT14, and H/myo-inositol transporter) and different pattern of protein expression or membrane translocation are observed in relation to tumor-specific glycolytic pathway modulation. Therefore, an increase in glucose uptake has been associated mainly with GLUT1 overexpression, but may also involve other GLUTs, including GLUT4 [14,15]. It has been demonstrated that an elevated AKT activity is involved in the high rate of glucose uptake in cancer cells by GLUT-1 [16–18] and that treatment with an AKT-specific inhibitor caused degradation of GLUT-1 in sustained AKT-activated breast cancer cells [19]. Human studies suggested that high levels of GLUT-1 expression in tumors are associated with poor survival [20,21]. The presence of CNTF and CNTFRα in the basal cell layer (in which reside stem cells) in the prostatic normal glandular epithelium has been previously demonstrated in our previous investigation [6]. Thus, the aim of this study is to detect the localization of CNTF and CNTFRα in androgen-sensitive and androgen-insensitive human PCa tissues and to analyze the possible role of CNTF in signaling-pathway modulation involved in glucose uptake, proliferation, migration and invasion processes using sensitive (LNCaP) and androgen-insensitive (22Rv1) cell lines.

2. Materials and Methods

2.1. Tissue Collection

In this study, we analyzed a total of 20 human prostate carcinoma samples: 10 untreated PCa were obtained from radical prostatectomy and 10 CRPC were obtained from transurethral resection of the prostate (TURP). A pathologist (R.M.) reviewed and selected the hematoxylin–eosin stained samples that were used in this study. All the samples were collected from the Pathology Services of the Polytechnic University of the Marche Region United Hospitals. The procedures followed for the collection of samples were in accordance with the Helsinki Declaration of 1975, as revised in 2013. The permission of the Human Investigation Committee of Marche Region (IT) was granted (protocol number 2020/395).

2.2. Immunohistochemistry

All prostate samples (PCa and CRPC) were fixed in 10% neutral buffered formalin and routinely processed for paraffin embedding. Immunohistochemical staining was per-formed as previously described [22,23]. Briefly, after dewaxing, paraffin sections were rinsed in phosphate-buffered saline (PBS), incubated with 3% hydrogen peroxide for 40 min to block endogenous peroxidase. Pre-treatment by heat in 10 mM citrate buffer, pH 6.0 for 5 min was used for CNTF while pre-treatment by 100 ng/mL Proteinase K (Sigma-Aldrich, St. Louis, MO, USA) for 5 min at 37 °C was used for CNTFRα. After pre-treatment, sections were rinsed with PBS and incubated with normal horse serum (Vector laboratories, Burlingame, CA, USA) diluted 1:75 in PBS for 1 h at room temperature (RT). Sections were then incubated overnight with the primary antibodies (listed in Table 1) diluted in PBS at 4 °C. After a thorough rinse in PBS, sections were incubated with the appropriate biotinylated secondary antibody (Vector laboratories, Burlingame, CA, USA) diluted 1:200 v/v solution for 30 min at room temperature. Vectastain ABC Kit (Vector Laboratories, Burlingame, CA, USA) for 1h at room temperature and 3′,3′- diaminobenzidine hydrochloride (Sigma-Aldrich, St. Louis, MO, USA) were used to develop the immunohistochemistry reaction. Sections were counterstained with Mayer's hematoxylin, dehydrated and mounted using Eukitt® solution (Kindler GmbH and Co., Freiburg, Germany). Negative controls were performed by omitting the first or secondary antibody for all the immunohistochemical reactions performed in this study as procedure control. Negative controls were performed by replacing the primary or the secondary antibody with an isotype antibody: monoclonal rabbit IgG (ab172730; Abcam, Cambridge, UK) for CNTF and monoclonal mouse IgG2a (ab18415; Abcam) for CNTFRα as control specificity antibody.

Table 1. Antibodies used in this study.

Antibody	IHC	WB	IF	Company
pAb Rabbit anti-human CNTF (#ab190985)	1:500	//	1:100	Abcam, Cambridge, UK
pAb Rabbit anti-human CNTFRα (#PA5-45053)	//	1:400	//	Thermo Fisher Scientific, Waltham, MA, USA
mAb Mouse anti-human CNTFRα (#ab89333)	1:150	//	1:100	Abcam, Cambridge, UK
mAb Rabbit anti-human pAKT (#4060)	//	1:1000	//	Cell signaling technology, Danvers, MA, USA
pAb Rabbit anti-human AKT (#9272)	//	1:1000	//	Cell signaling technology, Danvers, MA, USA
mAb Rabbit anti-human pERK1/2 (#4377)	//	1:800	//	Cell signaling technology, Danvers, MA, USA
mAb Rabbit anti-human ERK1/2 (#4695)	//	1:1000	//	Cell signaling technology, Danvers, MA, USA
mAb Mouse anti-human pSTAT3 (#4113)	//	1:800	//	Cell signaling technology, Danvers, MA, USA
mAb Rabbit anti-human STAT3 (#4904)	//	1:1000	//	Cell signaling technology, Danvers, MA, USA
pAb Rabbit anti-human GLUT1 (#PA1-46152)	//	1:500	//	Thermo Fisher Scientific, Waltham, MA, USA
pAb Rabbit anti-human GLUT4 (#PA5-23052)	//	1:500	//	Thermo Fisher Scientific, Waltham, MA, USA
pAb Rabbit anti-human MMP-9 (#10375-2-AP)	//	1:500	//	Proteintech Group, Manchester, UK
mAb Mouse anti-human MMP-2 (#436000)	//	1:500	//	Thermo Fisher Scientific, Waltham, MA, USA
mAb Mouse anti-human PCNA (#sc-56)	//	1:250	//	Santa Cruz Biotechnology, Inc, Dallas, TX, USA

mAb: monoclonal antibody; pAb: polyclonal antibody; IHC: Immunohistochemistry; IF: immunofluorescence; WB: Western blotting.

2.3. Cell Culture

Androgen-dependent (LNCaP) and androgen-independent (22Rv1) cell lines (ATCC/LGC Standards, Manassas, VA, USA) were grown in RPMI 1640 (Life technologies, Carlsbad, CA, USA) supplemented with 10% Fetal Bovine Serum (FBS; Gibco, Thermo Fisher Scientific, Waltham, MA, USA) and 100 U/mL penicillin and streptomycin (Gibco) at 37 °C, 95% humidity and 5% CO_2. The medium was changed every 2 to 3 days and cells were split 1:4 every 4 days.

2.4. Immunofluorescence

Immunofluorescence staining was performed as previously described [24,25]. Briefly, LNCaP and 22Rv1 cells were washed in Dulbecco's PBS (Life technology, Monza, Italy), fixed in 4% paraformaldehyde in PBS for 10 min at room temperature (RT), and permeabilized in PBS 0.1 M added with 0.1% Triton X-100 (Sigma, Milano, Italy) for 5 min. After washing in PBS at RT, cells were blocked with 10% Normal Donkey Serum (Jackson ImmunoResearch, Pennsylvania, PA, USA) in PBS 0.1 M and incubated overnight at 4 °C with the primary antibodies listed in Table 1. Cells were then washed 3 times in PBS and incubated with the FITC-conjugated donkey anti-rabbit (for CNTF) and TRITC-conjugated anti-mouse (for CNTFRα) IgG secondary antibodies (both from Jackson ImmunoResearch, West Grove, PA, USA) 1:400 for 30 min at room temperature. TOTO3 (1:5000) probe was used for nuclear staining. Finally, the slides were cover-slipped with propyl gallate and evaluated with a Leica TCS-SL spectral confocal microscope. For negative controls, see the immunohistochemistry Section 2.2.

2.5. Western Blotting

Once LNCaP and 22Rv1 cells reached 80% confluence, cells were lysed by using the following lysis buffer: 0.1 M PBS, 0.1% (*w*/*v*) SDS, 1% (*w*/*w*) NONIDET-P40, 1 mM (*w*/*v*) Na orthovanadate, 1 mM (*w*/*w*) PMSF (phenyl methane sulfonyl fluoride), 12 mM (*w*/*v*) Na deoxycholate, 1.7 μg/mL Aprotinin, pH 7.5. Cell lysates were centrifuged at 20,000× *g* for 20 min at 4 °C and the supernatants were aliquoted and stored at −80 °C. Viable counts using the trypan blue dye exclusion test were routinely performed. All experiments were performed in duplicate and were repeated at least 3 times. The proteins' concentrations were determined by a Bradford protein assay (Bio-Rad Laboratories, Milan, Italy). Protein samples were fractionated on 10% SDS-polyacrylamide gels (SDS-PAGE) and electrophoretically transferred (Trans-Blot® Turbo™ Transfer System; Bio-Rad Laboratories Inc, Richmond, CA, USA) to nitrocellulose membranes. Non-specific protein binding was blocked with 5% (*w*/*v*) non-fat-dried milk (Bio-Rad Laboratories) in Tris-buffered saline (TBS/0.05% Tween 20 (TBS-T)) for 1h. Blots were incubated with the primary antibodies listed in Table 1 overnight at 4 °C. After washing, blots were incubated with anti-rabbit secondary antibody conjugated with horseradish peroxidase (Amersham Italia s.r.l., Milano, Italy) diluted 1:5000 in TBS-T. Detection of bound antibodies was performed with the Clarity Western ECL Substrate (Bio-Rad Laboratories) and images were acquired with Chemidoc (Bio-Rad Laboratories). Bands were analyzed using the ImageJ software (https://imagej.nih.gov/ij/download.html, accessed on 12 September 2022) for quantification, and normalization was completed using β-actin band intensities. Data represent the mean ± SEM, and were analyzed for statistical significance ($p < 0.05$) using the Student's *t*-test. All experiments were performed in triplicate and were repeated at least three times.

2.6. CNTF Treatment of LNCaP and 22Rv1 Cell Lines

After verifying the presence of the CNTFRα, a dose/responsive curve was performed to test the best CNTF concentration showing a significant response for cellular treatments of LNCaP and 22Rv1 cell lines. These cell lines were grown in RPMI 1640 (Life technologies, Waltham, MA, USA) supplemented with 10% Fetal Bovine Serum (FBS; Gibco, Thermo Fisher Scientific, Waltham, MA, USA). They were not serum-starved, counted and plated in six well plates. These cells were treated with 0, 2, 10 and 20 ng/mL by recombinant human CNTF (rhCNTF) for 15 min, to detect which of the following signaling pathways was trigged: pERK/ERK; pAKT/AKT; and pSTAT3/STAT3. These signaling pathways and the expression of MMP-2, MMP-9, GLUT-1 and -4 were analyzed by Western blotting as described above, using the primary antibodies shown in Table 1. pSTAT3, pAKT and pERK1/2 quantities were normalized using total STAT3, AKT and ERK1/2, respectively, while GLUT-1, GLUT-4, MMP-2, MMP-9 and PCNA quantitates were normalized by β-actin. Results were calculated in arbitrary units (AU).

2.7. Analysis of Glucose Uptake after CNTF Treatment

The 22Rv1 cells were seeded in 96-well black plates (2×10^4 cells per well), then the cells, untreated and treated with 10 ng/mL rhCNTF, were incubated at 37 °C for 24 h at 5% CO_2. The cells were then incubated for 45 min with 50 μM 2-nitrobenzodeoxyglucose (2-NBDG) in glucose-free Dulbecco Modified Eagle's Medium (DMEM) (Sigma, Milano,

Italy). The level of cellular fluorescent 2-NBDG was evaluated at 550/590 nm with Synergy Ht fluorescence microplate reader (Biotek Instruments, VT, USA). Fluorescence values have been normalized for the number of live cells (by MTT analysis) and reported as Mean Fluorescence Intensity (MFI).

2.8. Wound Healing Assay

The 22Rv1 cells showed MMP-2 downregulation if treated with rhCNTF. To investigate the role of CNTF in cell motility, these cells were grown to confluence and then scratch wounded with a sterile plastic micropipette tip. Then, the cells were rinsed 3 times with warm media to wash away scraped-off cells in the wound. The cells were untreated or treated with 10 ng/mL of rhCNTF in RPMI 1640 (Life technologies, Waltham, MA, USA) supplemented with 10% Fetal Bovine Serum (FBS; Gibco, Thermo Fisher Scientific, Waltham, MA, USA). Digital images of these cells were taken at 0, 4, 8, and 24 h. The area (Relative Wound Area) of the wound not occupied by cells was measured using a morphological imaging system ImageJ software (National Institutes of Health, http://imagej.nih.gov/ij accessed on 12 September 2022). The experiment was repeated at least 3 times.

2.9. Transwell Invasion Assay

Matrigel invasion assays were performed as previously reported [26]. Briefly, 24-well transwell inserts with 8.0 μm pores (#353097; Corning, NY, USA) were coated with 100 μL of 250 μg/mL LDEV-free Matrigel (#356234; Corning, NY, USA) diluted in serum-free RPMI 1640 medium according to the manufacturer's instructions. The 22Rv1 cells were pretreated with 10 ng/mL CNTF for 24 h before being seeded in the transwell chambers. In the upper chambers, 5×10^4 cells were seeded on top of the Matrigel and supplemented with 200 μL serum-free RPMI 1640, with/without 10 ng/mL CNTF. Then, 10% FBS in 750 μL of the medium was used as a chemoattractant in lower chambers. After 24 h, non-invading cells were gently removed from the apical side of the membrane by scrubbing with a cotton swab, whereas the cells remaining on the lower surface were fixed with 100% methanol for 20 min and stained with 0.5% crystal violet for 15 min. Cells in four non-overlapping fields were counted using ImageJ software. Data were expressed as mean cell number in each of the four fields. All experiments were repeated 3 times in triplicate.

2.10. Statistical Analysis

The variables were not normally distributed in the Shapiro–Wilk test, and a non-parametric approach was used in the analysis. The distribution of fluorescent intensity (glucose uptake assay) and of invaded cells in control and CNTF groups were compared using the Wilcoxon rank-sum test. The nonparametric analysis of longitudinal data in factorial experiments [27] was used to compare the distribution of the relative wound area (dependent variable) between the control and CNTF groups (wound healing assay), to evaluate the distribution of the dependent variable over the four time points, and to analyze the interaction between the groups and the time points. Statistical analyses were carried out using R 4.0.5 (R Foundation for Statistical Computing; Vienna, Austria). The data are expressed as median and IQR. The significance level was set at 0.05.

3. Results

3.1. Human PCa and CRPC Tissues Expressed CNTF and CNTFRα

Neoplastic cells of PCa (Figure 1a,c) and CRPC (Figure 1b,d) samples were highly stained for CNTF (Figure 1a,b) and CNTFRα (Figure 1c,d). Stromal components (*) were weakly positive for CNTF and mainly negative for CNTFRα. Endothelial cells (arrow in d) in the majority of prostate vessels were positive or weakly stained for CNTFRα and CNTF, respectively, in all prostate samples analyzed.

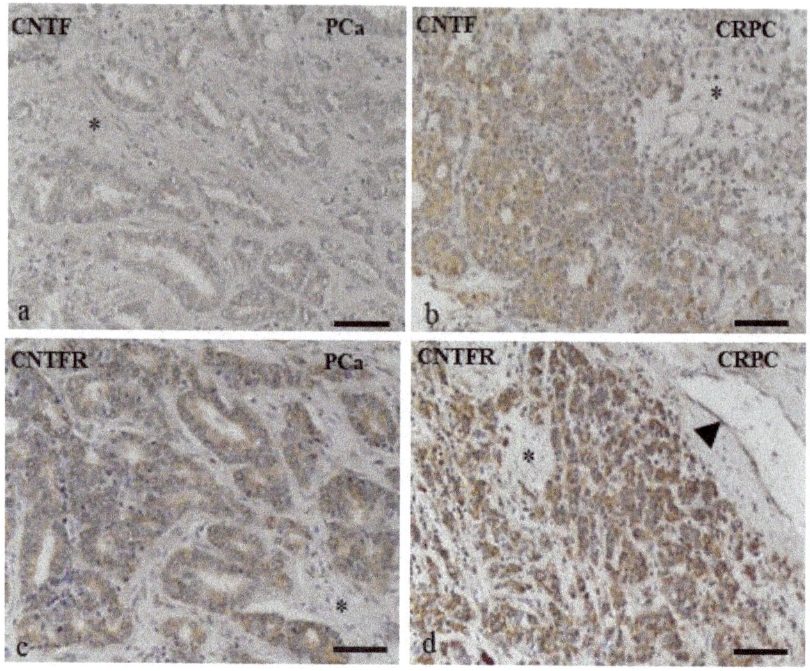

Figure 1. Immunohistochemistry localization of CNTF and CNTFRα in prostate cancer samples. CNTF (**a**,**b**) and CNTFRα (**c**,**d**) were expressed in PCa (**a**,**c**) and CRPC (**b**,**d**) tissues. Stromal components (*) were weakly positive for CNTF and mainly negative for CNTFRα. Endothelial cells (arrowhead in d) were positive or weakly stained for CNTFRα and CNTF, respectively. Scale bars: a,c = 100 μm; d,b = 200 μm.

3.2. LNCaP and 22Rv1 Cell Lines Expressed Both CNTF and CNTFRα

Immunopositivity for CNTF (Figure 2, green staining) was both nuclear (on chromatin; the nucleoli are negative) and cytoplasmatic in LNCaP and 22Rv1 cell lines (Figure 2b,h). The red signal for CNTFRα was present in the cytoplasm and cell membranes with a high intensity in the perinuclear region (Figure 2e,k) in both cell lines. Channels merging is shown in Figure 2c,f,i,l.

3.3. MAPK/ERK, AKT/PI3K and Jak/STAT Pathways Modulation after CNTF Treatments

rhCNTF at 2 ng/mL induced pSTAT3 phosphorylation, while at 10 ng/mL it induced pAKT and pERK de-phosphorylation in both LNCaP and 22Rv1 cell lines (Figure 3a,b, respectively). The histograms representing the quantitative analysis of pSTAT3, pAKT and pERK expression levels normalized by STAT3, AKT and ERK, respectively, were depicted on the right side of the representative Western blotting. The statistical analysis showed a significant pSTAT3 increase after rhCNTF treatments in both cell lines (Figure 3a,b). No significant decrease in pAKT and pERK after treatments with 2 ng/mL of rhCNTF were shown in both cell lines (Figure 3a,b) while treatments with 10 ng/mL of rhCNTF showed a significant decrease in pAKT ($p = 0.024$; $p = 0.0008$) and pERK ($p = 0.0007$; $p = 0.0009$) in theLNCaP and 22Rv1 cell lines, respectively. The effect of rhCNTF treatments on GLUT-1, GLUT-4, MMP-2, MMP-9 and PCNA regulation is depicted in Figure 4. The histograms of the quantitative analysis of these molecules normalized by β-actin were shown at the right side of the representative Western blotting. No significant difference was shown for all the molecules analyzed in the LNCaP cell line (Figure 4a) and for MMP-9, GLUT-4 and PCNA in the 22Rv1 cell line (Figure 4b). On the contrary, rhCNTF treatments (10 ng/mL) induced a significant decrease in GLUT-1 ($p = 0.0008$) and MMP-2 ($p = 0.03$) in the 22Rv1 cell line (Figure 4b).

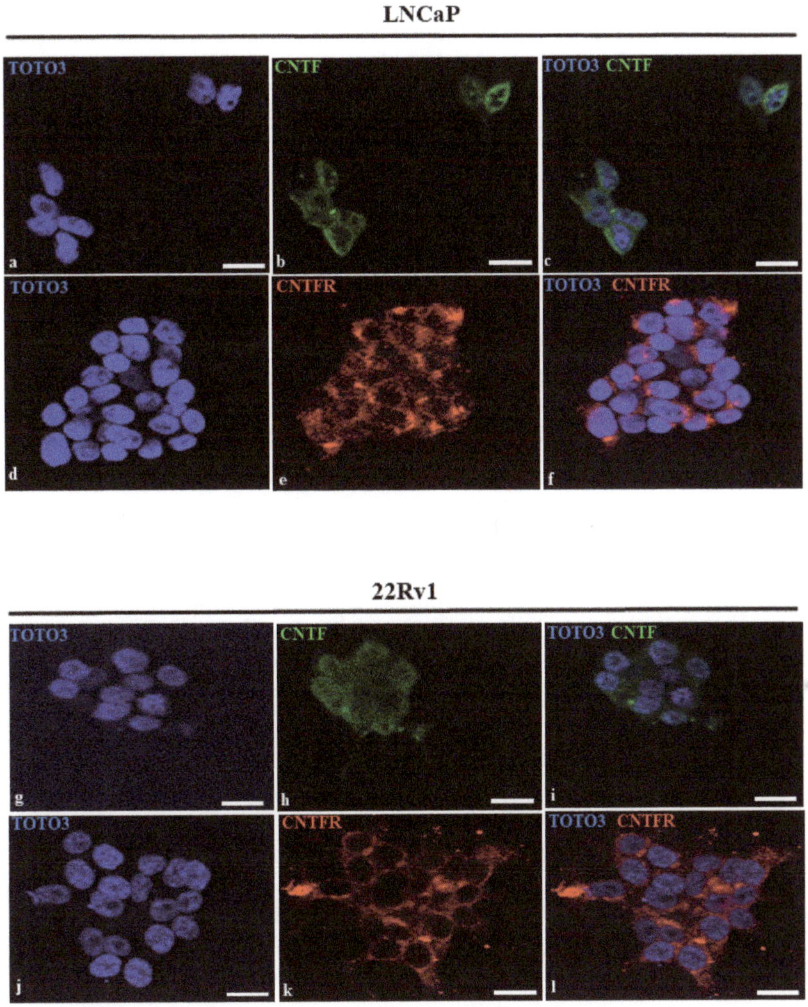

Figure 2. Immunofluorescence of CNTF and CNTFRα in LNCaP e 22Rv1cell lines. Nuclei (**a**,**d**,**g**,**j**) are stained in blue. CNTF (**b**,**h**) was stained in green and was mainly expressed in cytoplasm of both LNCaP and 22Rv1 cell lines. CNTFRα (**e**,**k**) was stained in red and was expressed in the cytoplasm and cell membranes with a high intensity in the perinuclear region of both cell lines. (**c**,**f**,**i**,**l**) represent merged channels. Scale bars: 30 μm.

3.4. CNTF Treatments Did Not Affect Cell Migration and Glucose Uptake but Reduce 22Rv1 Cells Invasiveness

The LNCaP cell line was not considered for the following experiments because rhCNTF treatments did not significantly modify GLUT-1, GLUT-4, MMP-2, MMP-9 and PCNA expressions. In order to investigate the role of CNTF in cell motility, the 22Rv1 cell lines were treated with 10 ng/mL hrCNTF for 0, 4, 8 or 24 h. As reported in Figure 5, no significant differences in cell migration were found at any times between the untreated and hrCNTF-treated 22Rv1 cells suggesting that CNTF is not involved in modulating cell migration as represented in the box plots below the wound healing photos.

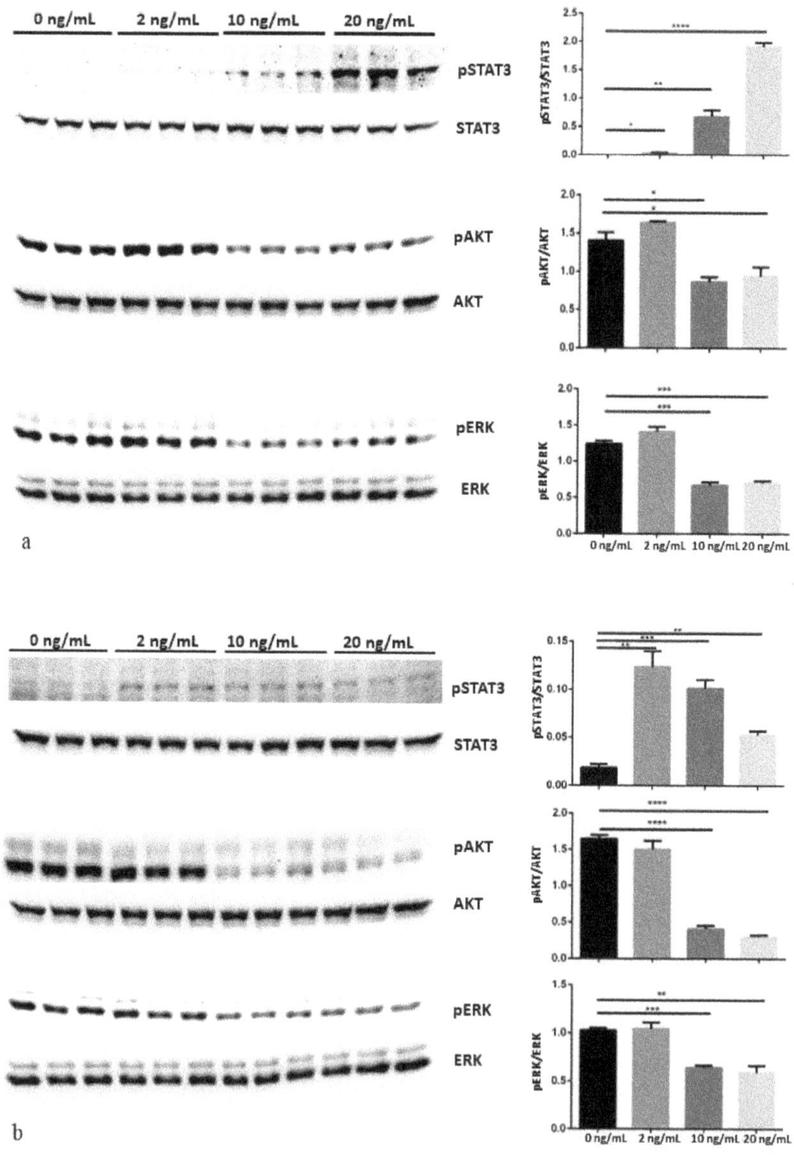

Figure 3. Dose-dependent response to rhCNTF of JAK/STAT3, PI3K/Akt and MAPK/ERK proteins pathways in LNCaP (**a**) and 22Rv1 (**b**) prostate cancer cell lines by Western blot analysis. LNCaP and 22Rv1 cell lines were treated with 0, 2, 10 and 20 ng/mL rhCNTF for 15 min. Bands were quantified and results were calculated in arbitrary units (AU) and reported as bars of a histogram. pSTAT3, pAKT and pERK1/2 quantities were normalized using total STAT3, AKT and ERK1/2, respectively. Data are represented as mean ± SD. * $p < 0.05$, ** $p < 0.01$, *** $p < 0.001$, **** $p < 0.0001$. Original images of immunoblotting data in the Figure S1.

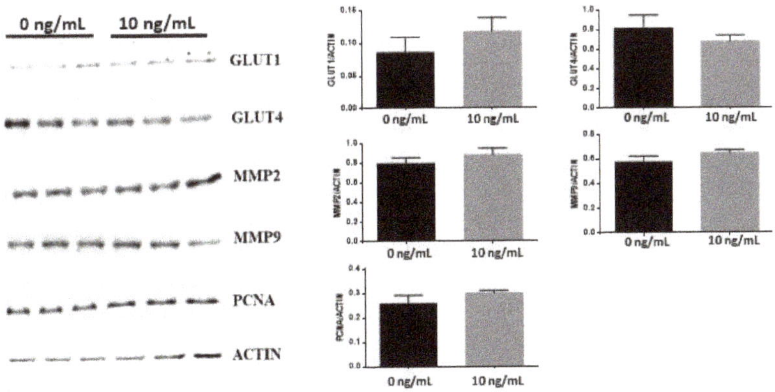

a

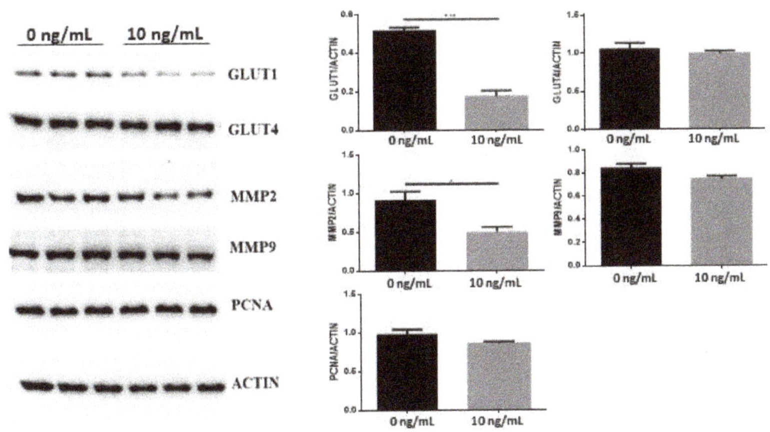

b

Figure 4. GLUT-1, GLUT-4, MMP-2, MMP-9 and PCNA expressions after treatment of LNCaP (**a**) and 22Rv1 (**b**) prostate cancer cell lines with 10 ng/mL rhCNTF for 24h. Bands were quantified and results were calculated in arbitrary units (AU) and reported as bars of a histogram. GLUT-1, GLUT-4, MMP-2, MMP-9 and PCNA expressions were normalized with β-actin expression. Data are represented as mean ± SD. * $p < 0.05$, *** $p < 0.001$. Original images of immunoblotting data in the Figure S1.

Moreover, in order to evaluate a possible role of CNTF in modulating glucose uptake, we treated the 22Rv1 cell lines with 10 ng/mL hrCNTF for 24 h and performed a glucose uptake assay. As reported in Figure 6, no changes in glucose uptake were found between CNTF-treated and -untreated cells proving that CNTF does not alter glucose uptake in the 22Rv1 cell line.

To evaluate the involvement of CNTF in 22Rv1 in invasiveness, we treated 22Rv1 cells with 10 ng/mL hrCNTF for 24 h. Interestingly, the distribution of the number of invaded cells is significantly ($p < 0.001$) lower in the cells treated with 10 ng/mL CNTF (median 89; IQR 71-98) compared to untreated cells (median 208; IQR 149-247) (Figure 7).

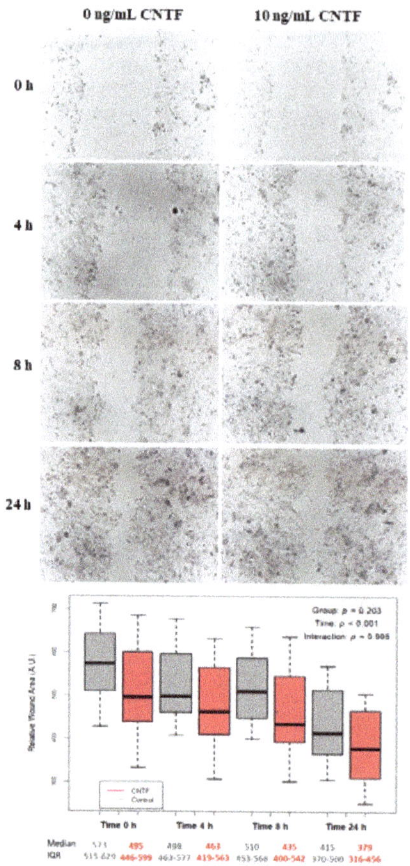

Figure 5. Wound healing assay after treatment with rhCNTF in 22Rv1 cell line. Representative photos of wound areas are shown; no significant difference in wound closure was found in cells treated with 10 ng/mL CNTF compared to untreated controls. Results are expressed in arbitrary units (A.U.) and reported as boxplots.

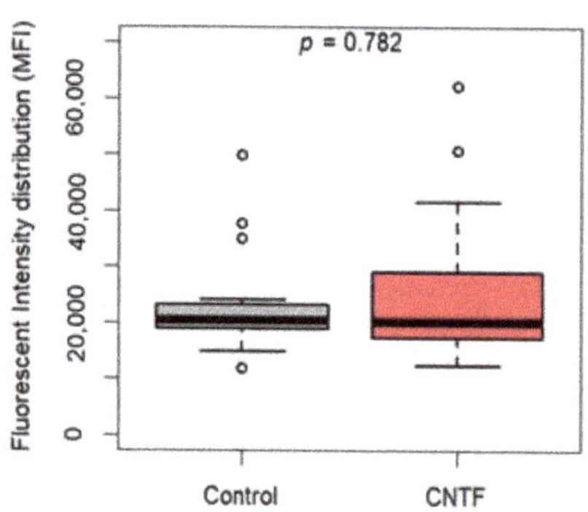

Figure 6. Glucose uptake after rhCNTF treatment of 22Rv1 cells. Boxplots show no differences in fluorescent intensity distribution (M.F.I.) between treated (median 20271, IQR 17416-28688) and untreated (median 20364, IQR 18784-23100) 22Rv1 cells.

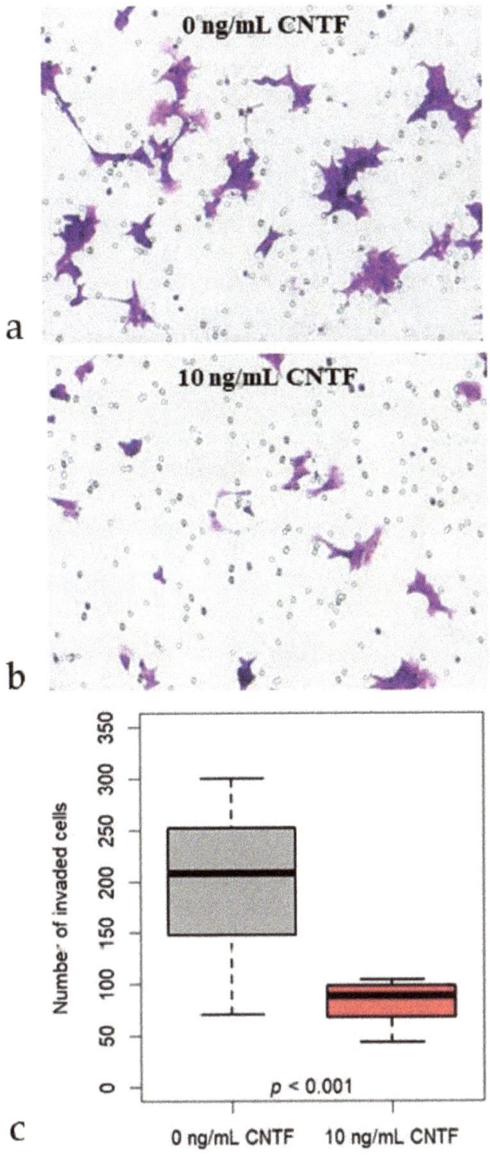

Figure 7. Transwell invasion assays. Representative images of 22Rv1 cells stained with crystal violet showing CNTF treatments effect: (**a**) untreated 22Rv1 cells; (**b**) 22Rv1 cells treated with 10 ng/mL rhCNTF; (**c**) Boxplot shows that the distribution of the number of 22Rv1 invaded cells is significantly ($p < 0.001$) lower in the group treated with 10 ng/mL CNTF (median 89; IQR 71-98) compared to the untreated controls (median 208; IQR 149-247).

4. Discussion

Despite many studies that have investigated CNTF and CNTFRα in various tissues and organs [26,28–31], there are no available data on the localization of these molecules and what role they have in PCa.

In the present study, we investigated these issues using both androgen-dependent (LNCaP) and androgen-independent (22Rv1) cell lines to mimic untreated PCa and CRPC, respectively. CNTF and CNTFRα were expressed in untreated PCa and CRPC, suggesting that CNTF can have a paracrine-constitutive role as previously suggested for normal prostate tissue [6]. In addition, our in vitro data showed that exogenous CNTF administration can modulate JAK/STAT3, MAPK/ERK and PI3K/AKT signaling pathways in both PCa cell models. In particular, CNTF upregulated pSTAT3 but downregulated pAKT and pERK in both cell culture models.

It is known that the activated ERK signaling pathway can facilitate the transformation of normal cells to malignant tumors [32] and that ERK1/2 activation is characteristic of CRPC [33]. In addition, Nickols and colleagues showed that frequent phosphorylation of ERK1/2 has been detected in metastatic CRPC [33]. In our hands, no modification was detected regarding PCNA expression, a proliferative marker. Therefore, we hypothesized that the upregulation of pSTAT3 could trigger apoptosis rather than a proliferative pathway by acting on Bcl-2/Bax balance as recently suggested [34,35]. In addition, we can speculate that the downregulation of pAKT and pERK due to CNTF can prevent cell proliferation and invasiveness, as our data demonstrated, masking the effect of pSTAT3 upregulation.

It is known that cell survival is also ensured by glucose supply and, particularly rapid glucose transport in cancer cells facilitated by GLUT proteins spanning across the cell membrane. Although there are several types of GLUT proteins, the GLUT-1 and GLUT-4 proteins have been pinpointed to be those involved in glucose transport through the cell membranes of neoplastic cells [14,36]. In addition, androgens are intimately associated with the development of PCa [37] by acting as important metabolic regulators stimulating the glycolytic flux in neoplastic prostate cells [12,38–40]. Both androgen-responsive and androgen non-responsive PCa cells display a distinct glycolytic profile [12]. In agreement, our experiments showed that GLUT-1 was decreased in 22Rv1 cells (androgen-non-responsive) after CNTF treatments by inhibiting MAPK/ERK and AKT/PI3K pathways while no significant modulation of GLUT-1 was present in LNCaP cells (androgen-responsive). Recently, it has been demonstrated that overexpression of GLUT-1 not only accelerates glucose metabolism but also protects cancer cells from glucose deprivation-induced oxidative stress [41]. Other studies involving human subjects suggested that high levels of GLUT-1 expression in tumors were associated with poor survival [20,21] and that knockdown of GLUT-1 inhibited cell glycolysis and proliferation and led to cell cycle arrest at G2/M phase in the 22RV1 cell line [42]. Moreover, these authors demonstrated that GLUT-1 knockdown inhibits the tumor growth in the 22Rv1 cell line and that GLUT-1 knockdown has no effect on cell migration in vitro. Interestingly, our data demonstrated that CNTF treatments, although decreasing GLUT-1 expression, did not affect glucose uptake and neoplastic cell motility, confirming in part the previous data [42].

Our data showing the downregulation of MAPK/ERK and AKT/PI3K pathways by CNTF treatments suggest that this molecule may be involved in MMP modulation that is downstream of these activation pathways. Generally, proteolytic degradation of extra-cellular matrix (ECM) is mediated by MMPs, with MMP-2 and MMP-9 playing a critical role in prostate cancer progression [43–45]. It was demonstrated that MMP-2 and MMP-9 inhibition suppressed the metastatic potential of PCa cells [46]. Moreover, MMP-2 overexpression by malignant prostatic epithelia was associated with a poor disease-free survival [47,48]. Interestingly, our data showed that CNTF treatments significantly inhibited MMP-2 expression and our invasion assay demonstrated a significant inhibition of cell invasiveness in the 22Rv1 cell model mimicking CRPC. A limit of this study is the definitive proof that MMP-2 downregulation is responsible for significant inhibition of 22Rv1 cell invasiveness. Therefore, our findings should be corroborated by MMP-2 overexpression and silencing assays in 22Rv1 cells treated with CNTF.

5. Conclusions

We affirm that CNTF has a pivotal role in the prostate-cancer environment remodeling and suggest that this cytokine may be viewed as a modulator of invasion processes. In addition, we demonstrated that CNTF downregulated MMP-2 expression which is the main MMP involved in cell cancer invasiveness (Figure 8).

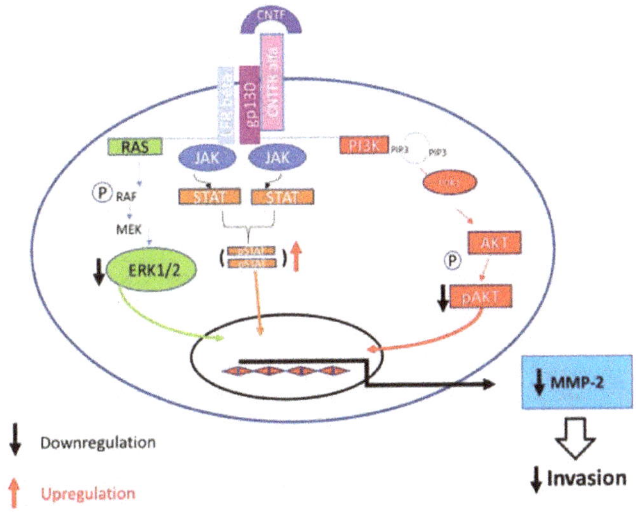

Figure 8. CNTF pathway in PCa. The binding of CNTF to the receptor actives JAK/STAT3 and downregulates MAPK/ERK, PI3K/AKT signaling pathways in CRPC cell model (22Rv1 cell line). Subsequently, MMP-2 is downregulated and cell invasion decreases.

More studies are clearly required to corroborate our data, as in vivo male murine studies concerning allographic transplantation of 22Rv1 cells in CNTF-treated and -untreated conditions to evaluate the CNTF effect in tumor growth and metastasis inhibition. These further studies could lead to a novel therapeutic approach in patients with CRPC.

Supplementary Materials: The following supporting information can be downloaded at: https://www.mdpi.com/article/10.3390/cancers14235917/s1, Figure S1: Original images of immunoblotting data.

Author Contributions: Conceptualization, G.T., R.M. and D.M.; methodology, G.T., S.F., R.G., M.S. and F.M.; software, R.G. and G.T.; investigation, G.T., S.F., R.G., M.S. and F.M.; data curation, G.T., R.G., D.M. and R.M.; writing—original draft preparation, G.T., D.M. and R.M.; writing—review and editing, G.T., D.M., R.M., G.G. and L.T.; funding acquisition, L.T., D.M. and R.M. All authors have read and agreed to the published version of the manuscript.

Funding: This research was funded by Università Politecnica delle Marche, grant number RSA 2022 to D.M and R.M.

Institutional Review Board Statement: The study was conducted in accordance with the Declaration of Helsinki, and approved by the Ethics Committee of Marche Region (IT) (protocol number 2020.395).

Informed Consent Statement: Informed consent was obtained from all subjects involved in the study according to the Ethics Committee.

Data Availability Statement: The data will be shared after the institutional approval.

Conflicts of Interest: The authors declare no conflict of interest.

References

1. Sung, H.; Ferlay, J.; Siegel, R.L.; Laversanne, M.; Soerjomataram, I.; Jemal, A.; Bray, F. Global Cancer Statistics 2020: GLOBOCAN Estimates of Incidence and Mortality Worldwide for 36 Cancers in 185 Countries. *CA Cancer J. Clin.* **2021**, *71*, 209–249. [CrossRef] [PubMed]
2. Marklund, M.; Schultz, N.; Friedrich, S.; Berglund, E.; Tarish, F.; Tanoglidi, A.; Liu, Y.; Bergenstrahle, L.; Erickson, A.; Helleday, T.; et al. Spatio-temporal analysis of prostate tumors in situ suggests pre-existence of treatment-resistant clones. *Nat. Commun.* **2022**, *13*, 5475. [CrossRef] [PubMed]
3. Culig, Z.; Puhr, M. Interleukin-6 and prostate cancer: Current developments and unsolved questions. *Mol. Cell. Endocrinol.* **2018**, *462*, 25–30. [CrossRef] [PubMed]
4. Pencik, J.; Wiebringhaus, R.; Susani, M.; Culig, Z.; Kenner, L. IL-6/STAT3/ARF: The guardians of senescence, cancer progression and metastasis in prostate cancer. *Swiss. Med. Wkly.* **2015**, *145*, w14215. [CrossRef]
5. Culig, Z. Proinflammatory cytokine interleukin-6 in prostate carcinogenesis. *Am. J. Clin. Exp. Urol.* **2014**, *2*, 231–238.
6. Fantone, S.; Tossetta, G.; Montironi, R.; Senzacqua, M.; Marzioni, D.; Mazzucchelli, R. Ciliary neurotrophic factor (CNTF) and its receptor (CNTFRalpha) signal through MAPK/ERK pathway in human prostate tissues: A morphological and biomolecular study. *Eur. J. Histochem.* **2020**, *64*, 3147. [CrossRef] [PubMed]
7. Jones, S.A.; Jenkins, B.J. Recent insights into targeting the IL-6 cytokine family in inflammatory diseases and cancer. *Nat. Rev. Immunol.* **2018**, *18*, 773–789. [CrossRef]
8. Johnson, D.E.; O'Keefe, R.A.; Grandis, J.R. Targeting the IL-6/JAK/STAT3 signalling axis in cancer. *Nat. Rev. Clin. Oncol.* **2018**, *15*, 234–248. [CrossRef]
9. Rascio, F.; Spadaccino, F.; Rocchetti, M.T.; Castellano, G.; Stallone, G.; Netti, G.S.; Ranieri, E. The Pathogenic Role of PI3K/AKT Pathway in Cancer Onset and Drug Resistance: An Updated Review. *Cancers* **2021**, *13*, 3949. [CrossRef]
10. Butler, D.E.; Marlein, C.; Walker, H.F.; Frame, F.M.; Mann, V.M.; Simms, M.S.; Davies, B.R.; Collins, A.T.; Maitland, N.J. Inhibition of the PI3K/AKT/mTOR pathway activates autophagy and compensatory Ras/Raf/MEK/ERK signalling in prostate cancer. *Oncotarget* **2017**, *8*, 56698–56713. [CrossRef]
11. Mulholland, D.J.; Kobayashi, N.; Ruscetti, M.; Zhi, A.; Tran, L.M.; Huang, J.; Gleave, M.; Wu, H. Pten loss and RAS/MAPK activation cooperate to promote EMT and metastasis initiated from prostate cancer stem/progenitor cells. *Cancer Res.* **2012**, *72*, 1878–1889. [CrossRef] [PubMed]
12. Vaz, C.V.; Alves, M.G.; Marques, R.; Moreira, P.I.; Oliveira, P.F.; Maia, C.J.; Socorro, S. Androgen-responsive and nonresponsive prostate cancer cells present a distinct glycolytic metabolism profile. *Int. J. Biochem. Cell Biol.* **2012**, *44*, 2077–2084. [CrossRef] [PubMed]
13. Gonzalez-Menendez, P.; Hevia, D.; Rodriguez-Garcia, A.; Mayo, J.C.; Sainz, R.M. Regulation of GLUT transporters by flavonoids in androgen-sensitive and -insensitive prostate cancer cells. *Endocrinology* **2014**, *155*, 3238–3250. [CrossRef] [PubMed]
14. Singh, S. Glucose decorated gold nanoclusters: A membrane potential independent fluorescence probe for rapid identification of cancer cells expressing Glut receptors. *Colloids Surf. B Biointerfaces* **2017**, *155*, 25–34. [CrossRef] [PubMed]
15. Szablewski, L. Expression of glucose transporters in cancers. *Biochim. Biophys. Acta* **2013**, *1835*, 164–169. [CrossRef]
16. Cairns, R.A.; Harris, I.S.; Mak, T.W. Regulation of cancer cell metabolism. *Nat. Rev. Cancer* **2011**, *11*, 85–95. [CrossRef]
17. Hong, S.Y.; Yu, F.X.; Luo, Y.; Hagen, T. Oncogenic activation of the PI3K/Akt pathway promotes cellular glucose uptake by downregulating the expression of thioredoxin-interacting protein. *Cell Signal.* **2016**, *28*, 377–383. [CrossRef]
18. Makinoshima, H.; Takita, M.; Saruwatari, K.; Umemura, S.; Obata, Y.; Ishii, G.; Matsumoto, S.; Sugiyama, E.; Ochiai, A.; Abe, R.; et al. Signaling through the Phosphatidylinositol 3-Kinase (PI3K)/Mammalian Target of Rapamycin (mTOR) Axis Is Responsible for Aerobic Glycolysis mediated by Glucose Transporter in Epidermal Growth Factor Receptor (EGFR)-mutated Lung Adenocarcinoma. *J. Biol. Chem.* **2015**, *290*, 17495–17504. [CrossRef]
19. Avanzato, D.; Pupo, E.; Ducano, N.; Isella, C.; Bertalot, G.; Luise, C.; Pece, S.; Bruna, A.; Rueda, O.M.; Caldas, C.; et al. High USP6NL Levels in Breast Cancer Sustain Chronic AKT Phosphorylation and GLUT1 Stability Fueling Aerobic Glycolysis. *Cancer Res.* **2018**, *78*, 3432–3444. [CrossRef] [PubMed]
20. Macheda, M.L.; Rogers, S.; Best, J.D. Molecular and cellular regulation of glucose transporter (GLUT) proteins in cancer. *J. Cell. Physiol.* **2005**, *202*, 654–662. [CrossRef]
21. Chandler, J.D.; Williams, E.D.; Slavin, J.L.; Best, J.D.; Rogers, S. Expression and localization of GLUT1 and GLUT12 in prostate carcinoma. *Cancer* **2003**, *97*, 2035–2042. [CrossRef]
22. Fantone, S.; Mazzucchelli, R.; Giannubilo, S.R.; Ciavattini, A.; Marzioni, D.; Tossetta, G. AT-rich interactive domain 1A protein expression in normal and pathological pregnancies complicated by preeclampsia. *Histochem. Cell Biol.* **2020**, *154*, 339–346. [CrossRef] [PubMed]
23. Tossetta, G.; Fantone, S.; Giannubilo, S.R.; Marinelli Busilacchi, E.; Ciavattini, A.; Castellucci, M.; Di Simone, N.; Mattioli-Belmonte, M.; Marzioni, D. Pre-eclampsia onset and SPARC: A possible involvement in placenta development. *J. Cell. Physiol.* **2019**, *234*, 6091–6098. [CrossRef]
24. Marinelli Busilacchi, E.; Costantini, A.; Mancini, G.; Tossetta, G.; Olivieri, J.; Poloni, A.; Viola, N.; Butini, L.; Campanati, A.; Goteri, G.; et al. Nilotinib Treatment of Patients Affected by Chronic Graft-versus-Host Disease Reduces Collagen Production and Skin Fibrosis by Downmodulating the TGF-beta and p-SMAD Pathway. *Biol. Blood Marrow Transplant.* **2020**, *26*, 823–834. [CrossRef] [PubMed]

25. Perugini, J.; Di Mercurio, E.; Tossetta, G.; Severi, I.; Monaco, F.; Reguzzoni, M.; Tomasetti, M.; Dani, C.; Cinti, S.; Giordano, A. Biological Effects of Ciliary Neurotrophic Factor on hMADS Adipocytes. *Front. Endocrinol.* **2019**, *10*, 768. [CrossRef] [PubMed]
26. Tossetta, G.; Fantone, S.; Busilacchi, E.M.; Di Simone, N.; Giannubilo, S.R.; Scambia, G.; Giordano, A.; Marzioni, D. Modulation of matrix metalloproteases by ciliary neurotrophic factor in human placental development. *Cell Tissue Res.* **2022**, *390*, 113–129. [CrossRef]
27. Noguchi, K.; Gel, Y.R.; Brunner, E.; Konietschke, F. nparLD: An R Software Package for the Nonparametric Analysis of Longitudinal Data in Factorial Experiments. *J. Stat. Softw.* **2012**, *50*, 1–23. [CrossRef]
28. Chen, J.; Chen, P.; Backman, L.J.; Zhou, Q.; Danielson, P. Ciliary Neurotrophic Factor Promotes the Migration of Corneal Epithelial Stem/progenitor Cells by Up-regulation of MMPs through the Phosphorylation of Akt. *Sci. Rep.* **2016**, *6*, 25870. [CrossRef]
29. Ding, J.; He, Z.; Ruan, J.; Ma, Z.; Liu, Y.; Gong, C.; Iqbal, K.; Sun, S.; Chen, H. Role of ciliary neurotrophic factor in the proliferation and differentiation of neural stem cells. *J. Alzheimers Dis.* **2013**, *37*, 587–592. [CrossRef]
30. McGregor, N.E.; Poulton, I.J.; Walker, E.C.; Pompolo, S.; Quinn, J.M.; Martin, T.J.; Sims, N.A. Ciliary neurotrophic factor inhibits bone formation and plays a sex-specific role in bone growth and remodeling. *Calcif. Tissue Int.* **2010**, *86*, 261–270. [CrossRef]
31. Johnson, R.W.; White, J.D.; Walker, E.C.; Martin, T.J.; Sims, N.A. Myokines (muscle-derived cytokines and chemokines) including ciliary neurotrophic factor (CNTF) inhibit osteoblast differentiation. *Bone* **2014**, *64*, 47–56. [CrossRef]
32. Xu, C.J.; Dong, L.L.; Kang, X.L.; Li, Z.M.; Zhang, H.Y. Leptin promotes proliferation and inhibits apoptosis of prostate cancer cells by regulating ERK1/2 signaling pathway. *Eur. Rev. Med. Pharmacol. Sci.* **2020**, *24*, 8341–8348. [CrossRef]
33. Nickols, N.G.; Nazarian, R.; Zhao, S.G.; Tan, V.; Uzunangelov, V.; Xia, Z.; Baertsch, R.; Neeman, E.; Gao, A.C.; Thomas, G.V.; et al. MEK-ERK signaling is a therapeutic target in metastatic castration resistant prostate cancer. *Prostate Cancer Prostatic Dis.* **2019**, *22*, 531–538. [CrossRef]
34. Rezende, A.C.; Vieira, A.S.; Rogerio, F.; Rezende, L.F.; Boschero, A.C.; Negro, A.; Langone, F. Effects of systemic administration of ciliary neurotrophic factor on Bax and Bcl-2 proteins in the lumbar spinal cord of neonatal rats after sciatic nerve transection. *Braz. J. Med. Biol. Res.* **2008**, *41*, 1024–1028. [CrossRef]
35. Guo, X.; Liu, X. Nogo receptor knockdown and ciliary neurotrophic factor attenuate diabetic retinopathy in streptozotocin-induced diabetic rats. *Mol. Med. Rep.* **2017**, *16*, 2030–2036. [CrossRef]
36. Hu, H.; Lee, H.J.; Jiang, C.; Zhang, J.; Wang, L.; Zhao, Y.; Xiang, Q.; Lee, E.O.; Kim, S.H.; Lu, J. Penta-1,2,3,4,6-O-galloyl-beta-D-glucose induces p53 and inhibits STAT3 in prostate cancer cells in vitro and suppresses prostate xenograft tumor growth in vivo. *Mol. Cancer Ther.* **2008**, *7*, 2681–2691. [CrossRef]
37. Wang, D.; Tindall, D.J. Androgen action during prostate carcinogenesis. *Methods Mol. Biol.* **2011**, *776*, 25–44. [CrossRef]
38. Emonds, K.M.; Swinnen, J.V.; van Weerden, W.M.; Vanderhoydonc, F.; Nuyts, J.; Mortelmans, L.; Mottaghy, F.M. Do androgens control the uptake of 18F-FDG, 11C-choline and 11C-acetate in human prostate cancer cell lines? *Eur. J. Nucl. Med. Mol. Imaging* **2011**, *38*, 1842–1853. [CrossRef]
39. Moon, J.S.; Jin, W.J.; Kwak, J.H.; Kim, H.J.; Yun, M.J.; Kim, J.W.; Park, S.W.; Kim, K.S. Androgen stimulates glycolysis for de novo lipid synthesis by increasing the activities of hexokinase 2 and 6-phosphofructo-2-kinase/fructose-2,6-bisphosphatase 2 in prostate cancer cells. *Biochem. J.* **2011**, *433*, 225–233. [CrossRef]
40. Massie, C.E.; Lynch, A.; Ramos-Montoya, A.; Boren, J.; Stark, R.; Fazli, L.; Warren, A.; Scott, H.; Madhu, B.; Sharma, N.; et al. The androgen receptor fuels prostate cancer by regulating central metabolism and biosynthesis. *EMBO J.* **2011**, *30*, 2719–2733. [CrossRef]
41. Shao, M.; Yu, Z.; Zou, J. LncRNA-SNHG16 Silencing Inhibits Prostate Carcinoma Cell Growth, Downregulate GLUT1 Expression and Reduce Glucose Uptake. *Cancer Manag. Res.* **2020**, *12*, 1751–1757. [CrossRef]
42. Xiao, H.; Wang, J.; Yan, W.; Cui, Y.; Chen, Z.; Gao, X.; Wen, X.; Chen, J. GLUT1 regulates cell glycolysis and proliferation in prostate cancer. *Prostate* **2018**, *78*, 86–94. [CrossRef]
43. Chen, Y.; Zheng, L.; Liu, J.; Zhou, Z.; Cao, X.; Lv, X.; Chen, F. Shikonin inhibits prostate cancer cells metastasis by reducing matrix metalloproteinase-2/-9 expression via AKT/mTOR and ROS/ERK1/2 pathways. *Int. Immunopharmacol.* **2014**, *21*, 447–455. [CrossRef]
44. Stearns, M.E.; Wang, M.; Stearns, M. IL-10 blocks collagen IV invasion by "invasion stimulating factor" activated PC-3 ML cells: Upregulation of TIMP-1 expression. *Oncol. Res.* **1995**, *7*, 157–163.
45. Nemeth, J.A.; Yousif, R.; Herzog, M.; Che, M.; Upadhyay, J.; Shekarriz, B.; Bhagat, S.; Mullins, C.; Fridman, R.; Cher, M.L. Matrix metalloproteinase activity, bone matrix turnover, and tumor cell proliferation in prostate cancer bone metastasis. *J. Natl. Cancer Inst.* **2002**, *94*, 17–25. [CrossRef]
46. Chien, C.S.; Shen, K.H.; Huang, J.S.; Ko, S.C.; Shih, Y.W. Antimetastatic potential of fisetin involves inactivation of the PI3K/Akt and JNK signaling pathways with downregulation of MMP-2/9 expressions in prostate cancer PC-3 cells. *Mol. Cell. Biochem.* **2010**, *333*, 169–180. [CrossRef]
47. Trudel, D.; Fradet, Y.; Meyer, F.; Harel, F.; Tetu, B. Significance of MMP-2 expression in prostate cancer: An immunohistochemical study. *Cancer Res.* **2003**, *63*, 8511–8515.
48. Trudel, D.; Fradet, Y.; Meyer, F.; Harel, F.; Tetu, B. Membrane-type-1 matrix metalloproteinase, matrix metalloproteinase 2, and tissue inhibitor of matrix proteinase 2 in prostate cancer: Identification of patients with poor prognosis by immunohistochemistry. *Hum. Pathol.* **2008**, *39*, 731–739. [CrossRef]

cancers

MDPI

Article

The Clinical Utility of Systemic Immune-Inflammation Index Supporting Charlson Comorbidity Index and CAPRA-S Score in Determining Survival after Radical Prostatectomy—A Single Centre Study

Piotr Zapała [1,*], Karolina Garbas [1], Zbigniew Lewandowski [2], Łukasz Zapała [1,*], Aleksander Ślusarczyk [1], Cezary Ślusarczyk [1], Łukasz Mielczarek [3] and Piotr Radziszewski [1]

1 Clinic of General, Oncological and Functional Urology, Medical University of Warsaw, Lindleya 4, 02-005 Warsaw, Poland
2 Department of Epidemiology, Medical University of Warsaw, Oczki 3, 02-007 Warsaw, Poland
3 Second Department of Urology, Centre of Postgraduate Medical Education, Cegłowska 80, 01-809 Warsaw, Poland
* Correspondence: zapala.piotrek@gmail.com (P.Z.); lzapala@wum.edu.pl (Ł.Z.)

Simple Summary: Radical treatment of prostate cancer (PCa) provides excellent oncological outcomes. However, curative treatment of primary PCa, as well as salvage treatment of biochemical recurrence after radical treatment, requires at least a 10-year life expectancy to be beneficial. To provide an accurate selection for active treatment, several tools evaluating individual life expectancy have been developed. Our retrospective study aimed to determine the utility of the systemic immune-inflammation index (SII) in predicting early survival when used as an adjunct to CAPRA-S and Charlson comorbidity index (CCI) scores in non-metastatic PCa. We confirmed the SII as an independent predictor of survival. We have also validated the SII as a supplement to scoring systems when stratifying the risk of early mortality. In the setting of patients that might require salvage treatment, supplementing comorbidity status with the SII provided accurate discrimination of survival. The SII seems then to be useful when estimating life expectancy in patients with non-metastatic PCa.

Abstract: The selection of candidates for the curative treatment of PCa requires a careful assessment of life expectancy. Recently, blood-count inflammatory markers have been introduced as prognosticators of oncological and non-oncological outcomes in different settings. This retrospective, monocentric study included 421 patients treated with radical prostatectomy (RP) for nonmetastatic PCa and aimed at determining the utility of a preoperative SII (neutrophil count × platelet count/lymphocyte count) in predicting survival after RP. Patients with high SIIs (≥900) presented significantly shorter survival ($p = 0.02$) and high SIIs constituted an independent predictor of overall survival [HR 2.54 (95%CI 1.24–5.21); $p = 0.01$] when adjusted for high (≥6) age-adjusted CCI (ACCI) [HR 2.75 (95%CI 1.27–5.95); $p = 0.01$] and high (≥6) CAPRA-S [HR 2.65 (95%CI 1.32–5.31); $p = 0.006$]. Patients with high scores (ACCI and/or CAPRA-S) and high SIIs were at the highest risk of death ($p < 0.0001$) with approximately a one-year survival loss during the first seven years after surgery. In subgroup of high CAPRA-S (≥6), patients with high ACCIs and high SIIs were at the highest risk of death ($p < 0.0001$). Our study introduces the SII as a straightforward marker of mortality after RP that can be helpful in pre- and postoperative decision-making.

Keywords: prostate cancer; life expectancy; systemic immune-inflammation index; Charlson comorbidity index; CAPRA-S; early survival

Citation: Zapała, P.; Garbas, K.; Lewandowski, Z.; Zapała, Ł.; Ślusarczyk, A.; Ślusarczyk, C.; Mielczarek, Ł.; Radziszewski, P. The Clinical Utility of Systemic Immune-Inflammation Index Supporting Charlson Comorbidity Index and CAPRA-S Score in Determining Survival after Radical Prostatectomy—A Single Centre Study. *Cancers* **2022**, *14*, 4135. https://doi.org/10.3390/cancers14174135

Academic Editors: José I. López and Claudia Manini

Received: 30 July 2022
Accepted: 24 August 2022
Published: 26 August 2022

1. Introduction

Prostate cancer (PCa) is the second most prevalent cancer in males, with 1.4 million patients diagnosed in 2020 worldwide [1] and the lifetime risk of PCa-specific death reaching 2.5–4% [2]. Due to favourable survival in patients with untreated localized disease, active treatment is considered beneficial only in patients with a life expectancy of at least 10 years [3]. The majority of comorbid, elderly patients will die from competing causes and the impact of their tumours' characteristics on their 10-year overall survival is unclear [4], which necessitates the prediction of life expectancy in patients with PCa as a part of clinical practice [5–8]. Simultaneously, stratifying survival in patients treated with curative intent remains challenging due to favourable prognoses and extensive censoring in observational studies, leaving the definition of the patient at risk of early death yet to be specified. The European Association of Urology mentions the commonly used Charlson comorbidity index (CCI) and the age-adjusted Charlson comorbidity index (ACCI) as convenient measures of non-cancer-specific death risk [8,9], whereas the Cancer of the Prostate Risk Assessment Postsurgical score (CAPRA-S) based on pathological characteristics has been extensively validated to predict biochemical recurrence [10] and cancer-specific mortality after radical prostatectomy (RP) [11].

In recent years, complete blood count inflammatory markers have been sequentially introduced as predictors of survival in localized renal cell cancer [12–14] and bladder cancer [15,16], as well as in radio-recurrent [17] and castration-resistant prostate cancer [18]. The systemic immune-inflammation index (SII) is a novel, recently developed blood count-derived marker that has shown particularly promising predictive properties in the cohort with non-metastatic prostate cancer [19], which constitutes a population with excellent early survival. In otherwise treatment-naïve, nonmetastatic patients, a high SII was associated with a risk of biochemical recurrence (BCR) after radical prostatectomy independently from postoperative staging and grading [20]. In a similar analysis of a large cohort by Rajwa et al., a high SII was an independent predictor of BCR both in preoperative and postoperative settings [19]. In the retrospective cohort of 214 patients with radio-recurrent PCa, a high SII was an independent prognostic factor for overall and cancer-specific survival both in the preoperative and postoperative adjustment settings [17]. Although data on radio-recurrent patients suggest an impact of SII on survival, the evidence of long-term outcomes in patients treated a priori with RP is limited to the association with biochemical recurrence.

The study aimed to determine the utility of the systemic immune-inflammation index in predicting early survival when used as an adjunct to CAPRA-S and CCI scores in patients with non-metastatic prostate cancer.

2. Materials and Methods

2.1. Patients

Patients treated with radical prostatectomy in the years 2012–2018 for clinically non-metastatic prostate cancer in a single tertiary centre were included in the analysis. All patients underwent routine metastatic screening as recommended by a relevant edition of the European Association of Urology Guidelines. This consisted of at least a bone scan in intermediate-risk disease and a bone scan combined with cross-sectional abdominopelvic imaging in high-risk disease. Patients treated with salvage intention or neoadjuvant hormonotherapy were excluded.

2.2. Data Collection

Clinical and pathological data were retrieved from the department database. Blood cell counts were obtained from routinely performed preoperative evaluations during the hospitalization. The systemic immune–inflammation index (SII) was calculated by multiplying the neutrophil count by the platelet count and dividing the score by the lymphocyte count as previously [17,19,20]. The Charlson comorbidity index was calculated based on its last update [21] and included 19 conditions with scores assigned based on their severity. For age-adjusted CCI, beginning at the age of 50 years, one point was added to the CCI for each subsequent decade. The cut-off for the high score was set at 4 or greater for CCI, or 6 or greater for age-adjusted CCI. Both calculations included prostate cancer (2 points) as a covariable. The cut-off selection for CCI and age-adjusted CCI (high vs. low) was based on cut-offs used in previous localized PCa cohorts, with a correction for PCa which was included in our calculations [22,23]. The CAPRA-S score included prostate-specific antigen (PSA), pathologic Gleason score, positive surgical margins, extracapsular extension, seminal vesicle invasion, and lymph node invasion. CAPRA-S groups were defined as previously with high risk defined as 6 or greater [10].

Survival follow-up data were retrieved from the Central Statistical Office. The study was approved by the Ethics Committee of the Medical University of Warsaw (nr AKBE/58/2022; 21 February 2022).

2.3. Statistical Analysis

All analyses were performed using SAS 9.4 software (SAS Institute, Cary, NC, USA). Continuous and qualitative variables were compared utilizing Mann–Whitney's U-test and Fisher's exact test, respectively. The cut-off for SII was set at a value maximizing the difference between overall survival probabilities, rounded to the nearest 100 units. Overall survival estimates were calculated using the Kaplan–Meier method. The log-rank test was used for comparisons of survival curves between predefined risk groups. For multiple comparisons, a log-rank test with an adjustment for multiple comparisons was implemented. Cox regression models were utilized for multivariate analysis issues to determine independent predictors of early death after RP. For clinical convenience, a quantitative evaluation of survival was performed utilizing restricted mean survival time. The threshold for significance was set at $p < 0.05$.

3. Results

3.1. Baseline Characteristics

A total of 421 patients were included in the analysis. All patients were white, non-Hispanic men. High SIIs (≥900) were present in 218 patients (51.80%). In general, a high SII was not associated with significantly higher pre- and postprostatectomy grading or adverse pathological features (APF). The baseline characteristics of the patients stratified by SII (high vs. low) are summarized in Table 1.

Table 1. Baseline pre- and postprostatectomy characteristics of patients stratified by SII (high vs. low).

Variable		Overall	SII < 900	SII ≥ 900	*p*
		Clinical baseline characteristics			
PSA (ng/mL, median, IQR)		7.40 (6.7)	7.5 (6.7)	7 (5.5)	0.44
cT (*n*%)	cT1	149 149 (37.44%)	101 (34.95%)	45 (43.27%)	0.17
	cT2	244 (61.31%)	183 (63.32%)	59 (56.73%)	
	≥cT3	5 (1.26%)	5 (1.73%)	0	
Biopsy grade group (*n*,%)	1	158 (38.73%)	115 (38.33%)	40 (38.83%)	0.73
	2	128 (31.37%)	90 (30%)	37 (35.92%)	
	3	54 (13.24%)	41 (13.67%)	13 (12.62%)	
	4	47 (11.52%)	37 (12.33%)	9 (8.74%)	
	5	21 (5.15%)	17 (5.67%)	4 (3.88%)	
CCI (*n*,%)	2	308 (73.33%)	224 72.49%)	79 (74.53%)	0.37
	3	78 (18.57%)	56 (18.12%)	22 (20.75%)	
	4	27 (6.43%)	24 (7.77%)	3 (2.83%)	
	5	4 (0.95%)	3 (0.97%)	1 (0.94%)	
	6	3 (0.71%)	2 (0.65%)	1 (0.94%)	
Age (years, median, IQR)		65 (8)	65 (8)	64 (8)	0.57
		Postprostatectomy specimen			
Prostatectomy grade group (*n*,%)	1	58 (13.94%)	46 (14.98%)	10 (9.62%)	0.30
	2	165 (39.66%)	116 (37.79%)	47 (45.19%)	
	3	89 (21.39%)	62 (20.20%)	26 (25%)	
	4	62 (14.90%)	50 (16.29%)	12 (11.54%)	
	5	42 (10.10%)	33 (10.75%)	9 (8.65%)	

Table 1. *Cont.*

Variable		Overall	SII < 900	SII ≥ 900	*p*
pT (*n*,%)	pT2	241 (57.93%)	171 (55.70%)	65 (62.50%)	0.41
	pT3	173 (41.59%)	134 (43.65%)	39 (37.50%)	
	pT4	2 (0.48%)	2 (0.65%)	0	
pN (*n*,%)	pN0	181 (43.20%)	134 (43.51%)	45 42.45%)	0.20
	pN1+	31 (7.40%)	27 (8.77%)	4 (3.77%)	
	pNx	207 (49.40%)	147 (47.73%)	57 (53.77%)	
EPE (*n*,%)		175 (42.37%)	136 (44.59%)	39 (37.86%)	0.25
SVI (*n*,%)		61 (14.52%)	48 (15.53%)	13 (12.26%)	0.52
PSM (*n*,%)		123 (29.50%)	91 (29.55%)	32 (30.77%)	0.81

SII—systemic immune-inflammation index; PSA—prostate-specific antigen [ng/mL]; cT—clinical staging; CCI—Charlson comorbidity index; pT—pathological local staging; pN—pathological nodal staging; EPE—extracapsular extension; SVI—seminal vesicles involvement; PSM—positive surgical margins.

3.2. *Survival Predictors*

The median follow-up was 69 months (95%CI 66.5–71.9 months). In the analysed period, a total of 33 deaths (7.86%) were reported.

To determine the optimal SII cut-off, several thresholds were tested (Table 2). The value of 900 provided the maximum difference between survival maintaining statistical significance and was used for stratification purposes.

In a univariate survival analysis using baseline clinical and pathological factors, high (≥4) CCI ($p = 0.041$), high (≥6) ACCI ($p = 0.015$), high (≥6) CAPRA-S score ($p = 0.0092$) and high (≥900) SII ($p = 0.021$) were associated with a higher risk of death. Positive surgical margin status ($p = 0.062$) and lymph node involvement ($p = 0.066$) also showed a tendency for statistical significance (Figure 1), whereas pre-and postoperative grading as well as local pathological staging, preoperative PSA, and CAPRA score revealed no significant association with survival (data not shown). CAPRA-S, CCI and ACCI revealed modest concordance in the univariate prediction of death (c-index 0.59, 0.55 and 0.58, respectively).

Table 2. Differences in survival probabilities derived using different SII cut-offs.

SII Cut-Off	Cut-Off Percentile	8-Year Survival Probability Difference	Log-Rank *p*-Value
600	50	7.28%	0.1525
700	60	9.93%	0.1259
800	68	12.75%	0.0789
900	75	16.58%	**0.0206**
1000	80	16.72%	0.056

SII—systemic immune-inflammation index. *p*-value in the bold is the only significant.

To provide a rationale for combining scoring systems with SII, two multivariate models were built (Table 3) confirming high CAPRA-S score, high ACCI or CCI and high SII as independent predictors of overall mortality. Stratification combining both scoring systems with an SII yielded significantly worse survival in patients presenting at least one high score simultaneously with a high SII (Figure 2A). CAPRA-S risk groups were then used for a subgroup analysis (Figure 2B), which revealed that patients with high SIIs coexisting with high ACCIs were at the highest risk of mortality in high CAPRA-S (≥6).

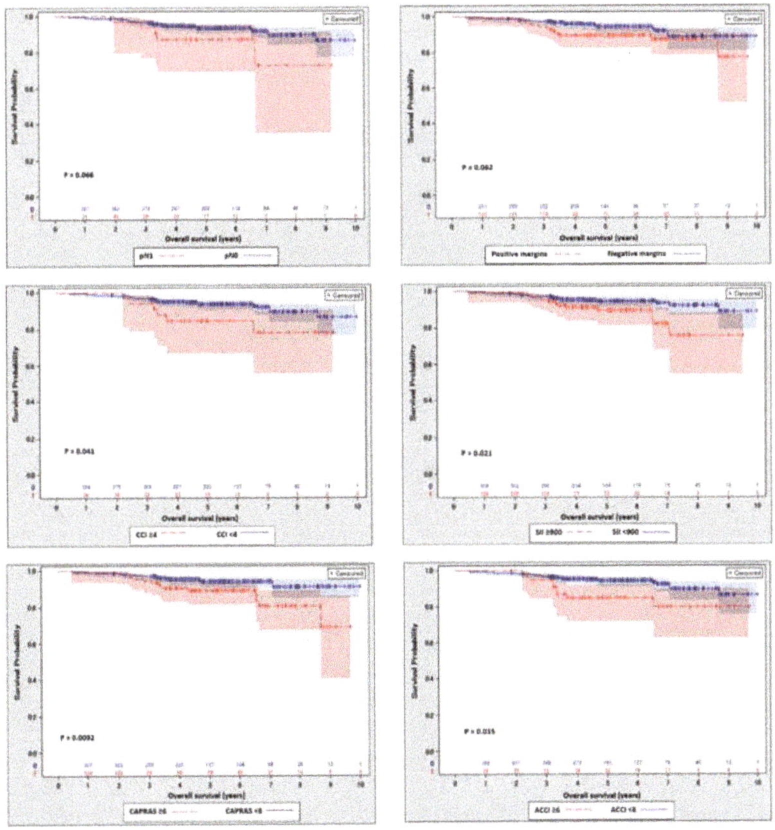

Figure 1. Kaplan-Meier curves with 95% confidence intervals for patients stratified with margin and nodal status, age-unadjusted Charlson comorbidity index (≥4) and age-adjusted Charlson comorbidity index (≥6), systemic immune-inflammation index (≥900) and CAPRA-S (≥6).

Table 3. The association of SII with overall survival in patients treated with radical prostatectomy for nonmetastatic prostate cancer-multivariable Cox regression analyses, including categorized SII, CAPRA-S and age-adjusted Charlson comorbidity index (ACCI) or age-unadjusted Charlson comorbidity index (CCI).

Variable	HR (95% CI)	*p*
	multivariate model 1 (c index = 0.67)	
CAPRA-S ≥ 6	2.67 (1.33–5.35)	0.006
CCI ≥ 4	2.79 (1.14–6.84)	0.025
SII ≥ 900	2.59 (1.26–5.31)	0.009
	multivariate model 2 (c index = 0.67)	
CAPRA-S ≥ 6	2.65 (1.32–5.31)	0.006
ACCI ≥ 6	2.75 (1.27–5.95)	0.01
SII ≥ 900	2.54 (1.24–5.21)	0.01

SII—systemic immune inflammation index; HR—hazard ratio; CI–confidence interval; CAPRA—The Cancer of the Prostate Risk Assessment score; CCI—Charlson comorbidity index; ACCI—age-adjusted Charlson comorbidity index.

(**A**)

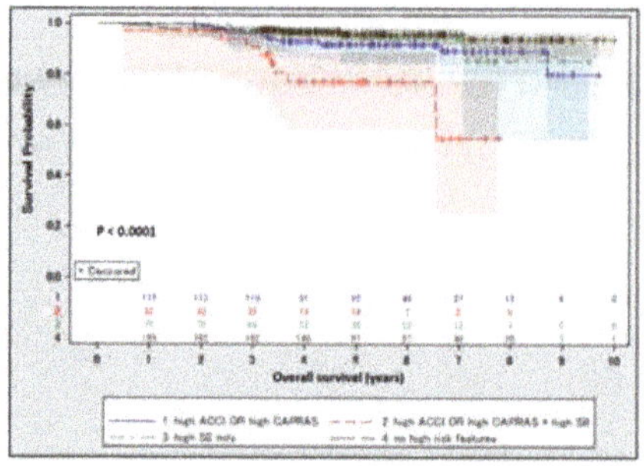

(**B**)

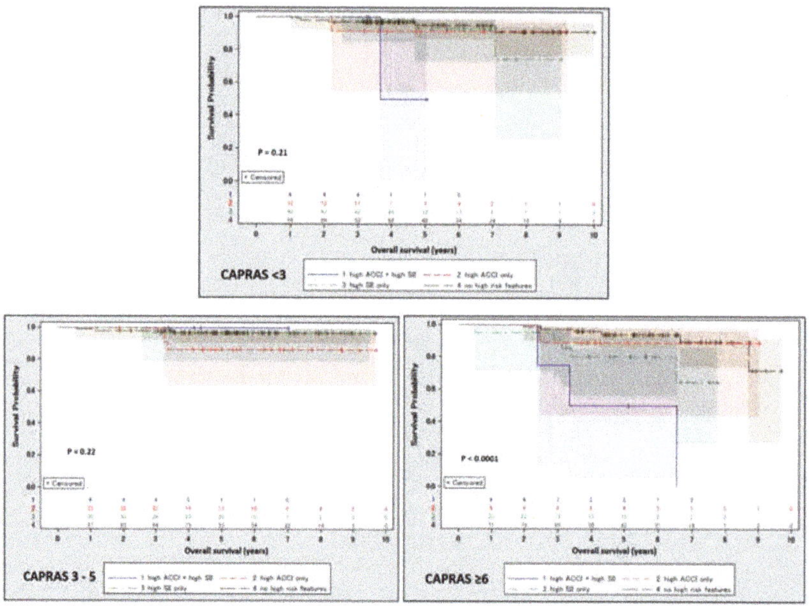

Figure 2. Kaplan-Meier curves with 95% confidence intervals depicting survival after radical prostatectomy depending on ACCI, CAPRA-S and SII. (**A**) Survival in patients stratified by the combination of either score with SII—entire cohort; (**B**) survival in patients stratified by the combination of ACCI with SII depending on the CAPRA-S risk group.

A quantitative analysis of survival yielded a clinically significant difference in survival between predefined subgroups, suggesting approximately a one-year survival loss in patients combining high SIIs with a single high score (CAPRA-S or CCI) achieved after 7 years when compared with other groups (Figure 3). In an explanatory analysis utilizing multiple comparisons, we found no significant differences in survival between patients presenting high SII only, a high score (ACCI or CAPRAS) and no risk features (comparing each of the groups with one another). We found however a significant difference in survival between patients presenting a high score (ACCI or CAPRAS) combined with high SII and patients from each of the three remaining groups—with high SII only ($p = 0.012$), with a high score (ACCI or CAPRAS) only (0.046) and with no risk features ($p = 0.013$).

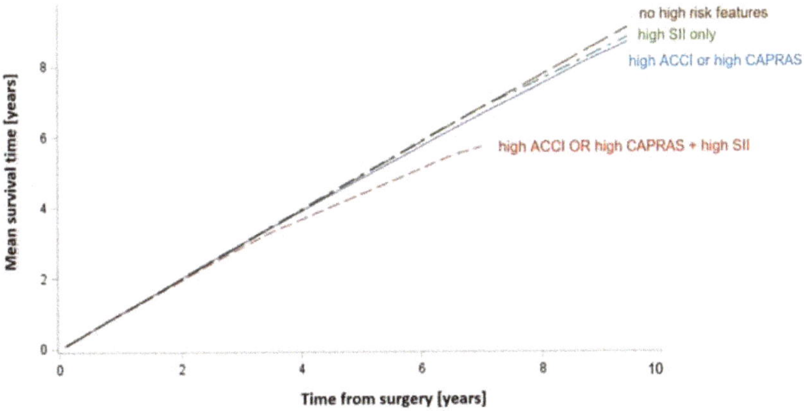

Figure 3. The mean survival time corresponding to the increasing length of follow-up among the four groups of patients.

4. Discussion

In this study, we present the systemic immune-inflammation index as a potential adjunct marker in predicting early death in patients treated with RP due to non-metastatic prostate cancer. We have validated the SII as an independent predictor of survival and developed a proposal for consecutive supplementation of the CAPRA-S score and Charlson comorbidity index with the SII.

The question of life expectancy in patients considered for radical treatment comes from existing evidence that men with less than 15 years of predicted survival are unlikely to benefit [24,25]. The clinical implementation of magnetic resonance imaging and novel markers, including next-generation PSA-based tests, improved the detection of clinically significant cancer and reduced overdiagnosis [26]. In turn, overtreatment reduction requires a deep insight into expected survival in a "risk and benefits" manner. Estimating individual life expectancy remains however a complex task which requires taking into account not only age [5] and co-morbidity [5,8,21], but also the risk profile of the cancer in itself [10,11], which, when neglected, can lead to undertreatment.

Increasing the utility of salvage treatment in recurrent PCa broadens the challenge of estimating life expectancy for patients who have already completed radical treatment. It has been shown that 20% of patients undergoing salvage RP die of other causes [27], whereas patients after RP with a favourable PSA doubling time (PSA-DT) and low postprostatectomy grading who experience biochemical recurrence might not benefit from salvage radiotherapy in a follow-up of 10 years [28]. Finally, even patients with postoperative PSA-DT shorter than 10 months who choose deferred radiotherapy can achieve an overall survival (OS) of more than 200 months [29]. Due to the following reasons, the paradigm of 10-year expected survival has been re-introduced, this time in patients considered for salvage treatment [30].

CAPRA-S has been developed as a straightforward calculator of biochemical recurrence after RP [10]. Extensive validation of the tool has yielded satisfactory accuracy and CAPRA-S has managed to outperform the Stephenson nomogram decision curve analysis when predicting BCR [11]. Interestingly, the good performance of CAPRA-S has also translated into the prediction of cancer-specific survival with a high concordance of 0.85 [11]. In our study, we have confirmed CAPRA-S as an independent predictor of OS. In the first place, however, we used CAPRA-S as primary stratification before adjusting for comorbidities and haematological markers to simulate the salvage treatment setting when validating the CCI, ACCI and SII.

The Charlson comorbidity index has been developed [9,21] and validated as a stratification of comorbidities influencing overall survival in multiple clinical settings [7,8,22,31].

Although the CCI has no value in prognosing PCa-specific mortality [31], it has proved to be a powerful predictor of overall mortality in nonmetastatic PCa, regardless of the D'Amico risk group [31]. In our study, patients with ACCIs ≥ 6 presented significantly worse OS and a high CCI/ACCI constituted an independent predictor of death in the multivariate analysis. Patients at postprostatectomy risk of receiving salvage therapy (CAPRA-S score higher than 5) were significantly less likely to survive if presenting a high ACCI at baseline. Since high comorbidity can translate into less apparent oncological benefits, we believe that the ACCI in the post-RP setting might tip the balance towards deferred salvage treatment or watchful waiting.

The systemic immune-inflammation index is a novel blood count-derived marker that has been introduced as a prognosticator in metastatic castration-resistant patients [32–34]. In the study of Lolli et al., an SII ≥ 535 predicted OS independently from the ECOG status, previous enzalutamide treatment and the presence of visceral metastases with HR = 1.8 [32]. The study of Fan et al. yielded similar outcomes, although with a lower threshold of SII ≥ 200 (overall mortality HR = 4.7; $p = 0.001$) [33].

Although in a preprostatectomy setting, the SII has not been confirmed as a predictor of unfavourable biopsy outcomes [35], recently the SII has been extensively validated in patients who have already undergone radical treatment. Similarly to other haematological markers, the SII can be easily retrieved and calculated from a full blood count, which is an obligatory part of preoperative laboratory evaluations. In a large multicenter cohort of patients treated with RP [19], high preoperative SIIs (≥620) were independently associated with extracapsular extension (odds ratio [OR] 1.16, $p = 0.041$), non-organ confined disease (OR 1.18, $p = 0.022$), adverse pathology (OR 1.36, $p < 0.001$) and upgrading at RP (OR 1.23, $p < 0.001$). A high SII also constituted an independent predictor of BCR in the preoperative model (HR 1.34, $p < 0.001$), but not in the postoperative model ($p = 0.078$). Another multicenter study by Rajwa et al. [17] included 214 patients with radio-recurrent PCa treated with salvage prostatectomy. Both preoperative and postoperative models confirmed a high SII (≥730) as an independent prognostic factor for cancer-specific survival (HR = 10.7, $p = 0.039$ and HR = 22.11, $p = 0.036$, respectively) and overall survival (HR = 8.57, $p < 0.001$ and 5.98, $p = 0.006$, respectively). Implementing the SII as an adjunct to previous preoperative models might therefore aid decision-making on salvage prostatectomy, especially in comorbid patients who are concerned about SRP harms.

In our study, we failed to validate a high SII as a preoperative predictor of adverse pathology. The discrepancy with previous studies might come from different characteristics of PCa in our cohort. The majority of the patients from the benchmark study by Rajwa et al. presented low pre- (61% grade group [GG] 1, 23% GG2) and postprostatectomy grading (32% GG1, 36% GG2). Nodal involvement (<2%), as well as an extracapsular extension (23%) and seminal vesicles invasion (6%), were uncommon in the development set [19]. In contrast, our patients were less likely to present with GG1 or GG2 in biopsy (38.73% and 31.37%, respectively) and post-RP specimen (13.94% and 39.66%, respectively), whereas the prevalence of pT3a and pT3b was considerably higher (42.37% and 14.52%, respectively). Noteworthy, previous cohorts with comparable prevalences of adverse pathology, such as the one analysed by Bravi et al., also yielded negative outcomes, failing to validate the association of NLR with adverse pathology [36]. This might suggest that the prognostic performance of particular haematological markers when predicting APF might depend on the risk profile of the cohort.

In our study, we managed to confirm a high SII as an independent predictor of overall survival when adjusted for adverse pathology using the CAPRA-S score and comorbidities using the CCI. Although isolated use of CAPRA-S and the CCI showed only limited value in predicting OS in short follow-ups, patients presenting with a high outcome of at least one of the scorings simultaneously with a high SII were particularly predisposed to early mortality. Finally, a high CCI supplemented with SII ≥ 900 provided genuine discrimination of overall mortality in patients from the high (>5) CAPRA-S risk group. The proposed stratifications might be easily utilised especially in a post-RP setting when deciding on salvage treatment.

The study has several limitations. Firstly, this is an observational analysis with retrospective data collection and limited follow-up. Although a detailed evaluation of comorbidities constitutes a routine workup of a patient before surgery, the CCI was evaluated based on reported comorbidities without a uniform survey procedure. Adjuvant or salvage treatment of the patients after RP was at a particular clinician's discretion and discrepancies in oncological follow-up might have theoretically confounded the survival outcome. Data regarding post-RP radiotherapy could not be collected, limiting bias detection and evaluation. Since this was a single-centre study, selection bias cannot be ruled out and some results might not be transferable to every clinical community. On the other hand, patient-tailored treatment and follow-up according to best local practice do not affect the major outcomes of this observational analysis. The primary aim of the study was to evaluate overall survival, which is described mainly by other-cause mortality in the post-RP setting; however, the lack of cancer-specific follow-up data prevented us from separating the contribution of cancer-dependent mortality.

5. Conclusions

Our study introduces the high systemic immune-inflammation index as a potential marker of survival in patients treated with radical prostatectomy for nonmetastatic prostate cancer. We have validated the SII as an adjunct to the commonly used CAPRA-S score and the Charlson comorbidity index. We estimated that patients combining high SIIs with high age-adjusted CCIs and/or high CAPRA-S survive approximately one year less during the first seven years after surgery. We have also confirmed the SII as useful when stratifying the risk of death in patients that might be potential candidates for salvage treatment (intermediate–high and high CAPRA-S). We believe that the implementation of the SII into already used prognostic tools might be of clinical value, especially when improving the selection of candidates for the radical treatment of prostate cancer.

Author Contributions: Conceptualization, P.Z.; data curation, P.Z.; formal analysis, P.Z.; investigation, P.Z., K.G., A.Ś., C.Ś., Ł.M. and Ł.Z.; methodology, P.Z. and Z.L.; project administration, P.Z.; resources, P.Z.; software, P.Z. and Z.L.; supervision, Z.L and P.R.; validation, P.Z. and Z.L.; visualization, P.Z.; writing—original draft, P.Z.; writing—review and editing, P.Z., Z.L., Ł.Z. and P.R. All authors have read and agreed to the published version of the manuscript.

Funding: This research received no external funding.

Institutional Review Board Statement: The study was conducted in accordance with the guidelines of the Declaration of Helsinki, and approved by the Ethics Committee of the Medical University of Warsaw.

Informed Consent Statement: Informed consent was obtained from all subjects involved in the study.

Data Availability Statement: The data presented in this study are available on request from the corresponding authors.

Conflicts of Interest: The authors declare no conflict of interest.

References

1. Culp, M.B.; Soerjomataram, I.; Efstathiou, J.A.; Bray, F.; Jemal, A. Recent Global Patterns in Prostate Cancer Incidence and Mortality Rates. *Eur. Urol.* **2020**, *77*, 38–52. [CrossRef]
2. Siegel, R.L.; Miller, K.D.; Jemal, A. Cancer Statistics, 2020. CA. *Cancer J. Clin.* **2020**, *70*, 7–30. [CrossRef] [PubMed]
3. Sandblom, G.; Dufmats, M.; Varenhorst, E. Long-Term Survival in a Swedish Population-Based Cohort of Men with Prostate Cancer. *Urology* **2000**, *56*, 442–447. [CrossRef]
4. Albertsen, P.C.; Moore, D.F.; Shih, W.; Lin, Y.; Li, H.; Lu-Yao, G.L. Impact of Comorbidity on Survival among Men with Localized Prostate Cancer. *J. Clin. Oncol. Off. J. Am. Soc. Clin. Oncol.* **2011**, *29*, 1335–1341. [CrossRef] [PubMed]
5. Boyle, H.J.; Alibhai, S.; Decoster, L.; Efstathiou, E.; Fizazi, K.; Mottet, N.; Oudard, S.; Payne, H.; Prentice, M.; Puts, M.; et al. Updated Recommendations of the International Society of Geriatric Oncology on Prostate Cancer Management in Older Patients. *Eur. J. Cancer Oxf. Engl. 1990* **2019**, *116*, 116–136. [CrossRef] [PubMed]
6. van Walree, I.C.; Scheepers, E.R.M.; van den Bos, F.; van Huis-Tanja, L.H.; Emmelot-Vonk, M.H.; Hamaker, M.E. Clinical Judgment versus Geriatric Assessment for Frailty in Older Patients with Cancer. *J. Geriatr. Oncol.* **2020**, *11*, 1138–1144. [CrossRef] [PubMed]

7. Casas Duran, F.; Valduvieco, I.; Oses, G.; Cortés, K.S.; Barreto, T.D.; Muñoz-Guglielmetti, D.; Ferrer, F. Spanish Validation of Charlson Index Applied to Prostate Cancer. *Clin. Transl. Oncol. Off. Publ. Fed. Span. Oncol. Soc. Natl. Cancer Inst. Mex.* **2020**, *22*, 1187–1192. [CrossRef]
8. Daskivich, T.J.; Kwan, L.; Dash, A.; Greenfield, S.; Litwin, M.S. Weighted versus Unweighted Charlson Score to Predict Long-Term Other-Cause Mortality in Men with Early-Stage Prostate Cancer. *Eur. Urol.* **2014**, *66*, 1002–1009. [CrossRef]
9. Charlson, M.E.; Pompei, P.; Ales, K.L.; MacKenzie, C.R. A New Method of Classifying Prognostic Comorbidity in Longitudinal Studies: Development and Validation. *J. Chronic Dis.* **1987**, *40*, 373–383. [CrossRef]
10. Cooperberg, M.R.; Hilton, J.F.; Carroll, P.R. The CAPRA-S Score: A Straightforward Tool for Improved Prediction of Outcomes after Radical Prostatectomy. *Cancer* **2011**, *117*, 5039–5046. [CrossRef]
11. Punnen, S.; Freedland, S.J.; Presti, J.C.; Aronson, W.J.; Terris, M.K.; Kane, C.J.; Amling, C.L.; Carroll, P.R.; Cooperberg, M.R. Multi-Institutional Validation of the CAPRA-S Score to Predict Disease Recurrence and Mortality after Radical Prostatectomy. *Eur. Urol.* **2014**, *65*, 1171–1177. [CrossRef] [PubMed]
12. Zapała, Ł.; Ślusarczyk, A.; Garbas, K.; Mielczarek, Ł.; Ślusarczyk, C.; Zapała, P.; Wróbel, A.; Radziszewski, P. Complete Blood Count-Derived Inflammatory Markers and Survival in Patients with Localized Renal Cell Cancer Treated with Partial or Radical Nephrectomy: A Retrospective Single-Tertiary-Center Study. *Front. Biosci. Sch. Ed.* **2022**, *14*, 5. [CrossRef]
13. Zapała, Ł.; Ślusarczyk, A.; Wolański, R.; Kurzyna, P.; Garbas, K.; Zapała, P.; Radziszewski, P. The Four-Feature Prognostic Models for Cancer-Specific and Overall Survival after Surgery for Localized Clear Cell Renal Cancer: Is There a Place for Inflammatory Markers? *Biomedicines* **2022**, *10*, 1202. [CrossRef] [PubMed]
14. Rajwa, P.; Życzkowski, M.; Paradysz, A.; Slabon-Turska, M.; Suliga, K.; Bujak, K.; Bryniarski, P. Novel Hematological Biomarkers Predict Survival in Renal Cell Carcinoma Patients Treated with Nephrectomy. *Arch. Med. Sci. AMS* **2020**, *16*, 1062–1071. [CrossRef] [PubMed]
15. Kool, R.; Marcq, G.; Shinde-Jadhav, S.; Mansure, J.J.; Saleh, R.; Rajan, R.; Aprikian, A.; Tanguay, S.; Cury, F.L.; Brimo, F.; et al. Role of Serum Lymphocyte-Derived Biomarkers in Nonmetastatic Muscle-Invasive Bladder Cancer Patients Treated with Trimodal Therapy. *Eur. Urol. Open Sci.* **2022**, *36*, 26–33. [CrossRef] [PubMed]
16. Sudoł, D.; Widz, D.; Mitura, P.; Płaza, P.; Godzisz, M.; Kuliniec, I.; Yadlos, A.; Cabanek, M.; Bar, M.; Bar, K. Neutrophil-to-Lymphocyte Ratio as a Predictor of Overall Survival and Cancer Advancement in Patients Undergoing Radical Cystectomy for Bladder Cancer. *Cent. Eur. J. Urol.* **2022**, *75*, 41–46. [CrossRef]
17. Rajwa, P.; Schuettfort, V.M.; Quhal, F.; Mori, K.; Katayama, S.; Laukhtina, E.; Pradere, B.; Motlagh, R.S.; Mostafaei, H.; Grossmann, N.C.; et al. Role of Systemic Immune-Inflammation Index in Patients Treated with Salvage Radical Prostatectomy. *World J. Urol.* **2021**, *39*, 3771–3779. [CrossRef]
18. Jiang, Z.-G.; Liao, S.-G. Baseline Neutrophil-Lymphocyte Ratio Is Associated with Outcomes in Patients with Castration-Resistant Prostate Cancer Treated with Docetaxel in South China. *Medicine (Baltimore)* **2021**, *100*, e27361. [CrossRef]
19. Rajwa, P.; Schuettfort, V.M.; D'Andrea, D.; Quhal, F.; Mori, K.; Katayama, S.; Laukhtina, E.; Pradere, B.; Motlagh, R.S.; Mostafaei, H.; et al. Impact of Systemic Immune-Inflammation Index on Oncologic Outcomes in Patients Treated with Radical Prostatectomy for Clinically Nonmetastatic Prostate Cancer. *Urol. Oncol.* **2021**, *39*, 785.e19–785.e27. [CrossRef]
20. Wang, S.; Yang, X.; Yu, Z.; Du, P.; Sheng, X.; Cao, Y.; Yan, X.; Ma, J.; Yang, Y. The Values of Systemic Immune-Inflammation Index and Neutrophil–Lymphocyte Ratio in Predicting Biochemical Recurrence in Patients with Localized Prostate Cancer After Radical Prostatectomy. *Front. Oncol.* **2022**, *12*, 907625. [CrossRef]
21. Quan, H.; Li, B.; Couris, C.M.; Fushimi, K.; Graham, P.; Hider, P.; Januel, J.-M.; Sundararajan, V. Updating and Validating the Charlson Comorbidity Index and Score for Risk Adjustment in Hospital Discharge Abstracts Using Data from 6 Countries. *Am. J. Epidemiol.* **2011**, *173*, 676–682. [CrossRef] [PubMed]
22. Park, J.W.; Koh, D.H.; Jang, W.S.; Lee, J.Y.; Cho, K.S.; Ham, W.S.; Rha, K.H.; Jung, W.H.; Hong, S.J.; Choi, Y.D. Age-Adjusted Charlson Comorbidity Index as a Prognostic Factor for Radical Prostatectomy Outcomes of Very High-Risk Prostate Cancer Patients. *PLoS ONE* **2018**, *13*, e0199365. [CrossRef] [PubMed]
23. Onderdonk, B.E.; Dorn, P.L.; Martinez, C.; Arif, F.; Cloutier, D.; Antic, T.; Golden, D.W.; Karrison, T.; Pitroda, S.P.; Szmulewitz, R.Z.; et al. A Prospective Clinical and Transcriptomic Feasibility Study of Oral-Only Hormonal Therapy with Radiation for Unfavorable Prostate Cancer in Men 70 Years of Age and Older or with Comorbidity. *Cancer* **2021**, *127*, 2631–2640. [CrossRef] [PubMed]
24. Wilt, T.J.; Brawer, M.K.; Jones, K.M.; Barry, M.J.; Aronson, W.J.; Fox, S.; Gingrich, J.R.; Wei, J.T.; Gilhooly, P.; Grob, B.M.; et al. Radical Prostatectomy versus Observation for Localized Prostate Cancer. *N. Engl. J. Med.* **2012**, *367*, 203–213. [CrossRef] [PubMed]
25. Schröder, F.H.; Hugosson, J.; Roobol, M.J.; Tammela, T.L.J.; Zappa, M.; Nelen, V.; Kwiatkowski, M.; Lujan, M.; Määttänen, L.; Lilja, H.; et al. The European Randomized Study of Screening for Prostate Cancer—Prostate Cancer Mortality at 13 Years of Follow-Up. *Lancet* **2014**, *384*, 2027–2035. [CrossRef]
26. Ferro, M.; Crocetto, F.; Bruzzese, D.; Imbriaco, M.; Fusco, F.; Longo, N.; Napolitano, L.; La Civita, E.; Cennamo, M.; Liotti, A.; et al. Prostate Health Index and Multiparametric MRI: Partners in Crime Fighting Overdiagnosis and Overtreatment in Prostate Cancer. *Cancers* **2021**, *13*, 4723. [CrossRef]
27. Wenzel, M.; Würnschimmel, C.; Nocera, L.; Collà Ruvolo, C.; Tian, Z.; Shariat, S.F.; Saad, F.; Briganti, A.; Graefen, M.; Becker, A.; et al. Salvage Radical Prostatectomy: Baseline Prostate Cancer Characteristics and Survival Across SEER Registries. *Clin. Genitourin. Cancer* **2021**, *19*, e255–e263. [CrossRef]

28. Matsumoto, K.; Niwa, N.; Hagiwara, M.; Kosaka, T.; Tanaka, N.; Takeda, T.; Morita, S.; Mizuno, R.; Shinojima, T.; Hara, S.; et al. Type of Patients in Whom Biochemical Recurrence after Radical Prostatectomy Can Be Observed without Salvage Therapy. *World J. Urol.* **2020**, *38*, 1749–1756. [CrossRef]
29. Marshall, C.H.; Chen, Y.; Kuo, C.; Cullen, J.; Jiang, J.; Rosner, I.; Markowski, M.; McLeod, D.G.; Trock, B.J.; Eisenberger, M.A. Timing of Androgen Deprivation Treatment for Men with Biochemical Recurrent Prostate Cancer in the Context of Novel Therapies. *J. Urol.* **2021**, *206*, 623–629. [CrossRef]
30. EAU Guidelines on Prostate Cancer—Uroweb. Available online: https://uroweb.org/guidelines/prostate-cancer (accessed on 30 May 2022).
31. Lee, J.Y.; Kang, H.W.; Rha, K.H.; Cho, N.H.; Choi, Y.D.; Hong, S.J.; Cho, K.S. Age-Adjusted Charlson Comorbidity Index Is a Significant Prognostic Factor for Long-Term Survival of Patients with High-Risk Prostate Cancer after Radical Prostatectomy: A Bayesian Model Averaging Approach. *J. Cancer Res. Clin. Oncol.* **2016**, *142*, 849–858. [CrossRef]
32. Lolli, C.; Caffo, O.; Scarpi, E.; Aieta, M.; Conteduca, V.; Maines, F.; Bianchi, E.; Massari, F.; Veccia, A.; Chiuri, V.E.; et al. Systemic Immune-Inflammation Index Predicts the Clinical Outcome in Patients with MCRPC Treated with Abiraterone. *Front. Pharmacol.* **2016**, *7*, 376. [CrossRef] [PubMed]
33. Fan, L.; Wang, R.; Chi, C.; Cai, W.; Zhang, Y.; Qian, H.; Shao, X.; Wang, Y.; Xu, F.; Pan, J.; et al. Systemic Immune-Inflammation Index Predicts the Combined Clinical Outcome after Sequential Therapy with Abiraterone and Docetaxel for Metastatic Castration-Resistant Prostate Cancer Patients. *Prostate* **2018**, *78*, 250–256. [CrossRef]
34. Stangl-Kremser, J.; Mari, A.; Suarez-Ibarrola, R.; D'Andrea, D.; Korn, S.M.; Pones, M.; Kramer, G.; Karakiewicz, P.; Enikeev, D.V.; Glybochko, P.V.; et al. Development of a Prognostic Model for Survival Time Prediction in Castration-Resistant Prostate Cancer Patients. *Urol. Oncol.* **2020**, *38*, 600.e9–600.e15. [CrossRef] [PubMed]
35. Rajwa, P.; Huebner, N.A.; Hostermann, D.I.; Grossmann, N.C.; Schuettfort, V.M.; Korn, S.; Quhal, F.; König, F.; Mostafaei, H.; Laukhtina, E.; et al. Evaluation of the Predictive Role of Blood-Based Biomarkers in the Context of Suspicious Prostate MRI in Patients Undergoing Prostate Biopsy. *J. Pers. Med.* **2021**, *11*, 1231. [CrossRef]
36. Bravi, C.A.; Rosiello, G.; Fallara, G.; Vertosick, E.; Tin, A.; Sjoberg, D.; Bianchi, M.; Mazzone, E.; Martini, A.; Dell'oglio, P.; et al. Predictive Value of Preoperative Neutrophil-to-Lymphocyte Ratio in Localized Prostate Cancer: Results from a Surgical Series at a High-Volume Institution. *Minerva Urol. Nephrol.* **2021**, *73*, 481–488. [CrossRef] [PubMed]

cancers

MDPI

Article

Identification and Validation of a Novel Ferroptotic Prognostic Genes-Based Signature of Clear Cell Renal Cell Carcinoma

Zhiyuan Shi [1,†], Jianzhong Zheng [1,†], Qing Liang [2,3], Yankuo Liu [1], Yi Yang [1], Rui Wang [2,3], Mingshan Wang [1], Qian Zhang [2,3], Zuodong Xuan [1], Huimin Sun [4], Kejia Wang [2,3,*] and Chen Shao [1,*]

1 Department of Urology, Xiang'an Hospital of Xiamen University, School of Medicine, Xiamen University, Xiamen 361101, China
2 Fujian Provincial Key Laboratory of Organ and Tissue Regeneration, Organ Transplantation Institute of Xiamen University, School of Medicine, Xiamen University, Xiamen 361101, China
3 Xiamen Key Laboratory of Regeneration Medicine, Organ Transplantation Institute of Xiamen University, School of Medicine, Xiamen University, Xiamen 361101, China
4 Central Laboratory, Xiang'an Hospital of Xiamen University, School of Medicine, Xiamen University, Xiamen 361101, China
* Correspondence: wangkejia@xmu.edu.cn (K.W.); cshao@xah.xmu.edu.cn (C.S.)
† These authors contributed equally to this work.

Simple Summary: Clear cell renal cell carcinoma (ccRCC) is one of the leading types of kidney malignancy and is closely related to ferroptosis that is an iron-dependent regulated cell death with lipid peroxide accumulation. A signature of nine ferroptotic genes was identified as an independent prognostic factor via construction in The Cancer Genome Atlas (TCGA) database and validation in the ArrayExpress database. This signature could successfully divide patients into low- and high-risk groups to predict survival rate. Compared with the other eight genes, glutaminase 2 (GLS2) played a crucial role during erastin-induced ferroptosis in ACHN and Caki-1 cells. It was discovered for the first time that GLS2 might be a ferroptotic suppressor in ccRCC.

Abstract: Renal cell carcinoma (RCC), as one of the primary urological malignant neoplasms, shows poor survival, and the leading pathological type of RCC is clear cell RCC (ccRCC). Differing from other cell deaths (such as apoptosis, necroptosis, pyroptosis, and autophagy), ferroptosis is characterized by iron-dependence, polyunsaturated fatty acid oxidization, and lipid peroxide accumulation. We analyzed the ferroptosis database (FerrDb V2), Gene Expression Omnibus database, The Cancer Genome Atlas database, and the ArrayExpress database. Nine genes that were differentially expressed and related to prognosis were involved in the ferroptotic prognostic model via the least absolute shrinkage and selection operator Cox regression analysis, which was established in ccRCC patients from the kidney renal clear cell carcinoma (KIRC) cohort in TCGA database, and validated in ccRCC patients from the E-MTAB-1980 cohort in the ArrayExpress database. The signature could be an independent prognostic factor for ccRCC, and high-risk patients showed worse overall survival. The Gene Ontology and Kyoto Encyclopedia of Genes and Genomes were utilized to investigate the potential mechanisms. The nine genes in ccRCC cells with erastin or RSL3 treatment were validated to find the crucial gene. The glutaminase 2 (GLS2) gene was upregulated during ferroptosis in ccRCC cells, and cells with GLS2 shRNA displayed lower survival, a lower glutathione level, and a high lipid peroxide level, which illustrated that GLS2 might be a ferroptotic suppressor in ccRCC.

Keywords: clear cell renal cell carcinoma; ferroptosis; prognostic model; TCGA; E-MTAB-1980; GLS2

Citation: Shi, Z.; Zheng, J.; Liang, Q.; Liu, Y.; Yang, Y.; Wang, R.; Wang, M.; Zhang, Q.; Xuan, Z.; Sun, H.; et al. Identification and Validation of a Novel Ferroptotic Prognostic Genes-Based Signature of Clear Cell Renal Cell Carcinoma. *Cancers* **2022**, *14*, 4690. https://doi.org/10.3390/cancers14194690

Academic Editors: José I. López and Claudia Manini

Received: 8 September 2022
Accepted: 23 September 2022
Published: 27 September 2022

1. Introduction

Renal cell carcinoma (RCC) is one of the main urological malignant tumors, and there were more than 430 thousand new cases and nearly 180 thousand deaths worldwide in 2020 [1]. The ratio of men to women diagnosed with RCC was nearly 1.7:1 and 90% of cases of RCC were in patients over 50 years of age [2]. The risk factors include hypertension, tobacco smoking, and obesity [3]. The leading pathological type of RCC is clear cell RCC (ccRCC) that accounts for approximately 75% [3], and the percentage of metastatic ccRCC is 83–88% [4]. Patients diagnosed at an early stage possess a better survival rate of 80–90% [5], however, only 30% of patients could be diagnosed according to symptoms and 40% of patients whose 5-year survival rate is under 11.2% develop distant metastases [6,7]. In summary, early diagnosis and precise management of ccRCC is significant because of their correlation with prognosis.

Ferroptosis, as one of the regulated cell deaths, was a term coined by Scott J. Dixon and Brent R. Stockwell in 2012 [8]. Differing from other cell deaths (such as apoptosis, necroptosis, pyroptosis, and autophagy), ferroptosis is dependent on the oxidization of polyunsaturated fatty acid (PUFA) and subsequent lipid peroxide accumulation [9]. When ferroptosis occurs in cells, there are several distinguishing features including the rupture of cellular membranes and a series of mitochondrial changes (mitochondria shrink, higher mitochondrial membrane density, reduced mitochondria cristae, and outer mitochondrial membrane rupture), however, there is no obvious change in the nucleus [10]. The molecular compound erastin and Ras selective lethal 3 (RSL3) could induce ferroptosis through different mechanisms. The erastin could inhibit solute carrier family 7 member A11 (SLC7A11) to prevent the import of cystine, which limits glutathione (GSH) synthesis [8]. Erastin could also bind with voltage-dependent anion channel 2 (VDAC2) to produce reactive oxygen species (ROS) in a NADH-dependent manner, which induces mitochondrial damage [11]. Different from erastin, RSL3 could directly repress the enzymatic activity of GSH peroxidase 4 (GPX4) that reduces lipid peroxides to corresponding alcohols via the inactivation of its enzymatic active site, and the reduction of GPX4 needs GSH to recover this site [12–14]. On the other hand, the intracellular level of iron is also significant because it plays a crucial role in participating in lipid peroxidation. Fe^{2+} could generate the hydroxyl radical (HO•) via the Fenton reaction [15]. The HO•, as the most active ROS, could initiate nonenzymatic lipid peroxidation including PUFA oxidation [10]. Therefore, the metabolism progress associated with GSH synthesis, iron metabolism, and lipid peroxidation might affect ferroptosis, especially in protein-coding genes, and the expression of some ferroptotic genes might change to reduce ferroptosis [16].

In previous research, several studies have explored prognostic markers in ccRCC, and a few ferroptosis-related prognostic models have been constructed [17–22]. However, previous studies only took into consideration 60 genes from references [23,24]. In this study, we made full use of the ferroptosis database (FerrDb V2), Gene Expression Omnibus (GEO) database, The Cancer Genome Atlas (TCGA) database, and the ArrayExpress database to explore the ferroptotic prognostic gene-based model of ccRCC. From the twenty genes that were differentially expressed and associated with prognosis, nine genes were involved in the ferroptotic prognostic model (FPM) via the least absolute shrinkage and selection operator (LASSO) Cox regression analysis, which was established in ccRCC patients from the kidney renal clear cell carcinoma (KIRC) cohort in TCGA database, used to establish a nomogram for conducting individualized prognosis assessment, and validated in ccRCC patients from the E-MTAB-1980 cohort in the ArrayExpress database. The signature could be an independent prognostic factor for ccRCC, and patients in high-risk groups showed worse prognosis. Then the Gene Ontology (GO) and Kyoto Encyclopedia of Genes and Genomes (KEGG) were employed to explore the potential mechanisms, and we tried to validate the nine genes in ACHN and Caki-1 cells with erastin or RSL3 treatment. The glutaminase 2 (GLS2) gene was upregulated during ferroptosis in ccRCC cells, and the cells treated with GLS2 shRNA displayed lower survival, a lower glutathione (GSH) level, and a high lipid peroxide level. Here, we discovered for the first time that GLS2 might be a ferroptotic suppressor in ccRCC.

2. Materials and Methods

2.1. Collection and Processing of Data

The GSE53757 (72 normal kidney tissues and 72 ccRCC tissues), GSE66272 (27 normal kidney tissues and 27 ccRCC tissues), and GSE71963 (16 normal kidney tissues and 32 ccRCC tissues) databases were downloaded from the GEO database (https://www.ncbi.nlm.nih.gov/geo/, accessed on 31 May 2022). The ferroptosis-related genes (including drivers, suppressors, and markers) were obtained from the FerrDb V2 database (http://www.zhounan.org/ferrdb, accessed on 31 May 2022). The datasets of ccRCC were obtained from the KIRC cohort (72 normal kidney samples and 539 ccRCC samples) in TCGA database (https://portal.gdc.cancer.gov, accessed on 31 May 2022) and the E-MTAB-1980 cohort (101 ccRCC samples) in the ArrayExpress database (https://www.ebi.ac.uk/arrayexpress, accessed on 31 May 2022). The proteomic dataset of ccRCC was downloaded from the Clinical Proteomic Tumor Analysis Consortium (CPTAC, https://proteomic.datacommons.cancer.gov/pdc, accessed on 31 May 2022).

The following data processing was handled by R language software. The gene expression of the above databases was normalized with the "edgeR" package [25]. The differentially expressed genes (DEGs) between normal kidney tissue and ccRCC tissue in the GSE53757, GSE66272, GSE71963, and KIRC cohort were obtained using the "limma" package [26], and DEGs were selected with the $|\log_2 (FC)| > 1$ and the false discovery rate (FDR) < 0.05. The tumor prognostic genes (TPGs) were obtained using the "survival" package and identified by univariate Cox analysis ($p < 0.05$) of overall survival (OS) [27].

2.2. Establishment and Validation of the FPM

The LASSO Cox regression analysis was used to select ferroptotic prognostic DEGs (FPEDGs) using the "glmnet" package, and the optimal lambda (λ) was identified as the optimal value through a tenfold cross-validation process [28,29]. The risk score was computed by summarizing the product of each normalized FPEDG expression and its corresponding multivariate Cox regression coefficient (β). In the KIRC cohort, the risk score of every patient was computed using the above formula, and ccRCC patients were divided into two groups (low-risk group and high-risk group) by the media risk score. In the E-MTAB-1980 cohort, ccRCC patients were divided into two groups (low-risk group and high-risk group) according to the media risk score of the KIRC cohort. The distribution patterns of patients were performed by uniform manifold approximation (UMAP) using the "umap" package and t-distributed stochastic neighbor embedding (t-SNE) with "Rtsne" package. The "survminer" package was employed to conduct Kaplan–Meier (K–M) survival analysis. The "time" ROC package was employed to conduct time-dependent receiver operating characteristic (ROC) analysis.

2.3. Independent Prognostic Value of FPM

The risk score and clinical factors including age, gender, grade, and stage were analyzed via Pearson's chi-square test and displayed using a heatmap. The univariate/multivariate Cox regression analysis was employed to estimate the independent prognostic value of FPM and traditional clinical characteristics, and the results were summarized using hazard ratios (HRs) and 95% confidence intervals (CIs).

2.4. Establishment and Validation of the Nomogram

According to the results of the multivariate Cox regression analysis, the factors were used to establish a nomogram for predicting survival rates using the "rms" package and the "survival" package. Time-dependent ROC analysis was utilized to evaluate the predictive performance of FPM using the "time" ROC package. Calibration curves were utilized to estimate the consistency between actual survival rates and predicted survival rates.

2.5. Molecular Functional Analysis

The GO and KEGG of the nine genes were displayed using the "clusterProfiler" package to screen the potential biological processes (BPs), cellular components (CCs), molecular functions (MFs), and pathways. These results were presented using the "ggplot2" package and the "G0plot" package [30].

2.6. Regents and Assay Kits

Fetal bovine serum (10100147), DMEM high glucose media (11965092), RPMI 1640 media (11875119), and penicillin-streptomycin (15140163) were purchased from Gibco (Foster, CA, USA). Erastin (S7242), ferrostatin-1 (Fer-1, S7243), liproxstatin-1 (Lip-1, S7699), hydroxychloroquine sulfate (HCQ, S4430), necrosulfonamide (NSA, S8251), Z-VAD-FMK (Z-VAD, S7023), and RSL3 (S8155) were purchased from Selleck (Houston, TX, USA). GLS2 rabbit polyclonal antibody (ab113509) was purchased from Abcam (Cambridge, UK). Actin mouse monoclonal antibody (A5441) and monobromobimane (MBB, B4380) were purchased from Sigma-Aldrich (Saint Louis, MO, USA). MTT (S19063) was purchased from Shanghai yuanye Bio-Technology Co., Ltd. (Shanghai, China). Lipofectamine 2000 (11668019), Hoechst 33,342 (H1398) and BODIPY 665/676 (B3932) were purchased from Thermo (Waltham, MA, USA). The SPARKeasy cellular RNA extraction kit (AC0205) was bought from Shandong Sparkjade Biotechnology Co., Ltd. (Jinan, China). The Evo M-MLV RT Kit (AG11711) was purchased from Accurate Biotechnology (Hunan) Co., Ltd. (Changsha, China). The puromycin (60210ES25) and Hieff®qPCR SYBR Green Master Mix (11201ES08) were purchased from Yeasen Biotechnology (Shanghai) Co., Ltd. (Shanghai, China). RIPA buffer (R0010), a GSH assay kit (BC1175) and a malondialdehyde (MDA) assay kit (BC0025) were bought from Beijing Solarbio Science & Technology Co., Ltd. (Beijing, China).

2.7. Cell lines and Cell Culture

ACHN and Caik-1 cells were provided by Cell Bank/Stem Cell Bank, Chinese Academy of Sciences. ACHN was cultivated in DMEM high glucose media with 1% penicillin-streptomycin and 10% fetal bovine serum, whereas Caki-1 cells were cultivated in RPMI 1640 media with 1% penicillin-streptomycin and 10% fetal bovine serum. All cells were cultured in a humidified incubator at 37 °C with 5% CO_2.

2.8. Cell Viability Assay

A total of 5×10^3 ACHN or Caik-1 cells were seeded into each well of a 96-well plate and were incubated overnight, then the cells were respectively treated with various concentrations of erastin (12 h) or RSL3 (6 h) with or without Fer-1 [1 µmol/L (µM)], Lip-1 (1 µM), HCQ (20 µM), NSA (1 µM), and Z-VAD (20 µM) [31]. Then, 5 mg/mL MTT was added to each well and incubated for 4 h at 37 °C. Then, the medium was carefully removed from the wells and 150 µL DMSO was added into each well. After shaking for 10 min, the 96-well plate was put into a multiskan sky high microplate reader (Thermo, Waltham, MA, USA) and the absorbance was detected at 570 nm wavelength.

2.9. Real Time-PCR Assay

Total RNA from cells were extracted by using a cellular RNA extraction kit, and reverse transcribed into cDNA by using the cDNA synthesis kit [32]. The real time PCR was performed using Bio-Rad CFX96 Real-time PCR systems (Bio-Rad, Hercules, CA, USA). The result was calculated by the comparative Ct method. All primers were designed by Primer Premier 6 and synthesized by Sangon Biotech (Shanghai, China). The primer sequences are shown in Table S1.

2.10. Lentiviral Infection

The GLS2 shRNA sequence was taken from references and synthesized by Shanghai Genechem. Viral vector and packaging vectors were transfected into HEK293T cells using Lipofectamine 2000. The medium was replaced after 6 h, and viral particles were harvested after 24 h. After ACHN and Caki-1 cells had been infected for 24 h, the medium was replaced, and cells were cultured for another 48 h. Then puromycin (2.0 μg/mL) was employed to select cell lines.

2.11. Western Blot

When cells were cultured to 90% confluence, they were harvested and lysed with RIPA buffer containing protease inhibitor cocktail (HY-K0010, MedChemExpress, Monmouth Junction, NJ, USA). The protein concentration was measured by using a BCA protein assay kit (23227, Thermo). Then, samples were separated by 10% SDS polyacrylamide gel, transferred onto polyvinylidene difluoride (PVDF) membrane (WBKLS0500, Millipore, Billerica, MA, USA), blocked with 5% skimmed milk for 1 h at room temperature, and blotted with the corresponding antibodies (actin 1:10,000; GLS2, 1:1000) in 5% skimmed milk overnight at 4 °C. The PVDF membranes were washed with TBST three times and incubated with HRP-conjugated secondary antibody for 1 h at room temperature. After being washed with TBST three times, the PVDF membranes were detected by enhanced chemiluminescence reagents (WBKLS0050, Millipore, Billerica, MA, USA) with C300 (Azure Biosystems, Dublin, CA, USA).

2.12. Detection of MDA and GSH Level

The levels of MDA and GSH in cells were measured using MDA and GSH assay kits and detected using a multiskan sky high microplate reader (Thermo, Waltham, MA, USA).

2.13. Lipid Peroxidation Assay

The cells were seeded on glass-bottomed culture dishes and incubated for 24 h. The cells were incubated with DMSO, erastin or erastin+Lip-1 solutions for 12 h and washed with PBS. Next, the cells were treated with 10 μM BODIPY 665/676 dissolved in PBS for 30 min at 37 °C and washed with PBS. Then the cells were treated with 2 μg/mL Hoechst 33,342 for 15 min at room temperature, washed with PBS, and examined using a Zeiss LSM 880+ Airyscan confocal microscope (Zeiss, Oberkochen, Germany).

2.14. MBB Staining

MBB dissolved in PBS was used to treat cells for 15 min at 37 °C after the removal of culture medium [33]. They were photographed using an Olympus IX51 inverted fluorescence microscope (Olympus, Tokyo, Japan).

2.15. Statistical Analysis

All data were displayed by mean plus or minus standard deviation. Statistical analysis was managed using Prism 9 and SPSS 13. The value of $p < 0.05$ was considered significant (* $p < 0.05$, ** $p < 0.01$, *** $p < 0.001$).

3. Results

3.1. Identification of the FPDEGs

In this study (Figure 1), we identified 1519 DEGs (Table S2) from three GEO databases (GSE53757, GSE66272 and GSE71963), 449 genes (Table S3) from the FerrDb V2 database, and 5865 DEGs (Table S4) from TCGA database, and then 41 significant DEGs (Table S5) were obtained. Next, we identified 207 tumor prognostic genes (TPGs) (Table S6) from the FerrDb V2 database and TCGA database. Finally, we harvested 20 FPEDGs that contained 11 upregulated genes and nine downregulated genes between normal tissues and ccRCC tissues in Figure 2A. The expression and HR (95% CI) of 20 FPEDGs in TCGA are displayed in Figure 2B,C.

3.2. Construction of the FPM in TCGA

The 20 FPEDGs were subjected to LASSO Cox regression analysis based on OS to screen key genes among FPEDGs (Figure 3A,B). A nine-gene signature with DPEP1, NOX4,

MT1G, GLS2, GLRX5, TIMP1, CA9, CDCA3, and CYBB was identified in the KIRC cohort based on the optimal λ, and their respective relative coefficients were calculated to establish the FPM. The risk score of each patient was computed using the following formula: Risk score = (−0.12661 × expression of DPEP1) + (−0.03457 × expression of NOX4) + (0.06429 × expression of MT1G) + (−0.06666 × expression of GLS2) + (−0.31540 × expression of GLRX5) + (0.21071 × expression of TIMP1) + (−0.17067 × expression of CA9) + (0.56213 × expression of CDCA3) + (−0.04120 × expression of CYBB).

After checking the clinical data and gene expression of the patients in TCGA, we deleted 13 cases without gene expression profiles, complete clinical data or patients with 0 days. Based on the media risk score, patients (n = 526) were divided into a low-risk group (n = 263) and a high-risk group (n = 263), and patients in the high-risk group possessed high mortality (Figure 3C). The two groups of ccRCC patients could be well distributed into two sets by using UMAP and t-SNE analysis (Figure 3D,E). The K–M survival curves of the two groups showed that patients in the high-risk group had a worse survival rate when compared with their counterparts (Figure 3F). Furthermore, the time-dependent ROC analysis was utilized to show the prognostic value and predictive performance of the FPM, and the area under curve (AUC) reached 0.751 at 1 year, 0.732 at 3 years, and 0.748 at 5 years (Figure 3G), which illustrated that the FPM could be a suitable prognostic predictor.

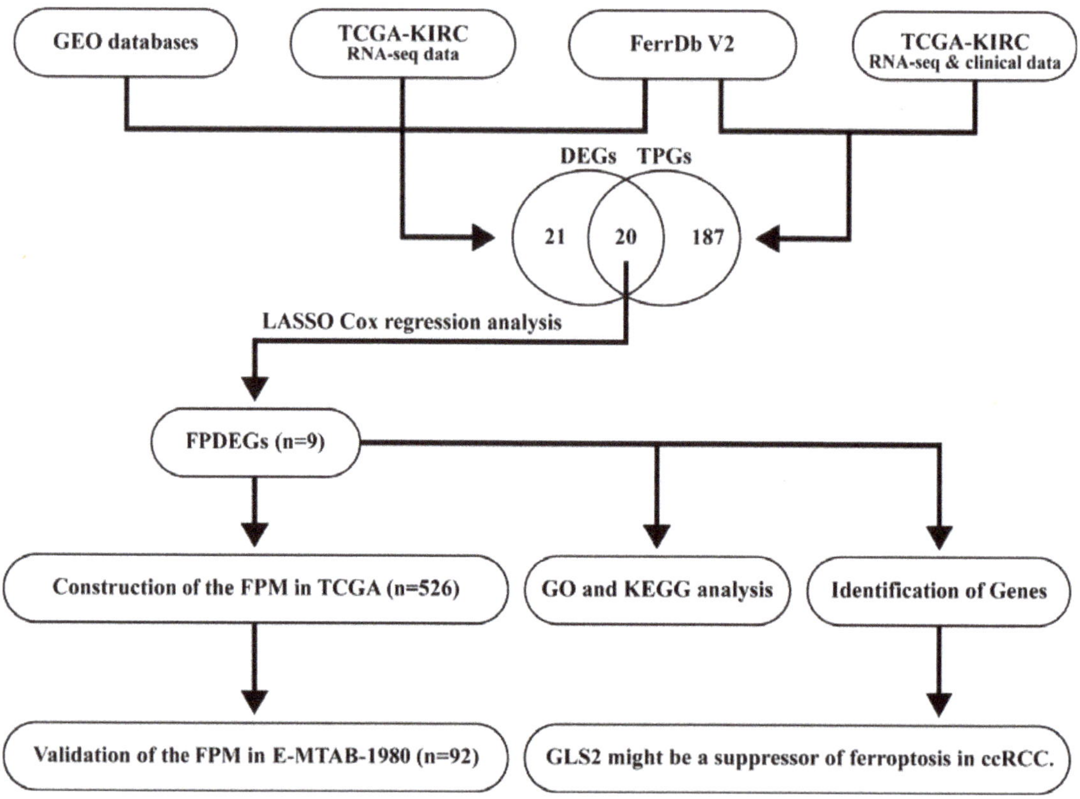

Figure 1. Flow chart of identification and validation.

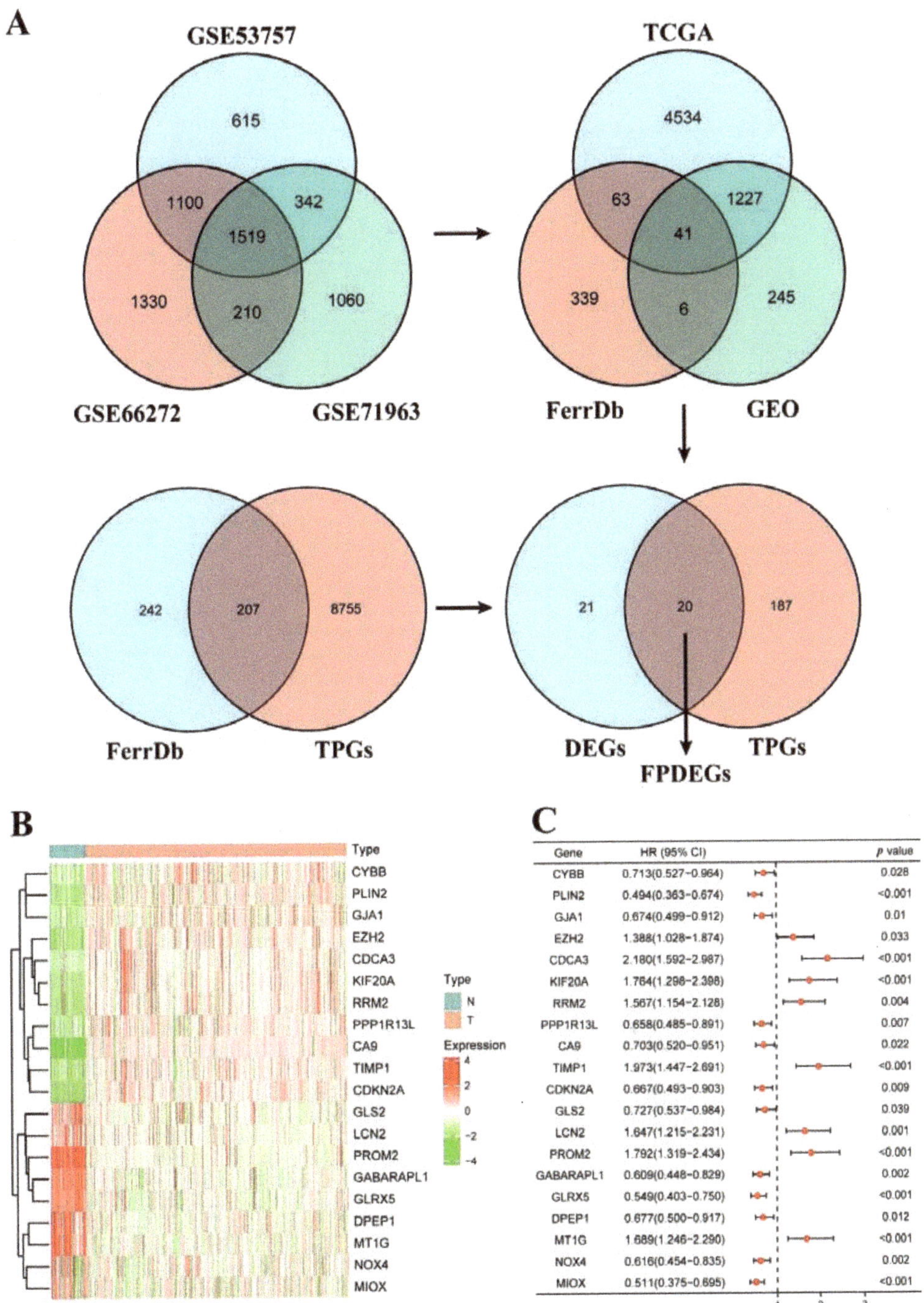

Figure 2. Identification of FPDEGs. (**A**) Process of identifying FPDEGs via Venn. (**B**) Heatmap showing expression of twenty FPDEGs in KIRC cohort of TCGA database. (**C**) Forest plot showing twenty FPDEGs identified by univariate Cox regression analysis based on OS.

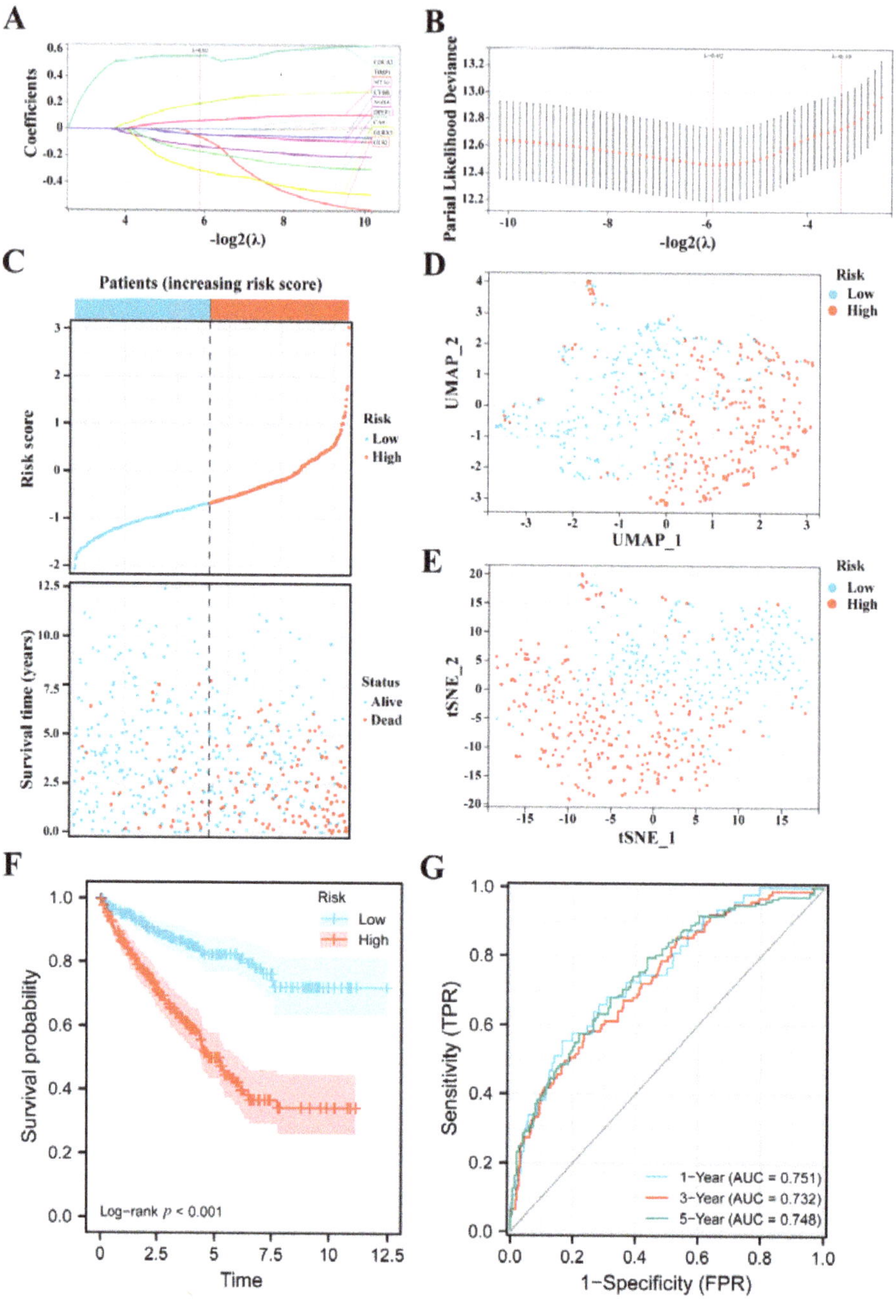

Figure 3. Construction of the FPM in TCGA. (**A**,**B**) LASSO Cox regression analyzed FPDEGs. (**C**) Distributions and media value of ccRCC patients with increasing risk score and distribution of ccRCC patients with corresponding survival status. (**D**) UMAP plot of the ccRCC patients showing the distribution in two risk groups. (**E**) t—SNE plot of the ccRCC patients showing the distribution in two risk groups. (**F**) K–M curves of patients in two risk groups. (**G**) AUC of the time-dependent ROC curves showing the prognostic value of the risk score.

3.3. FPM Could Be a Well Independent Prognostic Factor of ccRCC

The heatmap based on increasing risk score showed the relationship between risk group, basic clinical information, pathological feature, and nine-gene expression (Figure 4A). Based on Pearson's chi square test, the risk group had noticeable correlativity with the tumor grade ($p < 0.001$), tumor stage ($p < 0.001$), and patient status ($p < 0.001$) while it was not related to age ($p = 0.794$) or gender ($p = 0.066$).

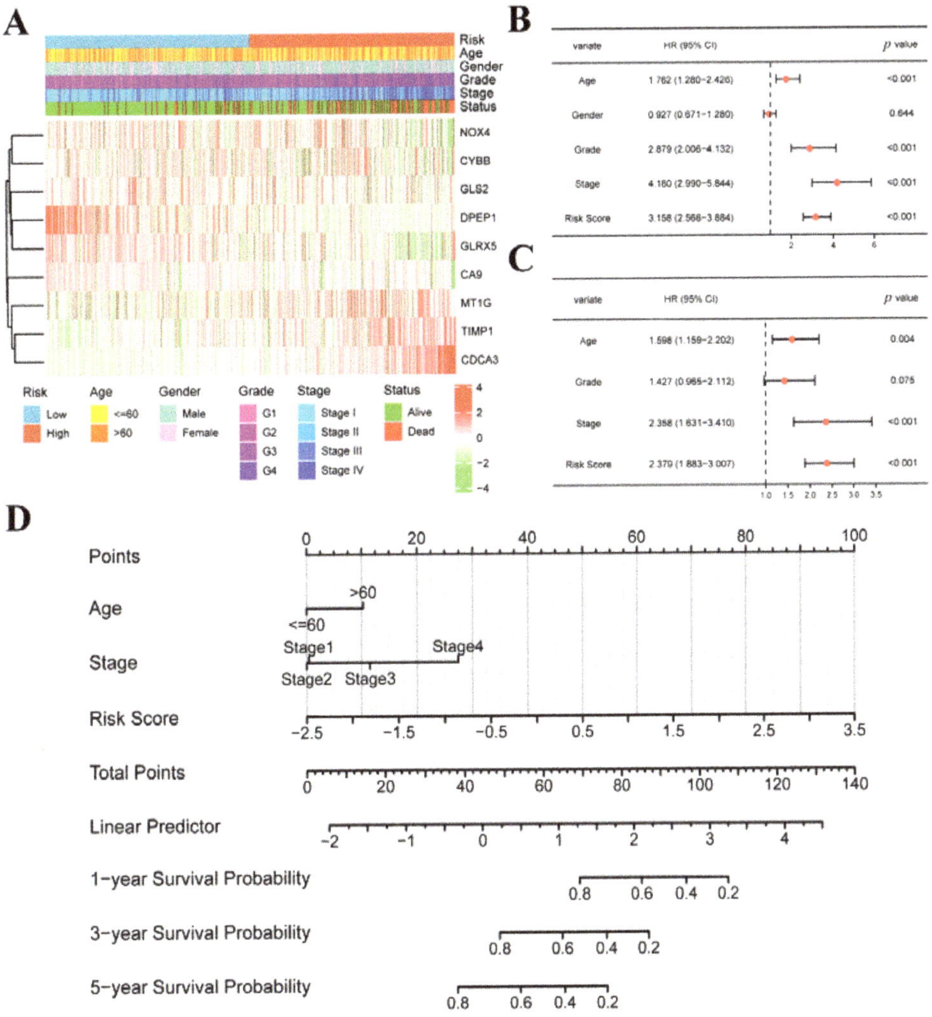

Figure 4. *Cont.*

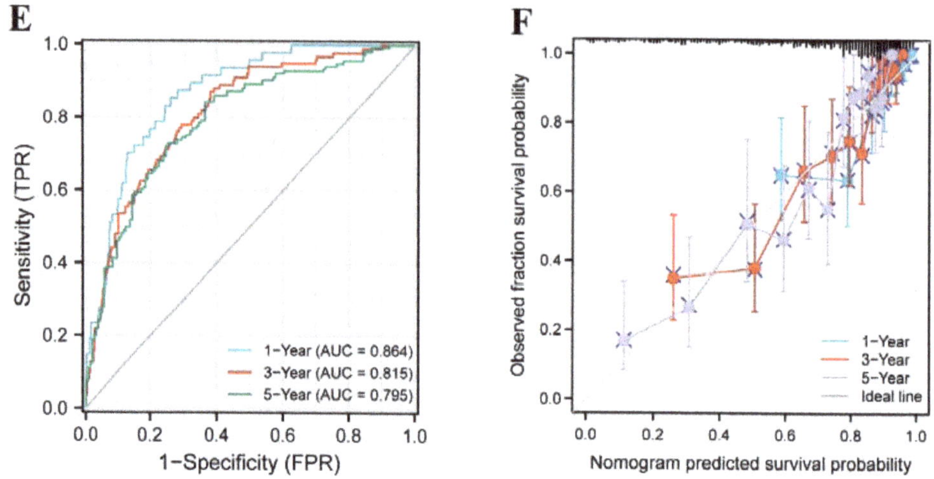

Figure 4. FPM could be an independent prognostic factor for ccRCC. (**A**) Heatmap showing the correlations between risk groups and clinical characteristics. (**B**) Univariate Cox regression analysis of the factors. (**C**) Multivariate Cox regression analysis of the factors. (**D**) A prognostic nomogram model for predicting 1-, 3-, and 5-year survival probability. (**E**) Time-dependent ROC curves of the nomogram. (**F**) Calibration curves of the nomogram for predicting 1-, 3-, and 5-year survival probability.

The age, tumor grade, tumor stage, and risk score possessed a strong relationship with OS through univariate Cox regression (Figure 4B). The age, tumor stage, and risk score showed a marked relationship with OS through multivariate Cox regression (Figure 4C). These results demonstrate that the risk score could be an independent prognostic predictor. The age, tumor stage, and risk score were employed to set up a nomogram to display the survival probability rates (Figure 4D). The AUC for 1, 3, and 5 years was 0.864, 0.815, and 0.795, respectively (Figure 4E). The calibration analysis displayed a good fitting effect between the actual survival probabilities and the predicted survival probabilities (Figure 4F).

3.4. The FPM Could Be a Convincing Independent Prognostic Predictor in E-MTAB-1980 Cohort

We checked the clinical data and gene expression of the patients in E-MATB-1980 and deleted the cases without complete information, then 92 cases were used to validate the prognostic value of the FPM. According to the media risk score of TCGA, 92 patients were divided into a low-risk group (n = 33) and a high-risk group (n = 59), and all dead patients were placed into the high-risk group, which illustrated the well-predicted value of FPM (Figure 5A). The two groups of ccRCC patients could be well distributed into two sets by using UMAP and t-SNE analysis (Figure 5B,C). The K–M survival curves of the two risk groups showed that patients in the high-risk group displayed a worse survival rate when no patient was in the low group (Figure 5D). The AUC for 1, 3, and 5 years was 0.888, 0.875, and 0.879, respectively, which showed the convincing prognostic value of the FPM (Figure 5E). The heatmap of the 92 patients in the E-MTAB-1980 cohort was also used to present the relationship between risk group, basic clinical information, pathological features, and nine-gene expression (Figure 5F).

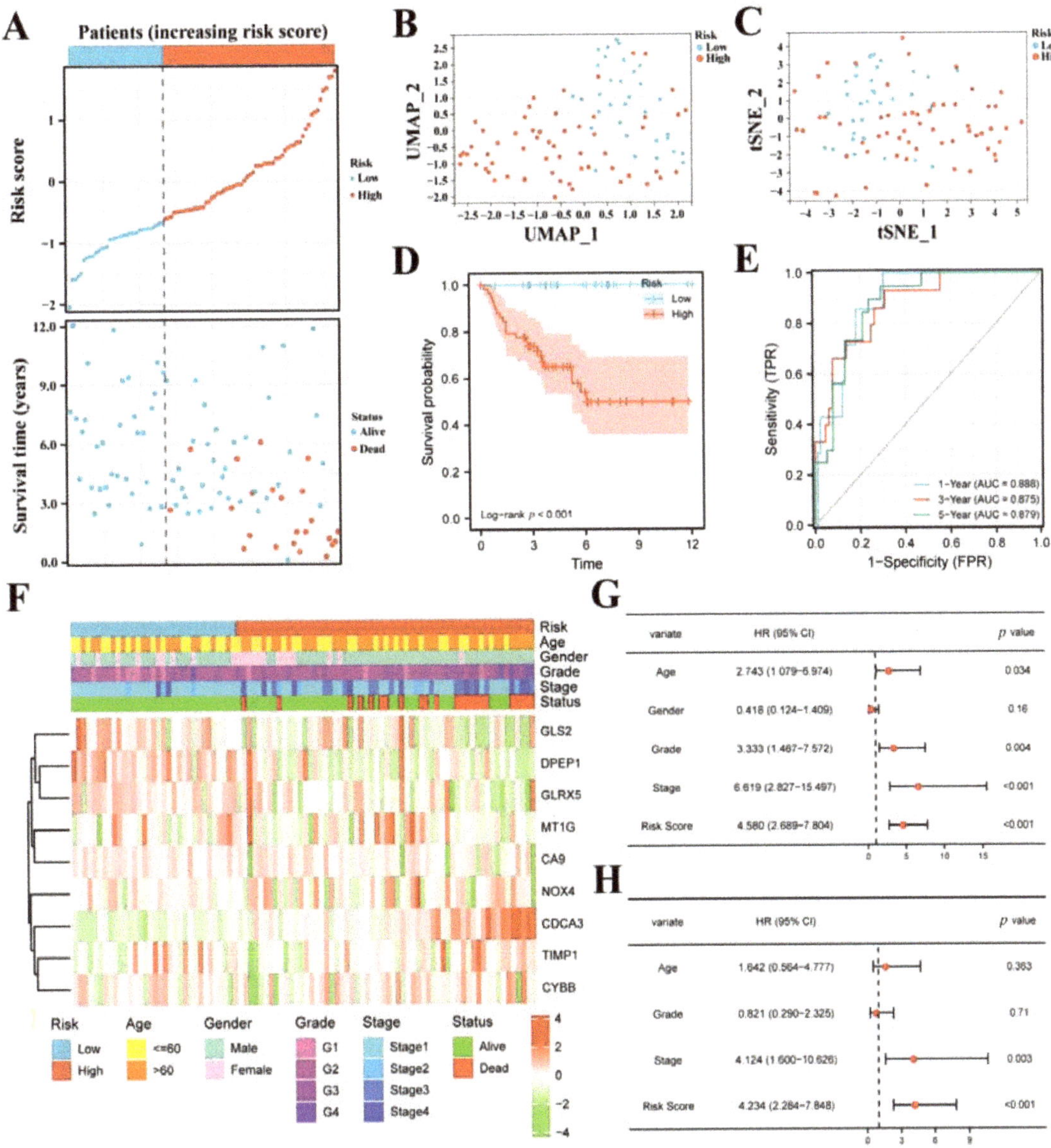

Figure 5. Validation of the FPM in E-MTAB-1980 cohort. (**A**) Distributions and media value of ccRCC patients with increasing risk score and distribution of ccRCC patients with corresponding survival status. (**B**) UMAP plot of the ccRCC patients showing the distribution in two risk groups. (**C**) t-SNE plot of the ccRCC patients showing the distribution in two risk groups. (**D**) K–M curves of patients in two risk groups. (**E**) AUC of the time-dependent ROC curves showing the prognostic value of the risk score. (**F**) Heatmap showing the correlations between risk groups and clinical characteristics. (**G**) Univariate Cox regression analysis of the factors. (**H**) Multivariate Cox regression analysis of the factors.

The age, tumor grade, tumor stage, and risk score possessed a strong relationship with OS through univariate Cox regression (Figure 5G). The tumor stage, and risk score showed a marked relationship with OS through multivariate Cox regression (Figure 5H). These results demonstrate that the risk score could be a convincing independent prognostic predictor.

3.5. Molecular Functional Analysis

GO and KEGG analysis were used to investigate the molecular function and potential signaling pathways of nine genes. The nine genes enriched in 132 BP, 22 CC, 36 MF, and 10 KEGG terms that possessed significant difference ($p < 0.05$). The top ten terms of BP, CC, and MF were chosen and are displayed in Figure 6A,B. The nine genes mainly focused on several amino acid metabolic processes, metal processes, and redox processes, such as the homocysteine metabolic process, α-amino acid metabolic process, sulfur amino acid metabolic process, ROS metabolic process, electron transport chain, electron transfer activity, oxidoreductase activity, and cellular response to metal iron. The 10 KEGG terms contained HIF-1 signaling pathway, ferroptosis, and amino acid metabolism (Figure 6C,D), however, DPEP1, GLRX5, and CDCA3 did not enrich in 10 KEGG terms.

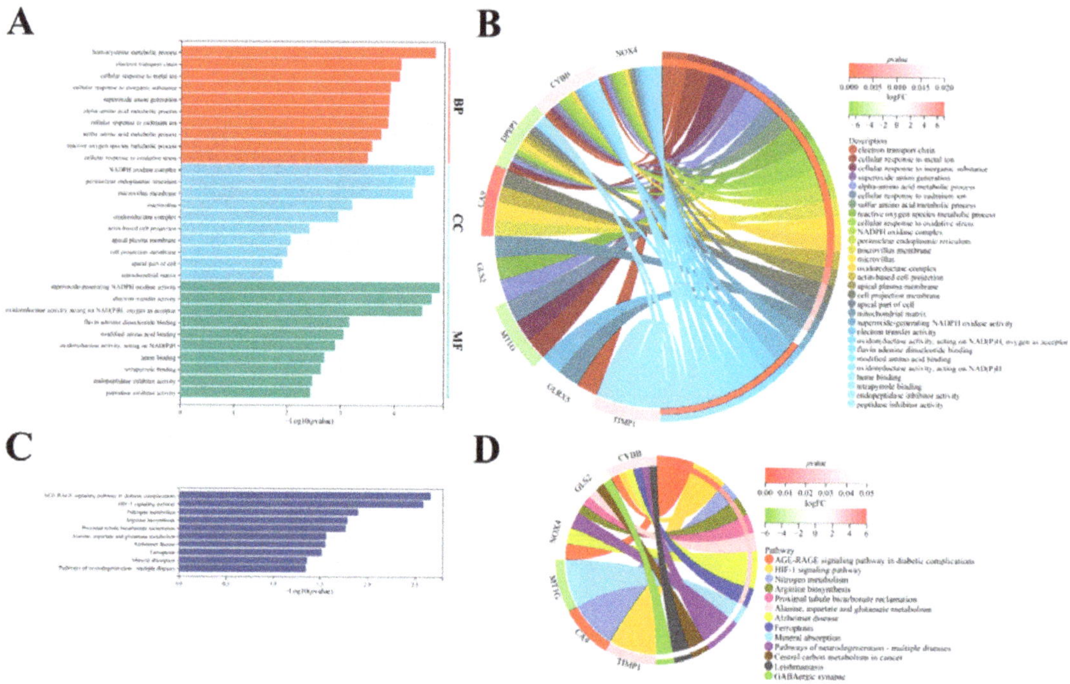

Figure 6. Enrichment analysis of nine FPDEGs. (**A**,**B**) Top ten enriched terms of nine FPDEGs in BP, CC, and MF. (**C**,**D**) Ten KEGG terms of nine FPDEGs.

3.6. GLS2 Was Upregulated during Ferroptosis

The erastin or RSL3 could dose-dependently induce the death of ACHN cells that was prevented by the ferroptosis inhibitors Fer-1 and Lip-1 in Figure 7A,B, and the cellular morphology is shown in Figure 7E. In contrast, the autophagy inhibitor HCQ, the necroptosis inhibitor NSA, and the apoptosis inhibitor Z-VAD could not repress erastin- or RSL3-induced cell death. Meanwhile, similar results also occurred in Caki-1 cells (Figure 7C–E). The mRNA expression of nine genes in ACHN and Caik-1 with different treatments are exhibited in Figure 7F,G. The mRNA expression of GLS2 was significantly upregulated in both ACHN and Caki-1 with erastin or RSL3 treatment, which indicated that GLS2 played a crucial role during ferroptosis.

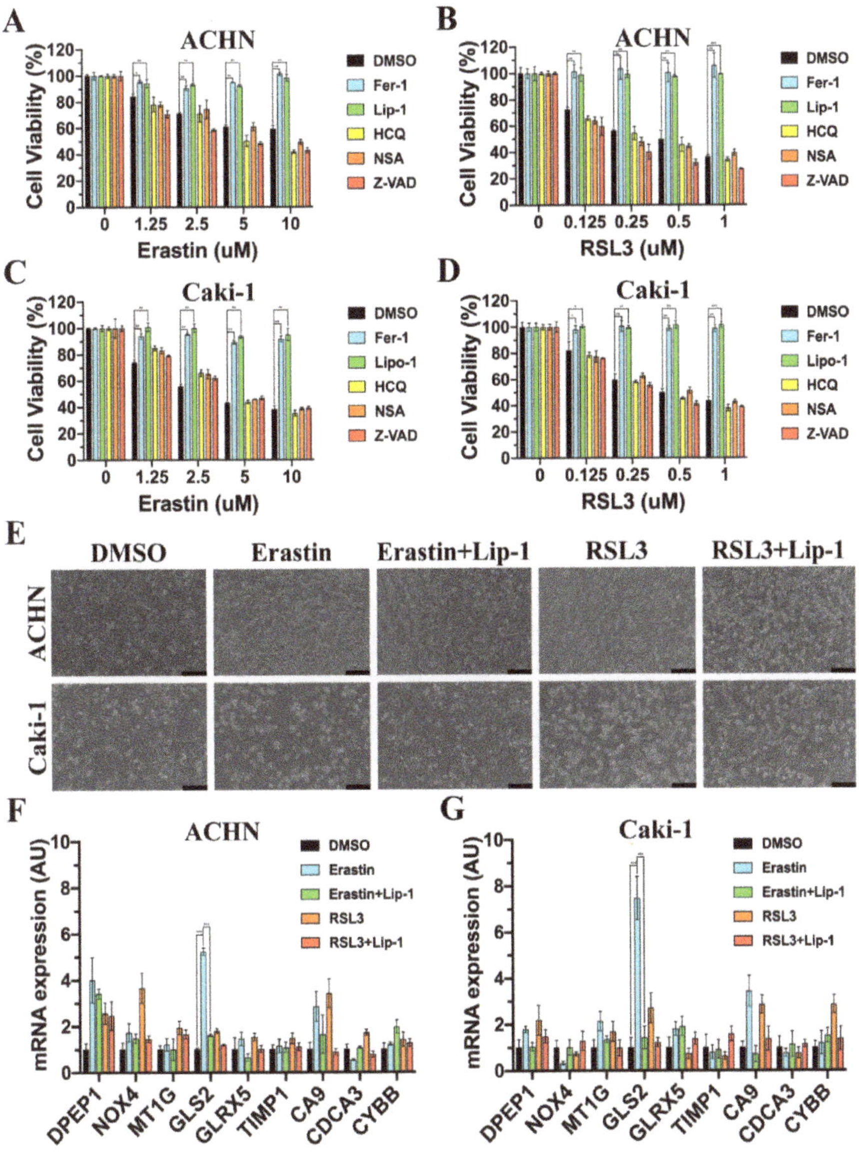

Figure 7. GLS2 was upregulated during ferroptosis. (**A**,**B**) ACHN cells were treated with various concentrations of erastin (12 h) or RSL3 (6 h) with or without Fer-1 (1 μM), Lip-1 (1 μM), HCQ (20 μM), NSA (1 μM), Z-VAD (20 μM). (**C**,**D**) Caki-1 cells were treated with various concentrations of erastin (12 h) or RSL3 (6 h) with or without Fer-1 (1 μM), Lip-1 (1 μM), HCQ (20 μM), NSA (1 μM), Z-VAD (20 μM). (**E**) The cellular morphology of AHCN and Caki-1 treated with erastin (10 μM, 12 h) or RSL3 (1 μM, 6 h) in the absence or presence of Lip-1 (1 μM). The scale bar represented 500 μm. (**F**) The mRNA expression of nine FPDEGs in ACHN treated with erastin (10 μM, 12 h) or RSL3 (1 μM, 6 h) in the absence or presence of Lip-1 (1 μM). (**G**) The mRNA expression of nine FPDEGs in Caki-1 treated with erastin (10 μM, 12 h) or RSL3 (1 μM, 6 h) in the absence or presence of Lip-1 (1 μM). (* $p < 0.05$, ** $p < 0.01$, *** $p < 0.001$).

3.7. GLS2 Was Low-Expressed in ccRCC Tissues and Closely Related with Prognosis

Compared with normal kidney tissues, the mRNA expression of GLS2 was low-expressed in ccRCC tissues from TCGA (Figure 8A,C), and the proteomic expression of GLS2 was also low-expressed in ccRCC tissues from CPTAC (Figure 8B). In the aspect of histopathological grade, the mRNA expression of GLS2 was lower in G3 and G4 than its counterparts (Figure 8D). In terms of clinical TNM stage from AJCC, the mRNA of GLS2 was low-expressed in Stage III and IV (Figure 8E). The OS, disease specific survival, and progress-free interval illustrated that patients with high GLS2 expression possessed high survival rates compared with their counterparts (Figure 8F–H). These results illustrate that GLS2 was significantly low-expressed in ccRCC tissues and is closely related to the prognosis of ccRCC patients.

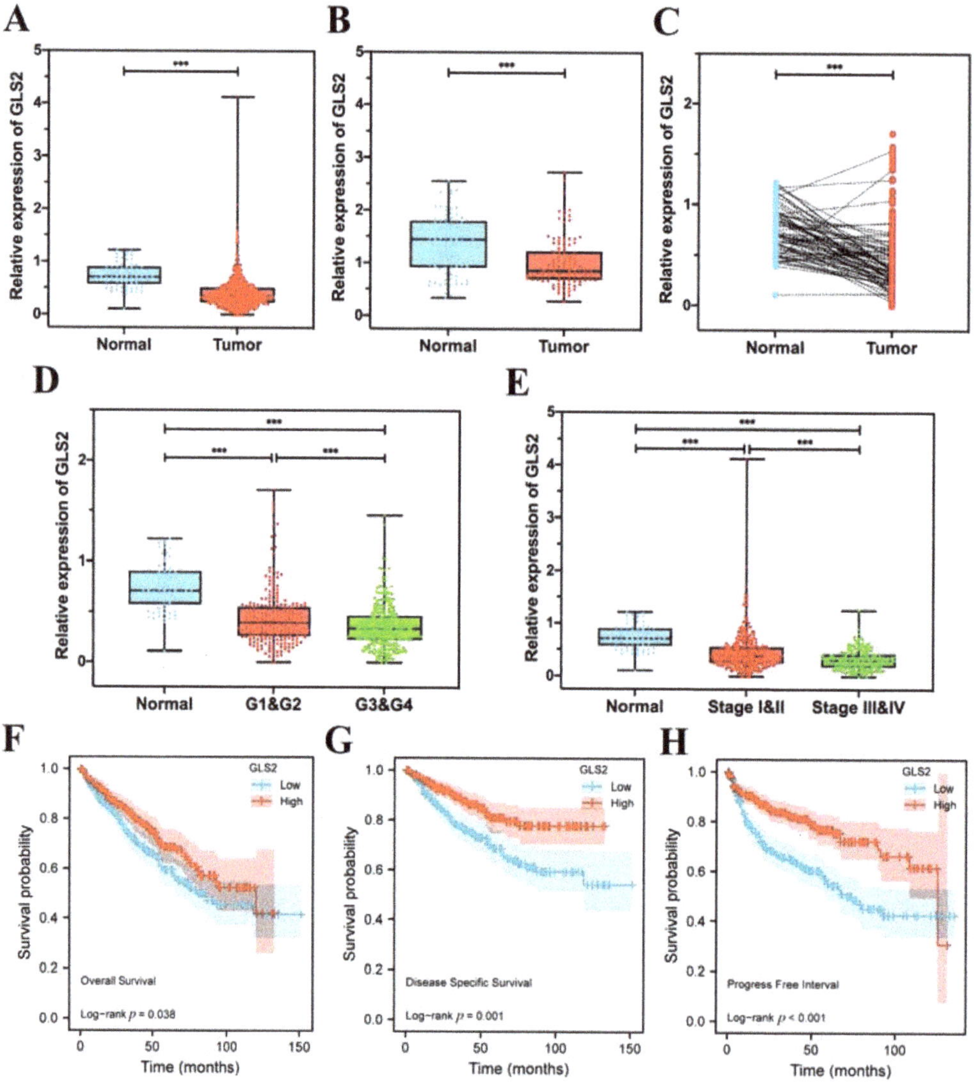

Figure 8. GLS2 was downregulated in ccRCC tissue and closely correlated with prognosis. (**A**) The mRNA expression of GLS2 between normal kidney tissues and ccRCC tissues in TCGA. (**B**) The

proteomic expression of GLS2 between normal kidney tissues and ccRCC tissues in CPTAC. (**C**) The mRNA expression of GLS2 between ccRCC tissues and paracancerous tissues in TCGA. (**D**) The mRNA expression of GLS2 in normal kidney tissues and different grade groups in TCGA. (**E**) The mRNA expression of GLS2 in normal kidney tissues and different stage groups in TCGA. (**F**) The OS of ccRCC patients in groups of high or low GLS2 expression. (**G**) The disease specific survival of ccRCC patients in groups of high or low GLS2 expression. (**H**) The progress-free interval of ccRCC patients in groups of high or low GLS2 expression. (*** $p < 0.001$).

3.8. GLS2 Might Be a Suppressor of Ferroptosis in ccRCC

As the expression of GLS2 obviously upregulated with treatment of erastin, we knocked down the mRNA expression of GLS2 using shRNA in ACHN and Caik-1 cells (Figures 9A–D and S3). After the knockdown of GLS2, the cell viabilities of ACHN and Caki-1 markedly decreased (Figure 9E,G). The cell viabilities of shRNA groups with erastin treatment were further decreased, and cell death could be repressed by Lip-1. The levels of MDA increased in both ACHN and Caki-1 cells after the knockdown of GLS2 and further reduced in knockdown groups with erastin treatment while the levels of MDA could downregulate by Lip-1 (Figure 9F,H). The BODIPY 665/676 could detect the lipid ROS in cells and the results were consistent with MDA results both in ACHN (Figure 9I) and Caki-1 (Figure S1). As GLS2 participated in the biosynthesis of GSH, the levels of GSH were tested in ACHN and Caki-1 cells (Figure 9J,K). The levels of GSH decreased in shRNA groups and further reduced in shRNA groups with erastin treatment, whereas Lip-1 treatment could not reverse the intracellular GSH level. The intracellular GSH level was detected by MBB staining because MBB could bind with GSH to emit blue fluorescence [31]. The fluorescence intensity of shRNA groups and erastin groups distinctly reduced and did not reverse with Lip-1 treatment in ACHN, which was consistent with GSH detection (Figure 9L). The MBB staining of Caki-1 with different treatments is displayed in Figure S2. These results show that GLS2 might be a suppressor of ferroptosis via affecting biosynthesis of GSH.

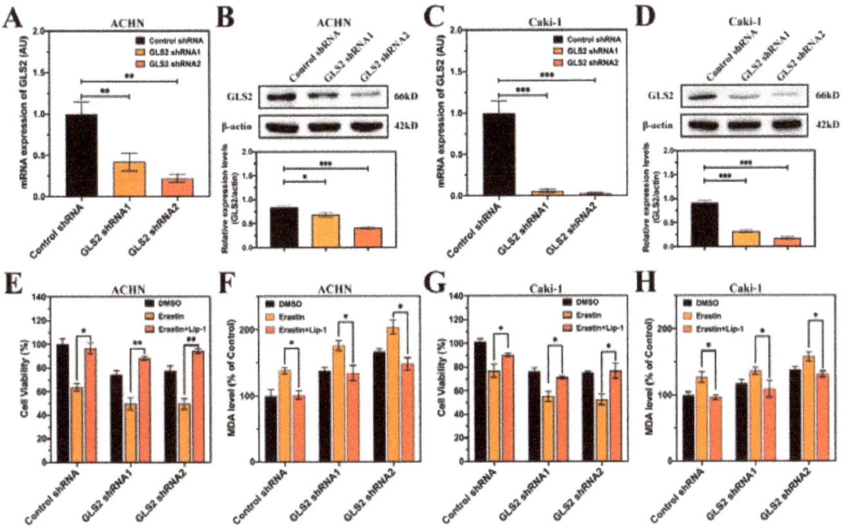

Figure 9. *Cont.*

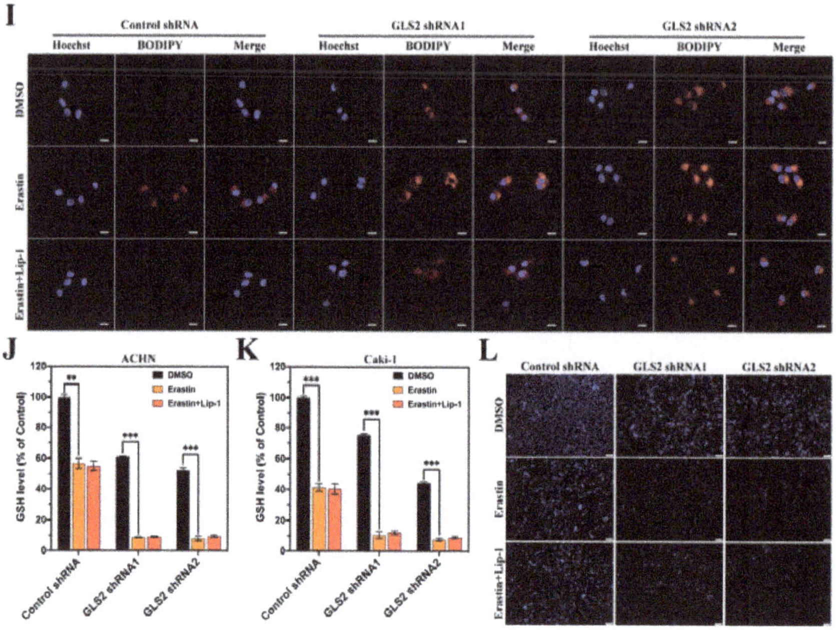

Figure 9. GLS2 might be a suppressor of ferroptosis in ccRCC. (**A**) The mRNA expression of GLS2 with different shRNAs treatments in ACHN. (**B**) The proteomic expression of GLS2 with different shRNAs treatments in ACHN. (**C**) The mRNA expression of GLS2 with different shRNAs treatments in Caki-1. (**D**) The proteomic expression of GLS2 with different shRNAs treatments in Caki-1. (**E**,**F**) The cell viabilities and MDA levels of ACHN treated with erastin (10 μM, 12 h) in the absence or presence of Lip-1 (1 μM). (**G**,**H**) The cell viabilities and MDA levels of Caki-1 treated with erastin (10 μM, 12 h) in the absence or presence of Lip-1 (1 μM). (**I**) Lipid peroxidation of ACHN with various treatments (Nuclei were stained with Hoechst 33342, and lipid peroxides were stained with BODIPY 665/676). The scale bar represented 20 μm. (**J**) GSH levels of ACHN treated with erastin (10 μM, 12 h) in the absence or presence of Lip-1 (1 μM). (**K**) GSH levels of Caki-1 treated with erastin (10 μM, 12 h) in the absence or presence of Lip-1 (1 μM). (**L**) The status of intracellular GSH levels was assessed by MBB staining in ACHN treated with erastin (10 μM, 12 h) in the absence or presence of Lip-1 (1 μM). The scale bar represents 100 μm. (* $p < 0.05$, ** $p < 0.01$, *** $p < 0.001$).

4. Discussion

Several methods of managing RCC have been utilized such as surgery, ablation, targeted therapy, chemotherapy, and immunotherapy. Although partial or radical nephrectomy and ablation could successfully be applied as ccRCC therapy, 30% of ccRCC patients still develop metastases, which is associated with higher mortality [4]. The 3-year survival rate of patients with nodal invasion is 20–30% after surgery without consideration of T stage [34]. Thermal ablation, cryoablation, and radiofrequency ablation could be taken into consideration for renal masses of less than 3 cm [3]. Targeted therapy, such as VEGF receptor inhibitors and tyrosine kinase inhibitors, has been used for RCC, however, many patients could develop drug resistance after treatment for 6–15 months, especially those with metastatic RCC [35]. These interventions might improve the OS of ccRCC patients, however, complete remission is rare because advanced RCC is a deadly disease.

Ferroptosis has been investigated in different cancers, such as breast cancer [36,37], glioblastoma [38,39], hepatocellular carcinoma [40,41], lung cancer [42,43], and pancreatic cancer [31,44]. A great number of genes, long non-coding RNAs (lncRNAs), and compounds have been studied and explored by researchers and different mechanisms have been reported. Different gene- or lncRNA-based signatures have been explored and possess prognostic value [18,19,45–47]. Prognostic models are fundamental to developing a personalized therapy, moreover, an early diagnosis is of paramount importance in these cases [48]. In a recent study, a signature containing eight ferroptotic lncRNAs was found to be accurate and reliable in predicting clinical outcomes, and the target genes (BNIP3, RRM2, and GOT) of three lncRNAs (LINC00460, LINC01550, and EPB41L4A-DT) were closely related to survival outcomes of ccRCC [45]. In the ferroptosis-related gene signature, seven genes were selected to set up a model that showed good prognostic value, and twelve genes were chosen to predict prognosis and reveal immune relevancy, however, the signatures only used a group of 60 ferroptosis-related genes and were only displayed in TCGA with or without simple validation in the E-MTAB-1980 cohort [18,19]. These signatures lack database validation or experimental validation and remain only at the level of computer operation and therefore need deeper investigation. Considering these deficiencies, we introduced three databases from GEO and obtained the whole gene expression, then we used these databases and the KIRC cohort from TCGA to find the common DEGs. In order to avoid the limitation of using only 60 genes, we downloaded the latest ferroptotic gene database (including drivers, suppressors, and markers) without unclassified genes, whose role in ferroptosis is unclear, from FerrDb V2. Based on the five databases, a signature including nine genes was employed to predict outcomes of ccRCC patients. Except MT1G and CA9, the other seven genes of the signature were not involved in previous signatures [16,18,19,46,49–51], which might be attributed to using a fragmentary gene set. Therefore, the nine-gene signature in this study could be more successful at distinguishing high-risk and low-risk patients and more accurate in predicting the prognosis of patients.

Except GLS2, the other eight genes involved in the signature could be divided into two groups, the drivers (DPEP1, NOX4, TIMP1, and CDCA3) and the suppressors (MT1G, GLRX5, CA9, and CYBB) of ferroptosis. Dipeptidase 1 (DPEP1) as a membrane-bound glycoprotein could hydrolyze a wide range of dipeptides, and it colocalized with clathrin (endocytic vesicle marker) to induce transferrin endocytosis [52]. Deficiency of DPEP1 could protect kidneys from cisplatin-induced ferroptosis. In kidney samples, DPEP1 expression is strongly related to SLC3A2 that combines with SLC7A11 to form a transport system for cystine. NADPH oxidase 4 (NOX4) could generate intracellular superoxide and promote ferroptosis via oxidative stress-induced lipid oxidation [53]. When NOX4 was inhibited by its inhibitor, cells could display resistance to erastin-induced ferroptosis [8]. Pseudolaric acid B could trigger ferroptosis by activating NOX4 in glioma, and the knockdown of NOX4 made it resistant to Pseudolaric acid B-induced cell death [54]. Metalloproteinase inhibitor 1 (TIMP1) targets and forms complexes with metalloproteinases to irreversibly inactivate the latter. Inhibition of TIMP1 could repress ferroptosis of CMEC cells by decreasing transferrin receptor 1 [55]. The cell division cycle associated protein 3 (CDCA3) mainly participates in drug resistance and cell cycle regulation in cancers and has not been studied in depth [56]. However, CDCA3 was just regarded as a driver of ferroptosis because of genome-wide CRISPR screens and has not been validated [57]. In the other group, metallothionein-1G (MT1G) has a high content of cysteine residues that bind various heavy metals and, as a transcriptional target of NRF2, could ameliorate heavy metals and free radicals to maintain cellular redox homeostasis while it is upregulated in sorafenib-resistant hepatocellular carcinoma cells [41]. Suppression of MT1G expression via shRNA or an inhibitor could significantly improve the sensibility of tumors to sorafenib. Glutaredoxin 5 (GLRX5) participates in iron–sulfur cluster biogenesis and regulates hemoglobin synthesis. GLRX5 knockdown could enhance intracellular lipid peroxidation and increase intracellular free iron, which is attributed to the upregulated transferrin and downregulated ferritin in head and neck cancer cells [58]. Carbonic anhydrase 9 (CA9), as one of the CAs that play a crucial role in equilibrating the reaction between CO_2, HCO_3^-, and H^+, is inductively expressed during hypoxia in various cancers [59,60]. Inhibition of CA9 could decrease the viability and migration of malignant mesothelioma cells, while Fe^{2+} is increased via upregulating transferrin receptor and downregulating ferritin [61]. The CA9 inhibition could be repressed by deferoxamine and ferrostatin-1, which indicated that CA9 might be a suppressor of ferroptosis. Cytochrome b-245 β chain (CYBB) is the terminal component of a respiratory chain [62]. However, its suppression of ferroptosis was only deduced and not entirely confirmed [8].

There are two types of glutaminase isoenzymes, GLS (encoded by *GLS* and regulated by regulated by c-Myc) and GLS2 (encoded by *GLS2* and regulated by p53), which are both significant enzymes participating in glutamine metabolism. The GLS-mediated deamination of glutamine results in ammonia release to maintain cell survival via biosynthesis with α-ketoglutarate and intermediates, meanwhile, glutamate coming from GLS2-mediated deamination of glutamine takes part in an antioxidant mechanism (GSH) [63]. GLS is correlated with tumor growth rate and malignancy [64], and it is high-expressed in various cancers including brain cancer, lung cancer, breast cancer, hepatocellular carcinoma, colorectal cancer [65,66]. In a recent study, the absence of exogenous glutamine induced glutamate level, which led GLS to convert from dimer to self-assembled filamentous polymer [67]. The catalytic activity of filamentous GLS increased and further depleted intracellular glutamine, which resulted in ROS-induced apoptosis that could be rescued by asparagine supplementation. GLS2 could increase the GSH level to enhance intracellular antioxidant function in HepG2, HCT116, and LN-2024 cells [68]. In our experiments, the viabilities of ACHN and Caki-1 decreased after GLS2 knockdown, and the GSH levels of ACHN and Caik-1 descended accompanied by increase of MDA. According to these findings and our experiments, GLS2 might be a negative regulator of ferroptosis. However, this conclusion was different from another study in which the knockdown of GLS2 repressed serum-dependent necroptosis in mouse embryonic fibroblasts through control of glutaminolysis [69]. The authors also considered that the results might be due to predominant expression of GLS2 in mouse embryonic fibroblasts, however, they did not further validate this. As there are various components in fetal bovine serum, the results lacked specific ferroptosis treatments and specific ferroptosis-related assays. In our study, we used erastin and RSL3 to induce ferroptosis of ACHN and Caki-1 and discovered the role of GLS2 in the ferroptosis of ccRCC cells.

5. Conclusions

In summary, this study identified a novel signature that could successfully distinguish patients with ccRCC on the basis of clinical and molecular characteristics. The novel nine-FPEDG signature could be an independent prognostic factor for ccRCC in TCGA and ArrayExpress databases. It was discovered for the first time that GLS2 might be a ferroptotic suppressor in ccRCC. The potential mechanisms of other FPEDGs remain unclear and need further investigation.

Supplementary Materials: The following are available online at https://www.mdpi.com/article/10.3390/cancers14194690/s1. Table S1: Primer sequences of genes. Table S2: List of DEGs from three GEO databases. Table S3: List of genes from FerrDb V2 database. Table S4: List of DEGs from three TCGA database. Table S5: List of 41 DEGs. Table S6: List of 207 TPGs. Figure S1: Lipid peroxidation of Caki-1 with various treatments (Nuclei were stained with Hoechst 33342, and lipid peroxides were stained with BODIPY 665/676). The scale bar represented 20 μm. Figure S2: The status of intracellular GSH levels was assessed by MBB staining in ACHN treated with erastin (10 μM, 12 h) in the absence or presence of Lip-1 (1 μM). The scale bar represented 100 μm. Figure S3: Original whole blot.

Author Contributions: Conceptualization, C.S. and Z.S.; Data curation, Q.L. and M.W.; Formal analysis, J.Z. and Q.L.; Investigation, Z.S.; Methodology, R.W.; Resources, J.Z. and Y.Y.; Software, Y.L. and Q.Z.; Supervision, Q.Z., Z.X., H.S. and K.W.; Validation, Z.S.; Visualization, Y.L. and Z.X.; Writing—original draft, Z.S.; Writing—review and editing, H.S., K.W. and C.S. All authors have read and agreed to the published version of the manuscript.

Funding: This work was supported by the National Natural Science Foundation of China (No. 81972373), the Scientific Research Foundation for Advanced Talents of Xiang'an Hospital of Xiamen University (No. PM201809170001), and the Medical Leading Talents of Xiamen City.

Institutional Review Board Statement: Not applicable.

Informed Consent Statement: Not applicable.

Data Availability Statement: The data used in this study are available from the corresponding authors.

Acknowledgments: Zhiyuan Shi wishes to thank his wife Meiying Huang and his parents, who have given him strong spiritual and financial support during his hard times. Ultimately, thanks to Jingru Huang, Xiang You, and Haiping Zheng from the Central Laboratory at the School of Medicine, Xiamen University for their kind technical support.

Conflicts of Interest: The authors declare that they have no conflicts of interest and that no commercial or financial relationships exist that affect this research.

References

1. Zheng, R.; Zhang, S.; Zeng, H.; Wang, S.; Sun, K.; Chen, R.; Li, L.; Wei, W.; He, J. Cancer Incidence and Mortality in China, 2016. *J. Natl. Cancer Cent.* **2022**, *2*, 1–9. [CrossRef]
2. Miller, D.K.; Nogueira, L.; Mariotto, A.B.; Rowland, J.H.; Yabroff, K.R.; Alfano, C.M.; Jemal, A.; Kramer, J.L.; Siegel, R.L. Cancer Treatment and Survivorship Statistics, 2019. *CA Cancer J. Clin.* **2019**, *69*, 363–385. [CrossRef] [PubMed]
3. Gray, E.R.; Harris, G.T. Renal Cell Carcinoma: Diagnosis and Management. *Am. Fam. Physician* **2019**, *99*, 179–184.
4. Hsieh, J.J.; Purdue, M.P.; Signoretti, S.; Swanton, C.; Albiges, L.; Schmidinger, M.; Heng, D.Y.; Larkin, J.; Ficarra, V. Renal Cell Carcinoma. *Nat. Rev. Dis Primers* **2017**, *3*, 17009. [CrossRef]
5. Campbell, S.; Uzzo, R.G.; Allaf, M.E.; Bass, E.B.; Cadeddu, J.A.; Chang, A.; Clark, P.E.; Davis, B.J.; Derweesh, I.H.; Giambarresi, L.; et al. Renal Mass and Localized Renal Cancer: Aua Guideline. *J. Urol.* **2017**, *198*, 520–529. [CrossRef]
6. Capitanio, U.; Montorsi, F. Renal Cancer. *Lancet* **2016**, *387*, 894–906. [CrossRef]
7. Rao, A.; Wiggins, C.; Lauer, R.C. Survival Outcomes for Advanced Kidney Cancer Patients in the Era of Targeted Therapies. *Ann. Transl. Med.* **2018**, *6*, 165. [CrossRef]
8. Dixon, S.J.; Lemberg, K.M.; Lamprecht, M.R.; Skouta, R.; Zaitsev, E.M.; Gleason, C.E.; Patel, D.N.; Bauer, A.J.; Cantley, A.M.; Yang, W.S.; et al. Ferroptosis: An Iron-Dependent Form of Nonapoptotic Cell Death. *Cell* **2012**, *149*, 1060–1072. [CrossRef]
9. Shi, Z.; Zheng, J.; Tang, W.; Bai, Y.; Zhang, L.; Xuan, Z.; Sun, H.; Shao, C. Multifunctional Nanomaterials for Ferroptotic Cancer Therapy. *Front. Chem.* **2022**, *10*, 868630. [CrossRef]
10. Shi, Z.; Zhang, L.; Zheng, J.; Sun, H.; Shao, C. Ferroptosis: Biochemistry and Biology in Cancers. *Front. Oncol.* **2021**, *11*, 579286. [CrossRef]
11. Yagoda, N.; Von Rechenberg, M.; Zaganjor, E.; Bauer, A.J.; Yang, W.S.; Fridman, D.J.; Wolpaw, A.J.; Smukste, I.; Peltier, J.M.; Boniface, J.J.; et al. Ras-Raf-Mek-Dependent Oxidative Cell Death Involving Voltage-Dependent Anion Channels. *Nature* **2007**, *447*, 864–868. [CrossRef] [PubMed]
12. Yang, W.S.; SriRamaratnam, R.; Welsch, M.E.; Shimada, K.; Skouta, R.; Viswanathan, V.S.; Cheah, J.H.; Clemons, P.A.; Shamji, A.F.; Clish, C.B.; et al. Regulation of Ferroptotic Cancer Cell Death by Gpx4. *Cell* **2014**, *156*, 317–331. [CrossRef] [PubMed]
13. Gaschler, M.M.; Stockwell, B.R. Lipid Peroxidation in Cell Death. *Biochem. Biophys. Res. Commun.* **2017**, *482*, 419–425. [CrossRef] [PubMed]
14. Weïwer, M.; Bittker, J.A.; Lewis, T.; Shimada, K.; Yang, W.S.; MacPherson, L.; Dandapani, S.; Palmer, M.; Stockwell, B.R.; Schreiber, S.L.; et al. Development of Small-Molecule Probes That Selectively Kill Cells Induced to Express Mutant Ras. *Bioorg Med. Chem. Lett.* **2012**, *22*, 1822–1826. [CrossRef]
15. Fenton, H.J.H. Oxidation of Tartaric Acid in Presence of Iron. *J. Chem. Soc. Trans.* **1894**, *65*, 899–910. [CrossRef]
16. Wu, Z.G.; Wang, Q.F.; Xu, Y.K.; Li, Q.L.; Cheng, L. A New Survival Model Based on Ferroptosis-Related Genes for Prognostic Prediction in Clear Cell Renal Cell Carcinoma. *Aging* **2020**, *12*, 14933–14948. [CrossRef]
17. Gao, S.; Ruan, H.; Liu, J.; Liu, Y.; Liu, D.; Tong, J.; Shi, J.; Yang, H.; Xu, T.; Zhang, X. A Novel Ferroptosis-Related Pathway for Regulating Immune Checkpoints in Clear Cell Renal Cell Carcinoma. *Front. Oncol.* **2021**, *11*, 678694. [CrossRef]
18. Hong, Y.; Lin, M.; Ou, D.; Huang, Z.; Shen, P. A Novel Ferroptosis-Related 12-Gene Signature Predicts Clinical Prognosis and Reveals Immune Relevancy in Clear Cell Renal Cell Carcinoma. *BMC Cancer* **2021**, *21*, 831. [CrossRef]
19. Zhao, J.G.; Wu, Z.; Ge, L.; Yang, F.; Hong, K.; Zhang, S.; Ma, L. Ferroptosis-Related Gene-Based Prognostic Model and Immune Infiltration in Clear Cell Renal Cell Carcinoma. *Front. Genet.* **2021**, *12*, 650416. [CrossRef]
20. Wu, J.; Sun, Z.; Bi, Q.; Wang, W. A Ferroptosis-Related Genes Model Allows for Prognosis and Treatment Stratification of Clear Cell Renal Cell Carcinoma: A Bioinformatics Analysis and Experimental Verification. *Front. Oncol.* **2022**, *12*, 815223. [CrossRef]
21. Cheng, G.; Liu, D.; Liang, H.; Yang, H.; Chen, K.; Zhang, X. A Cluster of Long Non-Coding Rnas Exhibit Diagnostic and Prognostic Values in Renal Cell Carcinoma. *Aging* **2019**, *11*, 9597–9615. [CrossRef] [PubMed]
22. Yang, F.; Liu, C.; Zhao, G.; Ge, L.; Song, Y.; Chen, Z.; Liu, Z.; Hong, K.; Ma, L. Long Non-Coding Rna Linc01234 Regulates Proliferation, Migration and Invasion Via Hif-2α Pathways in Clear Cell Renal Cell Carcinoma Cells. *PeerJ* **2020**, *8*, e10149. [CrossRef] [PubMed]
23. Chen, X.; Li, J.; Kang, R.; Klionsky, D.J.; Tang, D. Ferroptosis: Machinery and Regulation. *Autophagy* **2021**, *17*, 2054–2081. [CrossRef] [PubMed]

24. Elgendy, M.S.; Alyammahi, S.K.; Alhamad, D.W.; Abdin, S.M.; Omar, H.A. Ferroptosis: An Emerging Approach for Targeting Cancer Stem Cells and Drug Resistance. *Crit. Rev. Oncol. Hematol.* **2020**, *155*, 03095. [CrossRef] [PubMed]
25. Robinson, D.M.; McCarthy, D.J.; Smyth, G.K. Edger: A Bioconductor Package for Differential Expression Analysis of Digital Gene Expression Data. *Bioinformatics* **2010**, *26*, 139–140. [CrossRef]
26. Zhang, M.; Zhang, X.; Yu, M.; Zhang, W.; Zhang, D.; Zeng, S.; Wang, X.; Hu, X. A Novel Ferroptosis-Related Gene Model for Overall Survival Predictions of Bladder Urothelial Carcinoma Patients. *Front. Oncol.* **2021**, *11*, 698856. [CrossRef]
27. Liu, J.; Lichtenberg, T.M.; Hoadley, K.A.; Poisson, L.M.; Lazar, A.J.; Cherniack, A.D.; Kovatich, A.J.; Benz, C.C.; Levine, D.A.; Lee, A.V.; et al. An Integrated Tcga Pan-Cancer Clinical Data Resource to Drive High-Quality Survival Outcome Analytics. *Cell* **2018**, *173*, 400–416. [CrossRef]
28. Simon, N.; Friedman, J.; Hastie, T.; Tibshirani, R. Regularization Paths for Cox's Proportional Hazards Model via Coordinate Descent. *J. Stat. Softw.* **2011**, *39*, 1–13. [CrossRef]
29. Tibshirani, R. The Lasso Method for Variable Selection in the Cox Model. *Stat. Med.* **1997**, *16*, 385–395. [CrossRef]
30. Yu, G.; Wang, L.G.; Han, Y.; He, Q.Y. Clusterprofiler: An R Package for Comparing Biological Themes among Gene Clusters. *Omics* **2012**, *16*, 284–287. [CrossRef]
31. Kuang, F.; Liu, J.; Xie, Y.; Tang, D.; Kang, R. Mgst1 Is a Redox-Sensitive Repressor of Ferroptosis in Pancreatic Cancer Cells. *Cell Chem. Biol.* **2021**, *28*, 765–775. [CrossRef] [PubMed]
32. Sun, H.; Zheng, J.; Xiao, J.; Yue, J.; Shi, Z.; Xuan, Z.; Chen, C.; Zhao, Y.; Tang, W.; Ye, S.; et al. Topk/Pbk Is Phosphorylated by Erk2 at Serine 32, Promotes Tumorigenesis and Is Involved in Sorafenib Resistance in Rcc. *Cell Death Dis.* **2022**, *13*, 450. [CrossRef] [PubMed]
33. Deng, F.; Sharma, I.; Dai, Y.; Yang, M.; Kanwar, Y.S. Myo-Inositol Oxygenase Expression Profile Modulates Pathogenic Ferroptosis in the Renal Proximal Tubule. *J. Clin. Investig.* **2019**, *129*, 5033–5049. [CrossRef] [PubMed]
34. Capitanio, U.; Becker, F.; Blute, M.L.; Mulders, P.; Patard, J.J.; Russo, P.; Studer, U.E.; van Poppel, H. Lymph Node Dissection in Renal Cell Carcinoma. *Eur. Urol.* **2011**, *60*, 1212–1220. [CrossRef]
35. Xuan, Z.; Chen, C.; Tang, W.; Ye, S.; Zheng, J.; Zhao, Y.; Shi, Z.; Zhang, L.; Sun, H.; Shao, C. Tki-Resistant Renal Cancer Secretes Low-Level Exosomal Mir-549a to Induce Vascular Permeability and Angiogenesis to Promote Tumor Metastasis. *Front. Cell Dev. Biol.* **2021**, *9*, 689947. [CrossRef]
36. Wu, H.Z.; Tang, Y.; Yu, H.; Li, H.D. The Role of Ferroptosis in Breast Cancer Patients: A Comprehensive Analysis. *Cell Death Discov.* **2021**, *7*, 93. [CrossRef]
37. Li, R.; Zhang, J.; Zhou, Y.; Gao, Q.; Wang, R.; Fu, Y.; Zheng, L.; Yu, H. Transcriptome Investigation and in Vitro Verification of Curcumin-Induced Ho-1 as a Feature of Ferroptosis in Breast Cancer Cells. *Oxid Med. Cell Longev.* **2020**, *11*, 3469840. [CrossRef]
38. Yang, X.; Liu, J.; Wang, C.; Cheng, K.K.-Y.; Xu, H.; Li, Q.; Hua, T.; Jiang, X.; Sheng, L.; Mao, J.; et al. Mir-18a Promotes Glioblastoma Development by Down-Regulating Aloxe3-Mediated Ferroptotic and Anti-Migration Activities. *Oncogenesis* **2021**, *10*, 15. [CrossRef]
39. Xia, L.; Gong, M.; Zou, Y.; Wang, Z.; Wu, B.; Zhang, S.; Li, L.; Jin, K.; Sun, C. Apatinib Induces Ferroptosis of Glioma Cells through Modulation of the Vegfr2/Nrf2 Pathway. *Oxid Med. Cell Longev.* **2022**, *5*, 9925919. [CrossRef]
40. Asperti, M.; Bellini, S.; Grillo, E.; Gryzik, M.; Cantamessa, L.; Ronca, R.; Maccarinelli, F.; Salvi, A.; De Petro, G.; Arosio, P.; et al. H-Ferritin Suppression and Pronounced Mitochondrial Respiration Make Hepatocellular Carcinoma Cells Sensitive to Rsl3-Induced Ferroptosis. *Free Radic. Biol. Med.* **2021**, *169*, 294–303. [CrossRef]
41. Sun, F.X.; Niu, X.H.; Chen, R.C.; He, W.Y.; Chen, D.; Kang, R.; Tang, D.L. Metallothionein-1g Facilitates Sorafenib Resistance through Inhibition of Ferroptosis. *Hepatology* **2016**, *64*, 488–500. [CrossRef]
42. Koppula, P.; Lei, G.; Zhang, Y.; Yan, Y.; Mao, C.; Kondiparthi, L.; Shi, J.; Liu, X.; Horbath, A.; Das, M.; et al. A Targetable Coq-Fsp1 Axis Drives Ferroptosis- and Radiation-Resistance in Keap1 Inactive Lung Cancers. *Nat. Commun.* **2022**, *13*, 2206. [CrossRef]
43. Meng, C.; Zhan, J.; Chen, D.; Shao, G.; Zhang, H.; Gu, W.; Luo, J. The Deubiquitinase Usp11 Regulates Cell Proliferation and Ferroptotic Cell Death Via Stabilization of Nrf2 Usp11 Deubiquitinates and Stabilizes Nrf2. *Oncogene* **2021**, *40*, 1706–1720. [CrossRef] [PubMed]
44. Kremer, D.M.; Nelson, B.S.; Lin, L.; Yarosz, E.L.; Halbrook, C.J.; Kerk, S.A.; Sajjakulnukit, P.; Myers, A.; Thurston, G.; Hou, S.W.; et al. Got1 Inhibition Promotes Pancreatic Cancer Cell Death by Ferroptosis. *Nat. Commun.* **2021**, *12*, 4860. [CrossRef]
45. Zhou, Z.; Yang, Z.; Cui, Y.; Lu, S.; Huang, Y.; Che, X.; Yang, L.; Zhang, Y. Identification and Validation of a Ferroptosis-Related Long Non-Coding Rna (Frlncrna) Signature to Predict Survival Outcomes and the Immune Microenvironment in Patients with Clear Cell Renal Cell Carcinoma. *Front. Genet.* **2022**, *13*, 787884. [CrossRef]
46. Dong, Y.; Liu, D.; Zhou, H.; Gao, Y.; Nueraihemaiti, Y.; Xu, Y. A Prognostic Signature for Clear Cell Renal Cell Carcinoma Based on Ferroptosis-Related Lncrnas and Immune Checkpoints. *Front. Genet.* **2022**, *13*, 912190. [CrossRef] [PubMed]
47. Chen, X.; Tu, J.; Ma, L.; Huang, Y.; Yang, C.; Yuan, X. Analysis of Ferroptosis-Related Lncrnas Signatures Associated with Tumor Immune Infiltration and Experimental Validation in Clear Cell Renal Cell Carcinoma. *Int. J. Gen. Med.* **2022**, *15*, 3215–3235. [CrossRef] [PubMed]
48. Cochetti, G.; Cari, L.; Maulà, V.; Cagnani, R.; Paladini, A.; del Zingaro, M.; Nocentini, G.; Mearin, E.I. Validation in an Independent Cohort of Mir-122, Mir-1271, and Mir-15b as Urinary Biomarkers for the Potential Early Diagnosis of Clear Cell Renal Cell Carcinoma. *Cancers* **2022**, *14*, 1112. [CrossRef]

49. Zheng, B.; Niu, Z.; Si, S.; Zhao, G.; Wang, J.; Yao, Z.; Cheng, F.; He, W. Comprehensive Analysis of New Prognostic Signature Based on Ferroptosis-Related Genes in Clear Cell Renal Cell Carcinoma. *Aging* **2021**, *13*, 19789–19804. [CrossRef]
50. Chen, J.; Zhan, Y.; Zhang, R.; Chen, B.; Huang, J.; Li, C.; Zhang, W.; Wang, Y.; Gao, Y.; Zheng, J.; et al. A New Prognostic Risk Signature of Eight Ferroptosis-Related Genes in the Clear Cell Renal Cell Carcinoma. *Front. Oncol.* **2021**, *11*, 700084. [CrossRef]
51. Li, S.; Xu, X.; Zhang, R.; Huang, Y. Identification of Co-Expression Hub Genes for Ferroptosis in Kidney Renal Clear Cell Carcinoma Based on Weighted Gene Co-Expression Network Analysis and the Cancer Genome Atlas Clinical Data. *Sci. Rep.* **2022**, *12*, 4821. [CrossRef] [PubMed]
52. Guan, Y.; Liang, X.; Ma, Z.; Hu, H.; Liu, H.; Miao, Z.; Linkermann, A.; Hellwege, J.N.; Voight, B.F.; Susztak, K. A Single Genetic Locus Controls Both Expression of Dpep1/Chmp1a and Kidney Disease Development Via Ferroptosis. *Nat. Commun.* **2021**, *12*, 5078. [CrossRef] [PubMed]
53. Park, W.M.; Cha, H.W.; Kim, J.; Kim, J.H.; Yang, H.; Yoon, S.; Boonpraman, N.; Yi, S.S.; Yoo, I.D.; Moon, J.S. Nox4 Promotes Ferroptosis of Astrocytes by Oxidative Stress-Induced Lipid Peroxidation Via the Impairment of Mitochondrial Metabolism in Alzheimer's Diseases. *Redox Biol.* **2021**, *41*, 101947. [CrossRef]
54. Wang, Z.; Ding, Y.; Wang, X.; Lu, S.; Wang, C.; He, C.; Wang, L.; Piao, M.; Chi, G.; Luo, Y.; et al. Pseudolaric Acid B Triggers Ferroptosis in Glioma Cells Via Activation of Nox4 and Inhibition of Xct. *Cancer Lett.* **2018**, *428*, 21–33. [CrossRef] [PubMed]
55. Shi, P.; Li, M.; Song, C.; Qi, H.; Ba, L.; Cao, Y.; Zhang, M.; Xie, Y.; Ren, J.; Wu, J.; et al. Neutrophil-Like Cell Membrane-Coated Sirna of Lncrna Aabr07017145.1 Therapy for Cardiac Hypertrophy Via Inhibiting Ferroptosis of Cmecs. *Mol. Ther. Nucleic Acids* **2022**, *27*, 16–36. [CrossRef] [PubMed]
56. Liu, Y.; Cheng, G.; Huang, Z.; Bao, L.; Liu, J.; Wang, C.; Xiong, Z.; Zhou, L.; Xu, T.; Liu, D.; et al. Long Noncoding Rna Snhg12 Promotes Tumour Progression and Sunitinib Resistance by Upregulating Cdca3 in Renal Cell Carcinoma. *Cell Death Dis.* **2020**, *11*, 515. [CrossRef] [PubMed]
57. Zou, Y.; Henry, W.S.; Ricq, E.L.; Graham, E.T.; Phadnis, V.V.; Maretich, P.; Paradkar, S.; Boehnke, N.; Deik, A.A.; Reinhardt, F.; et al. Plasticity of Ether Lipids Promotes Ferroptosis Susceptibility and Evasion. *Nature* **2020**, *585*, 603. [CrossRef]
58. Lee, W.J.; You, J.H.; Shin, D.; Roh, J.L. Inhibition of Glutaredoxin 5 Predisposes Cisplatin-Resistant Head and Neck Cancer Cells to Ferroptosis. *Theranostics* **2020**, *10*, 7775–7786. [CrossRef]
59. Supuran, C.T. Carbonic Anhydrases: Novel Therapeutic Applications for Inhibitors and Activators. *Nat. Rev. Drug Discov.* **2008**, *7*, 168–181. [CrossRef]
60. Wykoff, C.C.; Beasley, N.J.; Watson, P.; Turner, K.J.; Pastorek, J.; Sibtain, A.; Wilson, G.; Turley, H.; Talks, K.L.; Maxwell, P.; et al. Hypoxia-Inducible Expression of Tumor-Associated Carbonic Anhydrases. *Cancer Res.* **2000**, *60*, 7075–7083.
61. Li, Z.; Jiang, L.; Chew, S.H.; Hirayama, T.; Sekido, Y.; Toyokuni, S. Carbonic Anhydrase 9 Confers Resistance to Ferroptosis/Apoptosis in Malignant Mesothelioma under Hypoxia. *Redox Biol.* **2019**, *26*, 101297. [CrossRef] [PubMed]
62. Chen, H.; Sun, Q.; Zhang, C.; She, J.; Cao, S.; Cao, M.; Zhang, N.; Adiila, A.V.; Zhong, J.; Yao, C.; et al. Identification and Validation of Cybb, Cd86, and C3ar1 as the Key Genes Related to Macrophage Infiltration of Gastric Cancer. *Front. Mol. Biosci.* **2021**, *8*, 756085. [CrossRef] [PubMed]
63. Masisi, K.B.; el Ansari, R.; Alfarsi, L.; Rakha, E.A.; Green, A.R.; Craze, M.L. The Role of Glutaminase in Cancer. *Histopathology* **2020**, *76*, 498–508. [CrossRef]
64. Katt, P.W.; Lukey, M.J.; Cerione, R.A. A Tale of Two Glutaminases: Homologous Enzymes with Distinct Roles in Tumorigenesis. *Future Med. Chem.* **2017**, *9*, 223–243. [CrossRef] [PubMed]
65. Cassago, A.; Ferreira, A.P.; Ferreira, I.M.; Fornezari, C.; Gomes, E.R.; Greene, K.S.; Pereira, H.M.; Garratt, R.C.; Dias, S.M.; Ambrosio, A.L. Mitochondrial Localization and Structure-Based Phosphate Activation Mechanism of Glutaminase C with Implications for Cancer Metabolism. *Proc. Natl. Acad. Sci. USA* **2012**, *109*, 1092–1097. [CrossRef] [PubMed]
66. Yu, D.; Shi, X.; Meng, G.; Chen, J.; Yan, C.; Jiang, Y.; Wei, J.; Ding, Y. Kidney-Type Glutaminase (Gls1) Is a Biomarker for Pathologic Diagnosis and Prognosis of Hepatocellular Carcinoma. *Oncotarget* **2015**, *6*, 7619–7631. [CrossRef] [PubMed]
67. Jiang, B.; Zhang, J.; Zhao, G.; Liu, M.; Hu, J.; Lin, F.; Wang, J.; Zhao, W.; Ma, H.; Zhang, C.; et al. Filamentous Gls1 Promotes Ros-Induced Apoptosis Upon Glutamine Deprivation Via Insufficient Asparagine Synthesis. *Mol. Cell* **2022**, *82*, 1821–1835. [CrossRef]
68. Hu, W.; Zhang, C.; Wu, R.; Sun, Y.; Levine, A.; Feng, Z. Glutaminase 2, a Novel P53 Target Gene Regulating Energy Metabolism and Antioxidant Function. *Proc. Natl. Acad. Sci. USA* **2010**, *107*, 7455–7460. [CrossRef]
69. Gao, M.; Monian, P.; Quadri, N.; Ramasamy, R.; Jiang, X. Glutaminolysis and Transferrin Regulate Ferroptosis. *Mol. Cell* **2015**, *59*, 298–308. [CrossRef]

cancers

MDPI

Review

Precision Targets for Intercepting the Lethal Progression of Prostate Cancer: Potential Avenues for Personalized Therapy

Max Christenson, Chung-Seog Song, Ya-Guang Liu and Bandana Chatterjee *

Department of Molecular Medicine, Long School of Medicine, University of Texas Health San Antonio, San Antonio, TX 78229, USA; christensonm@livemail.uthscsa.edu (M.C.); songc@uthscsa.edu (C.-S.S.); liuy9@uthscsa.edu (Y.-G.L.)

* Correspondence: chatterjee@uthscsa.edu; Tel.: +1-210-567-7218

Simple Summary: Metastatic prostate cancer is incurable and lethal. Tumor growth is initially reduced by radiation, surgery, or hormone therapy and later, by pairing them with chemotherapy for advanced cancer. Recent innovations have helped to develop prescription drugs against certain prostate cancer types, showing gene alterations that prevent the repair of damaged DNA or activate the body's anti-cancer natural immune defense. A panel of genes has been identified whose cancer genome alterations may predict whether non-metastatic prostate cancer would go on to metastasize. The activity of these genes may help to guide treatment decision for non-metastatic cancer with the choice for non-aggressive versus debilitating aggressive options. The probing of prostate cancer genome has uncovered hormonal abnormalities and genome changes specific to individual patients and studies are revealing how these changes can lead to treatment failure. The discovery of new druggable vulnerabilities of the cancer cells has presented opportunities to develop precision treatments of metastatic prostate cancer tailored to individual patients.

Abstract: Organ-confined prostate cancer of low-grade histopathology is managed with radiation, surgery, active surveillance, or watchful waiting and exhibits a 5-year overall survival (OS) of 95%, while metastatic prostate cancer (PCa) is incurable, holding a 5-year OS of 30%. Treatment options for advanced PCa—metastatic and non-metastatic—include hormone therapy that inactivates androgen receptor (AR) signaling, chemotherapy and genome-targeted therapy entailing synthetic lethality of tumor cells exhibiting aberrant DNA damage response, and immune checkpoint inhibition (ICI), which suppresses tumors with genomic microsatellite instability and/or deficient mismatch repair. Cancer genome sequencing uncovered novel somatic and germline mutations, while mechanistic studies are revealing their pathological consequences. A microRNA has shown biomarker potential for stratifying patients who may benefit from angiogenesis inhibition prior to ICI. A 22-gene expression signature may select high-risk localized PCa, which would not additionally benefit from post-radiation hormone therapy. We present an up-to-date review of the molecular and therapeutic aspects of PCa, highlight genomic alterations leading to *AR* upregulation and discuss AR-degrading molecules as promising anti-AR therapeutics. New biomarkers and druggable targets are shaping innovative intervention strategies against high-risk localized and metastatic PCa, including AR-independent small cell-neuroendocrine carcinoma, while presenting individualized treatment opportunities through improved design and precision targeting.

Keywords: high-risk prostate cancer; androgen receptor axis; castration resistance; neuroendocrine carcinoma; PARP inhibition; immunotherapy; cancer genome; precision targeting

Citation: Christenson, M.; Song, C.-S.; Liu, Y.-G.; Chatterjee, B. Precision Targets for Intercepting the Lethal Progression of Prostate Cancer: Potential Avenues for Personalized Therapy. *Cancers* **2022**, *14*, 892. https://doi.org/10.3390/cancers14040892

Academic Editor: José I. López

Received: 14 December 2021
Accepted: 8 February 2022
Published: 11 February 2022

1. Introduction

As the second most common male cancer worldwide, next only to lung cancer, prostate cancer affects roughly 1.3 million people and kills more than 360,000 people annually. In the U.S.A., the year 2020 saw nearly 192,000 new cases of prostate cancer (PCa) along with around 33,000 deaths [1]. Early stage PCa is treatable, with a 5-year survival rate of about 95%. Once PCa metastasizes, the 5-year survival rate falls to less than 30%. Conventional treatments for early stage PCa include watchful waiting (i.e., observation/no therapy), surgery and radiation, or active surveillance entailing routine monitoring of the blood level of prostate specific antigen (PSA), a circulatory biomarker for PCa cells, and repeat biopsy when deemed necessary. Cancer re-emergence occurs in about 30% of cases following radiation and/or surgery when hormone therapy, which reduces blood androgens to a castrate level, initially prevents androgen-dependent tumor growth. Hormone therapy is also considered for early stage cancer categorized as high risk for progression based on tumor grade and stage, and blood PSA. Hormone therapy and chemotherapy are the standard of care for PCa that has advanced locally or spread to distant tissues. No treatment course can offer permanent remission of advanced PCa. Technological advances are leading the way to find novel avenues for clinical interventions. In this review, we summarize progress in the molecular and genomic characterization of various stages of PCa—from high-risk localized and metastatic disease to neuroendocrine carcinoma, which is a highly aggressive form of advanced PCa—describe how insights gleaned from ongoing studies are revealing new vulnerabilities of cancer cells, and highlight several potential targets that are being pursued for future interventions tailored to individual patients.

In 1941, Charles Huggins discovered revolutionary aspects of PCa by demonstrating reduced cancer progression for the metastatic disease via androgen deprivation therapy (ADT) in the form of surgical castration. The Huggins discovery pioneered PCa management employing ADT and additional forms of hormone therapy, while also paving the way for a multitude of modern-day therapeutics [2]. Androgens play essential roles in the development of normal and malignant prostate, and the androgen dependence of PCa for cancer cell survival and proliferation has been amply documented. Androgens, synthesized predominantly in the testes, mediate biological effects via the androgen receptor (AR), which is a nuclear receptor and ligand-inducible transcription factor. Androgen-activated AR regulates normal prostate physiology and prostate cancer pathophysiology [3–6].

Androgen precursors in the form of androstenedione and dehydroepiandrosterone are secreted into circulation following biosynthesis in the adrenal glands. Adrenal androgens are converted to testosterone intratumorally in the prostate via multiple enzymatic steps, and 5α-dihydrotestosterone (DHT) is the reduced metabolite of testosterone, generated by the catalytic action of 5α-reductase. DHT and testosterone are high-affinity AR ligands. Cytosolic AR, when bound to androgens at the ligand-binding domain (LBD) of the receptor, dissociates from multiple AR-interacting proteins, including heat shock proteins hsp90 and hsp70, undergoes a conformational change that exposes a nuclear localization signal and facilitates transport of the androgen/AR complex to the nucleus [7,8]. Chromatin recruitment of this complex to androgen response element(s) in target genes and receptor interaction with coregulators lead to transcriptional induction or repression of AR-regulated genes (Figure 1). The ensuing impact on downstream effectors is a major driver in shaping prostate functions. Elevated androgen-inducible PSA in circulation under peripheral castration (characterized by low (<20 ng/dL) blood testosterone) is a biochemical signature of advanced PCa. Kinase-mediated receptor phosphorylation is an androgen-independent, alternative path to AR activation (Figure 1).

AR activity is upregulated in advanced PCa, principally from increased AR expression due to *AR* gene amplification. Continued AR activity also occurs due to AR mutants with ligand affinity for steroids beyond androgens; AR splice variants (AR-Vs) lacking LBD; and altered AR coregulator(s) abundance/activity. Hormone therapy, chemotherapy, and emerging genome targeted therapy are principal treatments for metastatic castration-resistant prostate cancer (mCRPC). Genome alterations impacting non-AR pathways can also lead to late-stage, treatment-resistant prostate cancer.

Recent insights into genome alterations in prostate cancer that impact signaling pathways beyond the androgen receptor axis, subtypes of early stage and advanced PCa, and the dynamics of tumor-promoting and tumor-suppressive effects of immune cells are revealing new approaches to disease intervention. A generally immunosuppressive tumor microenvironment makes most prostate cancer cases unresponsive to current immunotherapy except for genome instability in two contexts. New laboratory results are suggesting that the reach of immunotherapy for advanced PCa management may be extended by targeting selected pro-tumorigenic pathways in combination therapy. This review is a synthesis of key findings on genetic, epigenetic, and molecular changes impacting PCa aggressiveness and elaborates how the information has contributed to innovations in treatment strategies and has led to genome-targeted precision medicine and cancer immunotherapy. Finally, a future scenario built on current research is presented when routine personalized treatment decision for high-risk non-metastatic and advanced metastatic prostate cancer would be a reality.

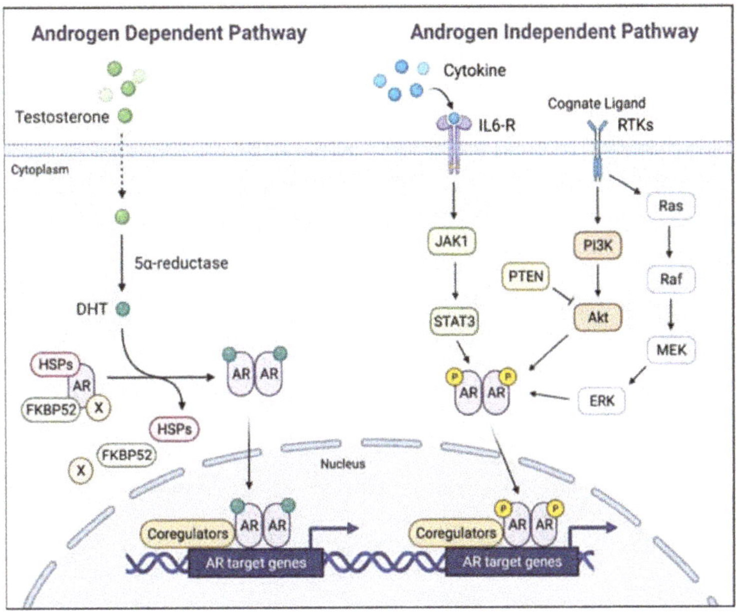

Figure 1. Simplified schema for AR-mediated target gene transcriptional response. DHT-bound AR dissociates from chaperone proteins (HSPs) and other interacting proteins, exposing AR's nuclear localization signal and facilitating receptor's nuclear import. Upon recruitment to androgen response elements of AR target genes, the hormone-bound AR interacts with various coregulators leading to target gene transcriptional response. An androgen-independent pathway may also drive AR activation and its cytoplasm to nucleus transfer subsequent to AR phosphorylation by various kinases. DHT: 5α-dihydrotestosterone; HSPs: heat shock proteins; FKBP52: forskolin-binding protein 52; IL6-R: interleukin 6 receptor; RTK: receptor tyrosine kinase.

2. Current Prostate Cancer Therapies

2.1. Low-Risk and High-Risk Non-Metastatic Prostate Cancer

The risks of localized prostate cancer for recurrence and metastasis are determined from the Gleason score, tumor stage and size, and blood PSA levels. Treatment options for localized PCa deemed low-to-moderate risk entail active surveillance, radical prostatectomy (RP) by surgery, and radiation therapy (RT) with external beam radiation or brachytherapy, while watchful waiting (i.e., observation/no therapy) is the norm for very low-risk cases [9]. Individual factors, such as patient's age and health, along with advantages and disadvantages of each option influence a treatment course [9,10] since regardless of the option selected, the low-risk disease shows a 10-year PCa-specific survival rate of nearly 99% [11]. Active surveillance helps to avoid health complications from RP and RT and therefore is recommended for younger and healthier patients. For high-risk localized PCa, surgery or external beam radiation is administered regardless of comorbidity-influenced life expectancy. RP and RT have similar efficacy. RP, however, is preferred over RT as primary monotherapy since it reduces local recurrence rates; potentially carries less risk for secondary cancer; allows the full-scale histopathological evaluation of tumor specimens; and also, potentially helps to avoid ADT [10]. Both RP and RT increase the risk for erectile dysfunction—with risk being greater for RP, and surgery leads to a higher risk of urinary incontinence, while the fecal incontinence risk is higher with RT [9]. Surgery is not an option when radiographic evidence indicates metastasis.

For histologically progressed non-metastatic PCa that is castration sensitive, a multimodal treatment design entailing ADT with or without RP and/or RT or combined with an AR inhibitor offers the greatest potential for improved long-term outcomes for patients who may harbor occult metastatic disease [9,10]. ADT induces the apoptosis of PCa cells, which leads to tumor shrinkage and inhibition of new tumor growth. The meta-analysis of a large dataset for high-risk prostate cancer showed that including external beam RT along with brachytherapy considerably shortens optimal duration of ADT and extends the period of metastasis-free survival [12]. This is a significant finding in view of the unpleasant side effects of ADT and its adverse impact on quality of life. A phase III trial (Stampede) has shown that ADT combined with an androgen synthesis blocking drug (abiraterone acetate plus prednisolone) provides improved benefits over ADT alone against non-metastatic castration sensitive PCa [13]. Therapeutic ADT includes agonists or antagonists for luteinizing hormone-releasing hormone (LHRH). The LHRH agonists, Leuprolide, Triptorelin, Histrelin, Goserelin, and two injectable LHRH antagonists, Degarelix (Firmagon®, Ferring Pharmaceuticals, Parsippany, NJ, USA) and orally deliverable Relugolix (Orgovyx®, Myovant Sciences, Brisbane, CA, USA), are FDA approved for ADT.

The development of non-metastatic CRPC is evidenced by rapidly rising blood PSA despite ADT. Disease progression in this case can be delayed by AR axis targeted therapy (ARAT) with a second-generation AR antagonist (enzalutamide, apalutamide or darolutamide) [14,15]. In a meta-analysis, a low proportion (16.5%) of non-metastatic CRPC patients was found to have avoided relapse within five years of initial therapy [12]. Of note, a 22-gene expression signature for high-risk localized PCa has recently been identified that can predict distant metastases and tumor responsiveness to second generation hormone therapy following radiation treatment in pretreatment biopsies [16]. The signature was identified based on the transcriptome analysis of archival pretreatment prostatectomy samples for which follow-up patient data over two to three decades was available. This genetic score-based predictive tool identified a subgroup of patients who did not additionally benefit from hormone therapy compared to radiation treatment only. Studies are ongoing to further validate this gene signature as a potential biomarker panel for stratifying high-risk primary PCa patients who should not be advised for additional hormone therapy beyond RT as primary monotherapy [16].

2.2. Targeting Metastatic Prostate Cancer

Androgen sensitive metastatic PCa is managed by ADT monotherapy or with RT or RP. The disease invariably progresses on ADT to develop into metastatic castration resistant prostate cancer (mCRPC), when ADT together with ARAT, which employs an AR antagonist or an androgen biosynthesis inhibitor that prevents intratumoral steroidogenesis, is temporarily successful as a first line treatment. ADT plus chemotherapy (with docetaxel) is another widely used first line option. Immunotherapy using immune checkpoint inhibitors and therapy that targets the DNA repair enzyme PARP (poly (ADP-ribose) polymerase) are more recent additions to the medical armamentarium as a second line treatment against mCRPC. Common treatment practices for prostate cancer at different stages are summarized in Figure 2.

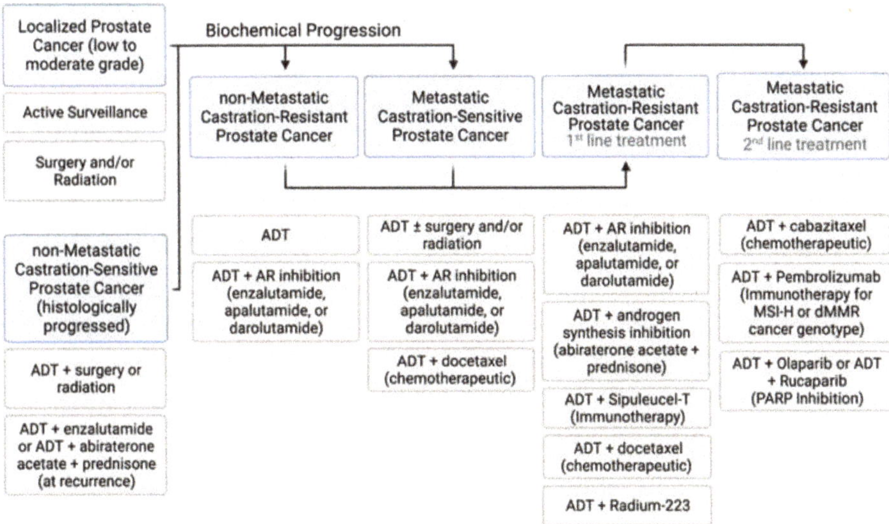

Figure 2. Schema for treatment modalities at progressively advancing prostate cancer stages. Boxes in blue indicate disease stages. Pink-colored boxes describe treatment choices. ADT: androgen deprivation therapy; MSI-H: microsatellite instability-high; dMMR: deficient mismatch repair.

Treatment choices for high-risk, localized PCa and advanced PCa, which has metastasized, are expected to grow in the future due to the discovery of new druggable targets, many of which are under clinical evaluation. The underlying molecular and biochemical principles for each therapy shown in Figure 2 are discussed below.

2.2.1. Hormone Therapy

Hormone therapy in prostate cancer implies the inhibition of androgen action in cancer cells by the systemic ablation of androgens through castration (medical or surgical), or by ARAT, which entails either inactivation of the AR due to its binding to an antagonist or suppression of intratumoral androgen biosynthesis by inactivating CYP17A1, a cytochrome P450 enzyme that catalyzes a rate limiting step in the steroidogenic pathway to testosterone and DHT production. Hormone therapy reduces cancer spread, prolongs life and controls disease-associated symptoms, such as skeletal related events. CRPC inhibition by ARAT is, however, short-lived, effective for about 4 to 5 months. Enhanced AR signaling in CRPC due to increases in AR expression and de novo androgen production leads to CRPC progression on hormone therapy. Treatment-induced stress may in part account for the upregulation of hormonal signaling in the ARAT setting. Elevated CYP17A1 in CRPC has been reported [17]. Whole genome sequencing of CRPC samples has revealed the recurrent amplification of genomic *AR* and its enhancer regions [18–22]. Resistance to ARAT develops also in response to functional changes in AR and is driven by mutation, coregulator alteration, and AR splice variants.

AR antagonists, upon binding to AR's LBD, promote corepressor recruitment to DNA-bound AR, which in turn triggers regulatory events that lead to androgen action suppression and tumor growth reduction [3]. Enzalutamide (Xtandi®, Astellas, Northbrook, IL, USA), and the newer drugs apalutamide (Erleada®, Janssen, Horsham, PA, USA) and darolutamide (Nubeqa®, Bayer, Whippany, NJ, USA), are second generation non-steroidal AR antagonists approved for CRPC management. Of the three antagonists, enzalutamide is deemed inferior due to certain adverse events. Among first generation AR antagonists (bicalutamide, flutamide and nilutamide), bicalutamide (Casodex, Astra Zeneca, Cambridge, UK) is still prescribed for ADT-unresponsive PCa. TAS3681 is a novel oral AR antagonist that inhibits the full-length AR and also AR splice variants. A first-in-human study with TAS3681 on mCRPC patients unresponsive to abiraterone, enzalutamide, and chemotherapy has been conducted [23].

Abiraterone acetate (Zytiga®, Janssen Biotech, Horsham, PA, USA) is a steroidal CYP17A1 inhibitor approved for blocking the adrenal production of androgen precursors, which lowers de novo androgen synthesis in tumor tissue. Prednisone co-administered with abiraterone acetate prevents corticosteroid deficiency. ADT paired with abiraterone-prednisone is approved for metastatic castration-resistant prostate cancer (mCRPC). A phase III trial (STAMPEDE) on non-metastatic castration-sensitive prostate cancer has shown superior patient outcomes with ADT plus abiraterone acetate/prednisone compared to stand-alone ADT [13]. In a phase III trial (PEACE-1) targeting patients with de novo metastatic castration-sensitive prostate cancer, adding abiraterone/prednisone to ADT and docetaxel (a chemotherapeutic) improved radiographic progression-free survival [24].

Ongoing studies are focusing on mechanisms of ARAT resistance, predictive biomarkers for resistance, and AR pathway independent targets toward the goal of new and improved treatments for high-risk non-metastatic PCa and advanced PCa that has progressed on ARAT. AR ablation as a next-generation therapy (discussed in Section 5) holds promise in this regard.

2.2.2. Chemotherapy

While not a standard treatment for early stage PCa, chemotherapy is used upon CRPC progression as combination therapy with ADT alone or ADT plus hormone therapy with second generation drugs. Docetaxel, cabazitaxel, mitoxantrone and estramustine are commonly used chemotherapy drugs. Docetaxel is a microtubules-binding taxane and a first-line chemotherapeutic that prevents nuclear translocation of AR. Castration-sensitive metastatic PCa can also benefit from ADT plus docetaxel [25]. Cabazitaxel is prescribed after disease progresses on docetaxel. Mitoxantrone and estramustine are used for palliation. Mitoxantrone is a type II topoisomerase inhibitor and estramustine is an estradiol-17β-phosphate conjugated to nor-nitrogen mustard, which is a genotoxic chemical and anti-mitotic. Estramustine disrupts the microtubular network, inhibits DNA replication and its estrogen component helps to lower the tumor issue testosterone level. A combination regimen of docetaxel plus estramustine showed a greater improvement in patient outcomes compared to docetaxel alone [26,27].

Adverse events are common with chemotherapy drugs. Docetaxel and cabazitaxel can cause severe allergic reactions and induce peripheral neuropathy, which leads to numbness, tingling or burning sensation in hands and feet. The use of mitoxantrone may lead to the development of leukemia a few years later, and estramustine increases the risk of thromboembolic events. Furthermore, since proliferation of rapidly dividing normal cells are also inhibited by chemotherapeutics similar to cancer cells, hair loss, sore throat, loss of appetite, fatigue due to low red blood count, and increased risk of infection from low count of white blood cells are commonly encountered examples of chemotherapy-induced morbidity.

Bone metastasis leading to severe morbidity, manifested as bone pain, spinal cord compression, hypercalcemia, and pathologic fractures, constitutes the most frequent bone-related complications in advanced PCa. The bisphosphonate class of drugs, such as zoledronic acid (Zometa®, Novartis, Cambridge, MA, USA) and denosumab (XGEVA®, Amgen, Thousand Oaks, CA, USA), which is a human monoclonal antibody against the RANK ligand (RANKL), are FDA approved for managing adverse skeletal events from bone metastasis [28]. Bisphosphonates prevent bone resorption by inhibiting osteoclast's resorptive activity and osteoclast formation, and by inducing osteoclast apoptosis. Bisphosphonates also suppress the tumor growth of bone metastases. Bisphosphonate therapy prevents bone loss caused by anti-androgen therapy. Denosumab inhibits osteoclasts by inactivating the RANK-RANKL signaling pathway, which plays an essential role in the differentiation, activity, and survival of osteoclasts. RANK activation is prevented when denosumab binds to its antigen, i.e., RANKL, the ligand for RANK (receptor activator of nuclear factor kappa B) [29]. In a phase III study, denosumab was found to be superior to bisphosphonate in delaying skeletal related events for mCRPC patients; however, denosumab causes osteonecrosis in the jaws [30].

Recombinant interleukin-24 (IL24) is potentially a novel therapeutic for preventing bone metastasis since IL24 suppressed colonization of CRPC cells to bone in a mouse prostate cancer metastasis model [31]. IL24-directed metastasis suppression was further enhanced by pharmacologic inhibition of the anti-apoptotic protein MCL-1, which is a member of the BCL2 protein family. The inactivation of the AKT survival signal also heightened metastasis suppression by IL24. IL24 inhibits the RANKL/RANK signaling axis, thus blocking osteoclast differentiation, which in turn prevents cancer cell dissemination in bone marrow. AKT and MCL-1 are downstream effectors of the RANK-RANKL pathway [31]. Beyond inhibiting osteoclast formation, IL24 induces apoptosis of PCa cells. In view of its inhibitory effect on osteoclasts, recombinant IL24 is potentially useful in the management of PCa bone metastasis, especially in combination with an MCL-1 or AKT inhibitor. Small molecule inhibitors of MCL-1 and AKT are in clinical trials in combination therapy [32,33].

2.2.3. PSMA-Targeted Radiation Therapy

Prostate specific membrane antigen (PSMA) is a cell surface protein expressed in PCa cells, not in normal prostate cells. PSMA is targeted for the delivery of a radioactive payload to metastasized PCa cells. ^{177}Lu-PSMA-617 is one such payload where the radioactive lutetium-177 is coupled to a ligand that binds PSMA with high selectivity and delivers radiation to PSMA-expressing cells. The outcome of ^{177}Lu-PSMA-617 radioligand therapy (RLT) combined with the standard of care was compared with the standard of care alone in a phase III trial when inclusion of RLT improved OS and progression-free survival (PFS) [34]. AR status may determine treatment outcome since early resistance to ^{177}Lu-PSMA-617 positively correlated to *AR* amplification detected in circulating DNAs [35]. In addition, a PSMA-targeting antibody (J591) linked to actinium-225, another radioactive element, reduced PSA for 70% of mCRPC cases in a phase I trial [36]. ^{225}Ac, an alpha particle emitter, is 1000-fold more potently radioactive than ^{177}Lu, which emits beta particles.

In contrast to PSMA-targeting radioactive agents, which destroy PCa cells anywhere in the body, radium-223 dichloride (Xofigo®, Bayer, Whippany, NJ, USA), an injectable radionuclide, is approved for targeting bone metastasis of ADT-resistant PCa [37]. Radium-223 is absorbed readily by bone and irradiates PCa cells in the bone environment.

2.2.4. Genome-Targeted Precision Therapy

The exploitation of the DNA repair gene mutation by targeting PARP (poly (ADP-ribose) polymerase) is an example of genome-targeted precision medicine. PARP inhibition therapy is approved for mCRPC with inactivating mutations of *BRCA2, BRCA1,* which regulate homologous recombination repair (HRR) of DNA double strand breaks (DSBs). Cases of mCRPC with alterations in other HRR pathway genes, such as *ATM, RAD51, MRE11, ATR, CHK2, MMR, MSH* and *CDK12*, are also known. Cancer genome exome sequencing identified somatically altered HRR genes in roughly 23% of mCRPC cases and germline events in 8% of cases [38,39]. Even primary PCa in the TCGA cohort showed HRR deficiency in 19% of cases [40].

PARP inhibition is synthetically lethal for HRR-mutant cells due to the buildup of DNA DSBs leading to stalled DNA replication forks. DSBs accumulate due to unrepaired single strand breaks (SSBs). It was initially thought that, without ADP-ribosylation, XRCC1 (a DNA repair enzyme) cannot be recruited to SSBs, which then prevents activation of base excision repair (BER), causing the failed repair of DNA SSBs. A more likely model based on additional results is that the inhibitor-bound PARP is trapped onto SSB repair intermediates, especially during BER, leading to DNA DSBs. Unresolved DSBs lead to replication block and cell death. Thus, PARP inhibition induces synthetic lethality in HRR-mutant cells

(Figure 3). PARP may also regulate DNA repair via an HRR-independent mechanism since hyper-activated PARP was detected in HRR-mutant cells [41].

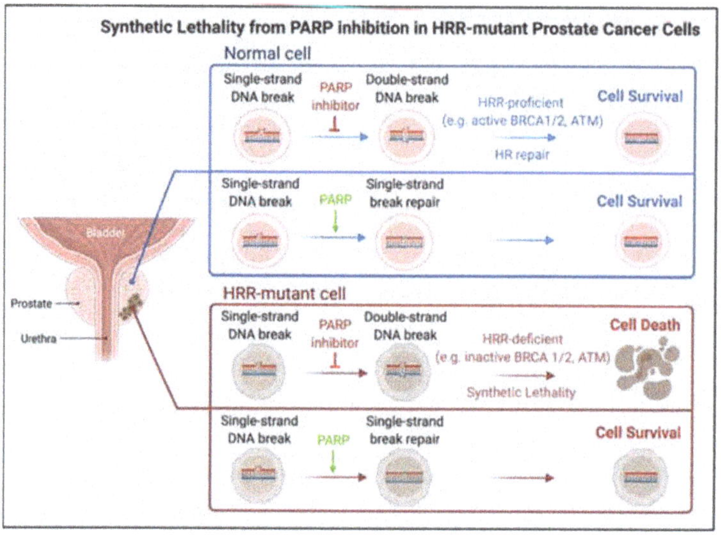

Figure 3. Schema for synthetic lethality from PARP inhibition in HRR mutant prostate cancer cells. Unlike HRR-proficient normal cells, HRR-mutant cells, upon PARP inactivation, accumulate DNA double-strand breaks, which cause replication block leading to cell death.

Two oral PARP inhibitors—Olaparib and Rucaparib—which inactivate PARP1 and other PARP family members, are approved, as of May 2020, for clinical targeting of mCRPCs harboring a mutant *BRCA2* or *BRCA1* genotype [42]. The inhibitors are used in combination with chemotherapy or ARAT. PARP inhibition may also induce synthetic lethality in PCa with alterations in other HRR-relevant genes. Breast and ovarian cancers respond to PARP inhibition when harboring HRR pathway deficiency. Talazoparib, a more potent PARP inhibitor than Olaparib and Rucaparib, is under study to test for efficacy against mCRPC with aberrant HRR [43]. Talazoparib is approved for HER2-negative breast cancer with germline *BRCA1*/*BRCA2* mutation.

The functional interplay of PARP with AR has been demonstrated. For example, PARP-1 is recruited to AR-occupied genomic sites in PCa cells [44]; PARP-2, a PARP-1 paralog, enhanced AR activity by interacting with chromatin-bound FOXA1, a pioneering factor that promotes AR-mediated target transcription [45]; and a functional HRR pathway requires AR activity [46–48]. Diminished DNA damage response and HRR activity was detected in biopsies from patients receiving ADT plus radiation compared to radiation alone. Furthermore, ADT led to enhanced PARP activity, evidenced by increased protein parylation (i.e., poly ADP-ribosylation) in biopsies from patients on ADT [46]. Concurrent PARP and AR inhibition caused synthetic lethality in PCa experimental models [47,48]. These results suggest that ADT-associated HRR downregulation precedes the development of castration resistance, and combined PARP and AR inactivation is possibly fruitful for targeting organ-confined high-risk PCa and castration-sensitive metastatic PCa [46–48]. Mutant DNA repair genes identified in primary PCa support this possibility. The clinical efficacy of combined ARAT and PARP inhibition against ADT-refractive mCRPC is under investigation [47].

2.2.5. Immunotherapy

Two FDA-approved immunotherapy options for mCRPC are the vaccine Sipuleucel-T (Provenge®, Dendreon, Seattle, WA, USA) and the immune checkpoint inhibitor Pembrolizumab (Keytruda®, Merck, Kenilworth, NJ, USA). Sipuleucel-T, approved in 2010 for mildly symptomatic mCRPC, is an autologous cellular immunotherapy entailing incubation of patient-derived dendritic cells, isolated from peripheral blood mononuclear cells, with the vaccine immunogen PA2024 and re-administering the immunogen/dendritic cell complex to patients [49]. PA2024 is a fusion of prostatic acid phosphatase (PAP), an immunogenic prostate-specific antigen elevated in PCa, with GM-CSF (granulocyte macrophage colony-stimulating factor). GM-CSF activates dendritic cells, which are antigen-presenting cells (APC). Activated APCs with the exposed PAP antigen stimulate cytotoxic T cells, which then recognize and kill PAP-positive PCa cells. Common adverse effects of Sipuleucel-T include fever, headache, influenza-like symptoms, elevated blood pressure, muscle ache and pain (myalgias), and abnormally excessive sweating.

Immune checkpoint inhibitors (ICIs) are approved against mCRPC and other solid tumor metastases that harbor genome alterations leading to high microsatellite instability (MSI-H) or deficient mismatch repair (dMMR) [50,51]. ICI Pembrolizumab is a human monoclonal antibody against PD-1 (Programmed Death-1), which is a cell surface protein expressed in cytotoxic CD8+T cells. PD-1 binds to tumor cell-expressed PD-L1 (Programmed Death Ligand-1), leading to T-cell suppression and immune escape of tumor cells [52]. PD-1 inhibition is regarded as more efficacious than PD-L1 inhibition since PD-L1 exists in two distinct forms [53]. Pembrolizumab is approved for mCRPC with MSI-H and/or dMMR. Pembrolizumab efficacy against mCRPCs, which show other categories of genome alterations, and which have progressed on ADT and enzalutamide, is being investigated [54]. In the KEYNOTE-365 trial, a small mCRPC cohort showed improved PSA response rate independent of HRR mutation status by the Pembrolizumab-Olaparib combination [55]. Pembrolizumab/radium-223 combination against mCRPC is under investigation (NCT03093428).

Another target for immune checkpoint suppression is CTLA-4 (cytotoxic T-lymphocyte antigen-4), which is a cell surface protein of CD8+T cells. Normally, immune blockade results from the inactivation of cytotoxic CD8+T cells following CTLA-4 binding to B7-1/B7-2 cell surface proteins that are expressed in APCs. Ipilimumab (Yervoy®, Bristol Myers Squibb, Princeton, NJ, USA) is a monoclonal antibody to CTLA-4 that disrupts the interaction between B7 and CTLA-4. T-cell receptors' recognition of exposed antigens and major histocompatibility complex proteins on APCs and the binding of CD28 to B7-1/B7-2 lead to T-cell activation and tumor cell killing. Ipilimumab plus Nivolumab (anti-PD-1 antibody) reduced mCRPC progression in a phase II study [56].

Somatic biallelic *CDK12* inactivation, identified so far in a limited number of mCRPC cases, may also present an opportunity for immune checkpoint blockade [57]. CDK12 is a cyclin dependent kinase, which partners with cyclin K to regulate various cellular processes. CDK12-deficient tumors have elevated neoantigen burden due to focal tandem genomic duplications that generate fusion-mediated chimeric open reading frames. CDK12-/-tumors showed increased tumor infiltration of T cells, and in a small-scale clinical study, anti-PD1 monotherapy was clinically effective against these tumors [57,58].

Prostate tumor is generally adept at immune evasion except when the cancer genome alterations lead to MSI-H or dMMR, as mentioned above. For a subgroup of patients, response to immunotherapy may be augmented by reducing neoangiogenesis, a hallmark of cancer progression. For example, anti-CTLA-4, anti-PD1 and other immune checkpoint inhibition therapies lead to elevated interferon-gamma (IFNγ) in the tumor microenvironment because of increased IFNγ expression in immune cells [59,60]. IFNγ has anti-angiogenic and tumor-suppressive activity [61]. The influence of IFNγ on immunotherapy is supported by the finding that deregulated IFNγ signaling leads to deficient immunotherapy responsiveness [62]. It has been suggested based on studies in PCa cell models that high-risk PCa patients, having tumors expressing low miR-221 microRNA and elevated VEGFR2 (i.e., vascular epithelial growth factor receptor 2, an miR-221 target), may benefit from initial anti-angiogenic therapy entailing VEGFR2 inhibition with a tyrosine kinase inhibitor, e.g., sunitinib, followed by immune checkpoint inhibition [63].

A sequential treatment modality as above may suppress PCa with high VEGFR2 and low miR-221 since sunitinib-treated PC3 prostate cancer cells showed increased miR-221 expression, which is likely an escape response to VEGFR2 inhibition given that sunitinib-mediated cell proliferation blockade was prevented by ectopic overexpression of miR-221 [63]. The upregulation of miR-221 may in part contribute to the failure of clinical trials of VEGFR2 inhibition for advanced PCa [64,65]. On the other hand, sunitinib treatment led to elevated immune-related signaling in PC3 cells, including an elevated IFN-related gene signature [66]. Notably, IRF2 and SOCS3, which are repressors of IFN signaling, are miR-221 targets. Increased pro-immunogenic traits also emerged in miR-221 overexpressed PC3 cells [63]. The interrogation of prostate cancer data bases has identified a subset of high-risk primary PCa cases with tumors having abundant expression of VEGFR2, but mostly undetectable miR-221. In contrast, primary PCa samples from the TCGA cohort showed very low VEGFR2 expression [63]. Collectively, in vitro studies and clinical specimen analysis of cancer genome database have revealed a microRNA-informed epigenetic signature that can identify a subset of high-risk PCa patients who are likely to show improved response to immunotherapy with a treatment plan that includes intercepting the VEGF/VEGFR regulated angiogenesis pathway with a drug such as sunitinib.

2.2.6. Summary of Approved Therapies

Current practices for managing prostate cancer are guided by the stage of malignancy, the nature of cancer genome alterations and the response of progressively advanced cancer to chemo-hormonal interventions. While low-to-moderate risk cancer is kept under active surveillance or subjected to surgery and/or radiation, a more aggressive intervention is needed for high-risk localized disease. ADT is the foundational hormone therapy and at biochemical recurrence, an AR antagonist or abiraterone acetate (plus prednisone), which inhibits intratumoral androgen biosynthesis, is added to ADT. ADT/anti-AR combination continues to be standard of care for CRPC—both non-metastatic and metastatic. Enzalutamide, apalutamide and darolutamide are second generation AR antagonists used interchangeably and their selection is based on adverse effects profiles. First-line treatment of mCRPC entails ADT plus AR axis targeted therapy—the latter employing abiraterone acetate/prednisone or an AR antagonist. Options also include combining ADT with docetaxel, a microtubule binding chemotherapy drug, or sipuleucel-T, an immunotherapy drug, or radium-223, which selectively targets PCa that has metastasized to bone. Skeletal-related events ensuing from bone metastasis of PCa are alleviated by osteoprotection with bisphosphonate therapy or denosumab-mediated antibody therapy. Both therapies inhibit osteoclast development and activity, albeit through different mechanisms. More recently, a genome-targeted precision treatment employing PARP inhibitors has been approved for second-line treatment along with ADT in the case of mCRPC, which shows deregulated DNA damage response leading to the deficient homologous recombinant repair (HRR) of DNA double strand breaks. Tumors of mCRPC patients carrying BRCA1 or BRCA2 gene mutation are approved for PARP inhibition therapy. PARP inhibition in a setting of aberrant DNA damage response, and thus deficient repair of DNA double-strand breaks, causes synthetic lethality that culminates in tumor cell death. Promising clinical responses to PARP inhibition have been observed for mCRPC patients with mutation in additional genes in the HRR pathway. This raises optimism that PARP inhibition will be an approved therapy for a wide variety of genomic mutations that interfere with HRR. Finally, immunotherapy with immune check point inhibitors (ICIs) is an approved second-line treatment for mCRPC presenting with genome alterations that cause microsatellite instability (MSI) and deficient mismatch repair (dMMR) and lead to excessive neoantigen burden and strongly immunogenic tumor cells as a consequence. These genomic alterations also promote increased PD-L1 expression and increased tumor infiltration of cytotoxic T cells.

3. Molecular Subtypes of Early Stage and Late-Stage Prostate Cancer

Genome-wide sequencing revealed distinct PCa subtypes and mechanistic studies uncovered affected signaling pathways. The TCGA cohort of 333 primary PCa cases identified 7 major molecular subtypes for 74% of early stage PCa and 26% of cases showed heterogeneous genome alterations [40]. The subtypes fall into two mutually exclusive categories—ETS fusion-positive and ETS fusion-negative (Figure 4). ETS family proteins are oncogenic transcription factors.

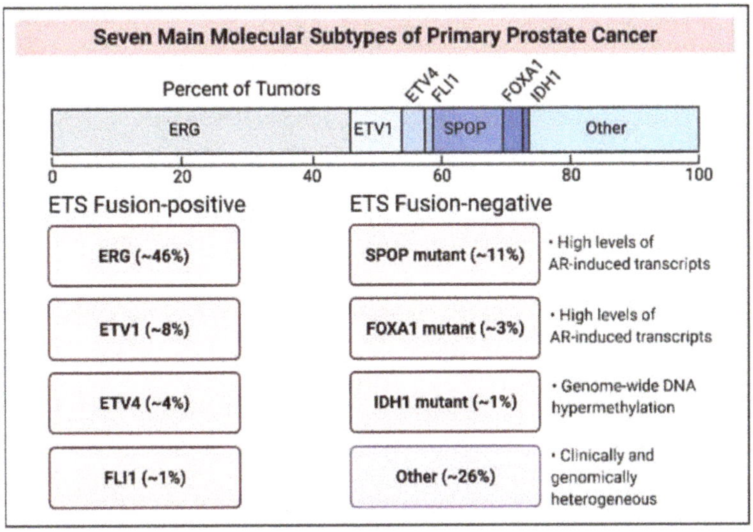

Figure 4. Molecular subtypes of prostate cancer. The ERG fusion subtype is most prevalent in primary PCa. The other category (~26%) associates with diverse genomic alterations, including enrichment for high copy-number alterations, DNA hypermethylation, mutations in TP53, KDM6A (encoding lysine demethylase 6A), KMT2D (encoding lysine methyltransferase 2D), and deletions of chromosomes 6 and 16. The amplification of chromosome 8 (spanning MYC oncogene) and chromosome 11 (spanning cyclin D1 encoding CCND1) is observed in tumors categorized as "other" (schema is based on Information in ref. [40]).

AR activity is variable across all cases and between seven subtypes. Sub-clonal heterogeneity within each subtype, arising from variable genome and epigenome signatures, is believed to set the disease course. Primary and metastatic PCa show a similar subtype distribution except that the IDH1-mutant subtype is absent in mCRPC. Actionable alterations affecting PI3 kinase or MAP kinase pathways were found in 20% of the TCGA cohort. The clinical translation of these findings holds promise for precision medicine.

3.1. ETS Fusion-Positive Subtypes

This category includes four subtypes of *ETS* gene fusion with an upstream androgen-inducible promoter. The most prevalent fusion involves the ETS protein ERG and the androgen-inducible promoter of *TMPRSS2* encoding a transmembrane serine protease. Less common promoter fusions involve other androgen-inducible genes, such as *SLC45A3*, encoding a sucrose: proton symporter [67]. Promoter fusions with the *ETS* members *ETV1*, *ETV4* and FLI1 also exist. ETS overexpression without fusion was detected in limited examples. ETS fusion-positive tumors show chromoplexy leading to extensive genome rearrangement. The deletion of the *PTEN* tumor suppressor gene, prevalent in TMPRSS2-ERG-positive PCa and a likely outcome of chromoplexy, activates the PI3K oncogenic axis. *TP53* alteration, leading to p53 tumor suppressor inactivation, is frequently detected in

ERG-rearranged PCa. In *PTEN*-deleted PCa, activating the mutation of *PIK3CB* (encoding the PI3K catalytic subunit, β isoform) is a dominant driver of PI3K signaling [68,69]. AR inhibition leading to the inactivation of the androgen-inducible promoter along with PIK3CB inactivation is expected to inhibit PCa with *PTEN* deletion and activating mutation of *PIK3CB* [70]. In transurethral resections of the prostate (TURP), ETS-rearranged PCa showed increased aggressiveness [71]. The association of TMPRSS2-ERG positive cancer with a higher risk of death from PCa is inconclusive since negative and neutral associations of ETS fusions with PCa aggressiveness were also found [40].

3.2. ETS Fusion-Negative Subtypes

ETS fusion-negative genetic subtypes involve mutations in *SPOP* (encoding a Speckle Type BTB/POZ domain protein), *FOXA1* (encoding a pioneering factor promoting AR-mediated transactivation), and *IDH1* encoding isocitrate dehydrogenase1, which catalyzes isocitrate conversion to 2-oxoglutarate.

SPOP mutations are most common (Figure 4, ref [40]). SPOP regulates the proteasomal degradation of selected proteins, including the AR coactivator SRC3, which is consistent with the high AR transcriptional activity in SPOP-mutant PCa [72]. SPOP also regulates ERG degradation. SPOP interacts with the E3 ubiquitin ligase CULLIN3 and E3 ligase substrates. A role of SPOP in HRR-directed DNA DSBs repair is consistent with the association of genome instability with *SPOP* mutation. PCa cells expressing SPOP mutants are sensitized to PARP inhibitors and other DNA damaging agents [73]. *SPOP* mutations frequently associate with the somatic deletion of the genomic locus containing *CDH1*, which encodes a chromatin remodeling protein. PCa with both *SPOP* mutation and *CDH1* deletion is elevated for SPINK1, a secreted serine protease inhibitor. In the tumor microenvironment, SPINK1 promotes tumor growth and survival, and SPINK1 overexpression correlated to increased PCa aggressiveness and shortened PFS [74]. The prognostic value of *SPOP* mutation is uncertain since in one study pathological parameters did not correlate to *SPOP* mutations, while another study linked shortened PFS to reduced SPOP [40].

FOXA1 mutation has been detected in 3% of primary PCa. The molecular features of *FOXA1*-mutant and *SPOP*-mutant tumors are similar—both showing high AR activity. Primary PCa cases with *FOXA1* and *SPOP* mutations in the same clonal subpopulation are known.

IDH1 mutation, detected in 1% of primary PCa, appears also in other cancers, and ongoing trials for *IDH1*-mutant malignancies highlight its clinical actionability. IDH1 mutation inactivates TET2, a DNA demethylating enzyme that converts methyl cytosine to hydroxymethyl cytosine leading to cytosine demethylation. IDH1 catalyzes isocitrate conversion to α-ketoglutarate, which is an essential TET2 cofactor. IDH1-mutant PCa shows robust genome-wide CpG hypermethylation. *ERG* fusion positive PCa shows a heterogeneous CpG methylation profile along with two clusters of elevated methylation. *ETV1* and *ETV4* fusions lack these clusters, showing genome-wide heterogeneous methylation. Homogeneously distributed genome-wide hypermethylation characterizes *SPOP* and *FOXA1* mutants [40].

4. Metastatic Castration-Resistant Prostate Cancer

The subtype distribution for mCRPC is similar to that for castration-sensitive primary, locally advanced, and metastatic PCa, except that *IDH1* mutation is absent and the mutation burden is higher. AR activity is more robust, primarily due to *AR* amplification or mutation—a feature not found prior to castration resistance. An upward of 65% mCRPC cases show actionable targets based on altered PI3K/AKT, WNT, and cell cycle and DNA repair pathways [22,75]. Other treatment-induced common genetic changes are *TP53* mutation or deletion, *PTEN* deletion, *RB1* (retinoblastoma1) loss, along with the loss-of-function mutation of *CDK12, BRCA2, BRCA1*. The PI3K/AKT pathway, which regulates tumor growth, survival, and therapy resistance [76], is activated in nearly 50% of mCRPC cases, frequently from deletion/mutation of *PTEN* [38,77]. The activation of *PIK3CA, PIK3CB,*

PIK3R1, and *AKT1* is less frequent. Beyond gene deletion, PTEN loss-of-function can occur from mutation, methylation, microRNA directed post-transcriptional regulation and post-translational modification [78,79].

The reciprocal feedback inhibition of AR and PI3K/AKT signaling, demonstrated in PCa experimental models, revealed the activation of AKT upon AR inhibition and activation of AR when AKT is inactivated [80]. In our study, we showed that the dual targeting of AR and the PI3K/AKT/mTORC1 axis by the antibiotic salinomycin caused PCa inhibition in cell and xenograft models [81,82]. Salinomycin is a human stem cell inhibitor and an anti-parasite antibiotic useful in veterinary medicine. The clinical translation of combined targeting of AR and PI3K/AKT pathways yielded promising results. Ipatasertib (a pan-AKT inhibitor) plus abiraterone-prednisone improved PFS for a subgroup of mCRPC cases with PTEN-deficient, AKT hyperactivated tumors [33]. Trials with combined Capivasertib, another pan-AKT inhibitor, and enzalutamide are ongoing (NCT02525068; NCT03310541).

The aberration of DNA repair genes is found in 20–30% of mCRPC cases. Among frequently altered genes are *BRCA2*, *BRCA1*, and *ATM*, which have central roles in HRR of DNA DSBs [39]. HRR gene defects appear as somatic or germline alteration. *BRCA2/BRCA1* mutated mCRPCs that progress on ARAT are approved for PARP inhibition therapy. Exceptional responses to platinum-based chemotherapy for HRR-deficient mCRPC cases are known [83,84].

Alteration in mismatch repair genes—*MSH2*, *MSH6*, *PMS2* and *MLH1*—leads to deficient mismatch repair (dMMR), genome hypermutation and microsatellite instability-high (MSI-H). A total of 3–12% of mCRPC cases showed dMMR and/or MSI-H, which associate with elevated tumor-expressed neoantigens and PD-L1, and increased tumor infiltration of cytotoxic T cells [50]. MSI-H is a pan-cancer biomarker for clinical response to immune checkpoint inhibition. PD-1 inhibition by Pembrolizumab is approved for mCRPC and other solid tumors with dMMR and/or MSI-H [51].

Biallelic somatic *CDK12* inactivating mutations are more prevalent in mCRPC than primary PCa [46]. No germline mutation was found. CDK12 promotes genome stability by regulating DNA damage response genes. *CDK12-/-*tumors, which are mutually exclusive with tumors bearing ETS fusions, SPOP mutations or DNA repair deficiency, show focal tandem duplications that lead to gene fusions, chimeric coding sequences and expression of neoantigens. *CDK12-/-*mCRPC responded to anti-PD-1 monotherapy [57,58].

5. AR Ablation for Intercepting Advanced Prostate Cancer

Heightened AR activity and AR pathway outputs—the outcomes of increased AR expression, AR activating mutations, and AR splice variants that are activated in an androgen-independent manner—are prominent features of mCRPC. Copy number gain at the genomic *AR* locus and at a far upstream enhancer locus occurs recurrently in mCRPC and correlates to a shorter time on ARAT without any influence on OS [50]. CRPC metastases from a cohort of 197 patients showed enhanced AR signaling in 80% cases, with ~57% cases showing amplification at the *AR* locus harboring a proximal enhancer [22]. In another cohort of 101 mCRPC patients, tumors in 81% cases amplified both the enhancer-containing *AR* locus and a far-away enhancer at a 66.94 mega base upstream of the *AR* locus [21]. Copy number gain exclusively at this far upstream enhancer without *AR* locus amplification was also detected. Increased AR mRNA scores strongly associated with the two amplified loci. The far upstream amplified locus independently associated with upregulation of AR mRNAs.

An alternative approach to AR pathway inhibition is to reduce the receptor in tumor cells by its degradation via the ubiquitin-proteasomal pathway, which can be initiated by a small-molecule SARD (Specific AR Degrader) or AR-selective PROTAC (proteolysis-targeting chimera) [58,85]. One class of SARDs, which activates the proteasome machinery by having a hydrophobic degron linked to an AR ligand, caused AR ablation in experimental models, and led to sensitization of enzalutamide resistant PCa cells to AR antagonism [86]. A PROTAC is a bifunctional molecule with affinity for an E3 ubiquitin ligase and E3 substrate. By bringing the ligase near the substrate, a PROTAC facilitates substrate ubiquitination and degradation. The SARD ASC-J9® degrades the wild-type AR and a clinically relevant mutant AR that confers enzalutamide resistance to PCa. ASC-J9® inhibited PCa growth/proliferation and prevented AR upregulation in docetaxel-resistant CRPC [87,88]. AZD-3514 is a moderately active SARD against mCRPC, causing a decline in PSA and soft tissue tumor burden, and disease stabilization in a subset of patients [89].

Several AR PROTACs are in clinical or pre-clinical testing. The ARCC-4 PROTAC ablated cellular AR inhibited PCa cell proliferation, and degraded AR point mutants that convert enzalutamide to an AR agonist [90]. The ARD-61 PROTAC showed anti-proliferative and pro-apoptotic effects on enzalutamide-resistant CRPC in cell and xenograft models [91]. ARV-110 can induce the degradation of the wild-type AR, gene-amplified AR, and certain AR mutants in preclinical enzalutamide-sensitive and enzalutamide-resistant models. Phase I study with ARV-110 showed PSA response in two mCRPC patients who progressed on ARAT—one patient even showing reduced soft tissue tumor burden. Responding tumors harbor AR mutations that lead to *ARAT* resistance [92]. The A031 PROTAC caused AR degradation in VCaP prostate cancer cells that contain a high copy number genomic *AR*. A031 inhibited VCaP xenograft growth [93]. AR-targeting PROTACs and SARDs are promising next-generation drugs for managing AR pathway-active PCa that stops responding to current anti-AR therapy.

Unlike full-length AR, the tumor levels of the splice variant AR-V7 did not correlate to time on ARAT, although the association of AR-V7 mRNA expression in circulating tumor cells with time to ARAT resistance and clinical outcome has been reported in other studies [94,95]. Of note, a long non-coding RNA (lncRNA, designated as NXTAR) is a negative regulator of AR/AR-V7 expression while NXTAR expression itself is suppressed by androgen-activated AR [96]. A NXTAR gene-specific oligonucleotide led to the reduction in AR/AR-V7 and proliferation of PCa cells, while in a xenograft study, a small-molecule AR inhibitor restored NXTAR expression and blocked the tumor growth of enzalutamide-resistant PCa.

6. Lineage Switch and mCRPC Progression

RB1 loss has emerged as a strong predictor of poor survival based on a study with taxane-naïve mCRPC patients receiving ARAT, whereas *Rb1*, *AR* and *TP53* alterations associated with a shorter time on ARAT [50]. The *RB1*-encoded tumor suppressor regulates the E2F transcription factor activity and G1→S checkpoint. A subset of ARAT-unresponsive mCRPC tumors bypasses AR pathway dependence due to lineage switch. This switch leads to a phenotype shift from luminal epithelial-like prostate adenocarcinoma to AR-independent, basal-like cells expressing neuroendocrine markers, such as synaptophysin and chromogranin A. *RB1* inactivation is the major driver of this lineage crossover to neuroendocrine-type prostate cancer (NEPC). SOX-2, a stem cell factor and regulator of neuronal progenitor differentiation, is a downstream effector of *RB1*-dependent lineage transition [97].

NEPC is a poorly differentiated mCRPC variant associated with low serum PSA and poor prognosis. A subset of NEPC histologically resembles small cell carcinoma. Variants with NEPC/small cell traits are diagnosed in up to 20% of mCRPC patients who progress on ARAT [98,99]. Prominent molecular features of NEPC/small cell carcinoma include loss of *RB1* and *TP53*, low or absent AR and AR pathway outputs, and induction of SOX2 and the epigenetic enzymes EZH2 and DNMT1. EZH2 is a lysine methyltransferase that confers a repressive epigenetic mark on histone H3 by trimethylating lysine-27 (H3K27-Me3) and DNMT1 mediates cytosine methylation at genomic context-specific CpG motifs. EZH2 is a component of polycomb repressive complex2 (PRC2). EZH2 can also coactivate AR in a PRC2-independent manner [100], which is consistent with our finding that methylation enhances AR's transcriptional activity [101]. EZH2 inhibition reversed the lineage switch for mCRPC variants and restored enzalutamide sensitivity in enzalutamide resistant PCa models [102,103].

Aurora kinase A (AURKA), a G2→M regulating cell cycle kinase, and delta-like protein 3 (DLL3), a cell surface notch signal inhibiting ligand, are other examples of proteins enriched in NEPC and its small-cell variant compared to adenocarcinoma [98]. N-MYC oncoprotein is elevated in NEPC—a sequela to AURKA-mediated N-MYC phosphorylation. AURKA inhibition induces synthetic lethality in CRPC variants exhibiting the combined loss of *RB1* and *TP53* [98]. EZH2, AURKA and DLL3 are promising targets for which clinically relevant inhibitors are in early phase trials against NEPC and small-cell carcinoma (NCT03480646; NCT04179864; NCT01799278; NCT04702737; NCT04471727). Of note, combined inhibition of PARP and the G1 phase cyclin dependent kinases CDK4/CDK6 induced apoptosis and suppressed neuroendocrine differentiation of PCa in preclinical models [104]. Platinum-based chemotherapy, the only treatment option for NEPC and its variants, has shown limited efficacy. Clinical development of targeted inhibition of NEPC-enriched proteins is paramount if sustained control is to be achieved against CRPC variants, such as NEPC and small-cell prostate carcinoma.

7. Concluding Perspectives and Future Directions

Precision medicine entails diagnosis and disease management after considering individual differences in genomics and gene interactions with environment and lifestyle. The recent finding that intestinal microbiota produces androgen precursors and promotes CRPC presents the exciting possibility for antibiotics therapy in future prostate cancer management [105]. Innovations in managing mCRPC, including small cell/neuroendocrine carcinoma variants, are facilitated by a growing list of actionable genomic changes revealed by next-generation genome sequencing. Furthermore, the identification of a 22-gene expression signature, serving as a potential blueprint to managing high-risk primary PCa [16], has led to a trial focused on transcriptome profiles of pretreatment biopsy samples as the guide to treatment decision for high-risk localized disease. The efficacy of the combination therapy entailing PARP inhibition and anti-PD1 guided immunotherapy for suppressing mCRPC with gene mutation affecting DNA damage response is an example of how the cancer genome landscape might fruitfully guide clinical decision.

Immune checkpoint inhibition in monotherapy normally fails to inhibit PCa, in part due to an immunosuppressive tumor environment—the exception being mCRPC with an MSI-H or dMMR genotype that leads to enhanced cancer cell immunogenicity from high neoantigen burden. For a subgroup of mCRPC patients, tumor-suppressive immune functions may be augmented by the prior inhibition of angiogenesis. A microRNA (miR-221) has been identified as a potential biomarker for selecting patients who would possibly benefit from the neoadjuvant use of an angiogenesis-blocking drug in a cancer immunotherapy setting [63]. The clinical testing of advanced PCa expressing low miR-221 and high VEGF receptor-2 (VEGFR2) for responses to sequential treatment at first with a VEGFR2 inactivating drug, such as a clinically active tyrosine kinase inhibitor, to reduce angiogenesis followed by immunotherapy will be an important future study [63].

The inevitable progression of prostate cancer on hormone therapy leaves other strategies and actionable targets open to experimental probing and clinical validation. AR degradation by small-molecule PROTACs and SARDs as an approach to overcome AR axis blockade is in early stage clinical investigation. A clinically viable approach may be in the horizon where AR expression would be disrupted epigenetically by a long non-coding RNA known as NXTAR. The reciprocal negative regulation of gene expression for NXTAR and AR has been demonstrated, and a xenograft study showed that a small-molecule AR inhibitor restored NXTAR expression and sensitized enzalutamide-resistant prostate tumor to growth inhibition [96].

Finally, progress in CRISPR/Cas-enzymes assisted gene editing in clinical settings has ushered optimism for a future arena of prostate cancer management utilizing this cutting-edge technology. Vast possibilities for genome-targeted interventions as well as other avenues of intervention that are coming to light from mechanistic studies are advancing clinical-translational research on prostate cancer through both personalized approach and precision targeting.

Author Contributions: M.C.: writing of the initial draft, preparing figures, literature review; C.-S.S.: generating certain data that was discussed, scientific comments, reading and editing; Y.-G.L.: scientific inputs, literature search, reading and editing; B.C.: conceptualization, funding acquisition, literature review, writing, supervision, and finalization. All authors have read and agreed to the published version of the manuscript.

Funding: The work was supported by grants W81XWH-21-1-0307 and W81XWH-14-1-0606 from the U.S. Department of Defense, and a Research Career Scientist Award (IK6 BX004207) from the U.S. Department of Veterans Affairs. We acknowledge BioRender.com for resources used to create Figures. Templates at BioRender.com (2021) retrieved from https://app.biorender.com/biorender-templates (accessed on 20 November 2021) were adapted partially to generate Figures 1 and 3.

Conflicts of Interest: The authors declare no conflict of interest.

References

1. Giona, S. The Epidemiology of Prostate Cancer. In *Prostate Cancer*; Bott, S.R.J., Ng, K.L., Eds.; Exon Publications: Brisbane, Australia, 2021; Chapter 1. [CrossRef]
2. Nelson, P. Beyond the Androgen Receptor: Targeting Actionable Drivers of Prostate Cancer. *JCO Precis. Oncol.* **2017**, *1*, 1–3. [CrossRef]
3. Watson, P.A.; Arora, V.K.; Sawyers, C.L. Emerging mechanisms of resistance to androgen receptor inhibitors in prostate cancer. *Nat. Rev. Cancer* **2015**, *15*, 701–711. [CrossRef] [PubMed]
4. Sartor, O.; de Bono, J.S. Metastatic prostate cancer. *N. Engl. J. Med.* **2018**, *378*, 645–657. [CrossRef] [PubMed]
5. Mohler, J.L. A brief history of intracrine androgen metabolism by castration-recurrent prostate cancer. *Am. J. Clin. Exp. Urol.* **2018**, *6*, 101–106. [PubMed]
6. Cunha, G.R.; Cao, M.; Franco, O.; Baskin, L.S. A comparison of prostatic development in xenografts of human fetal prostate and human female fetal proximal urethra grown in dihydrotestosterone-treated hosts. *Differentiation* **2020**, *115*, 37–52. [CrossRef] [PubMed]
7. Tyagi, R.K.; Lavrovsky, Y.; Ahn, S.C.; Song, C.S.; Chatterjee, B.; Roy, A.K. Dynamics of Intracellular Movement and Nucleocytoplasmic Recycling of the Ligand-Activated Androgen Receptor in Living Cells. *Mol. Endocrinol.* **2000**, *14*, 1162–1174. [CrossRef] [PubMed]
8. Fujita, K.; Nonomura, N. Role of Androgen Receptor in Prostate Cancer: A Review. *World J. Men's Health* **2019**, *37*, 288–295. [CrossRef]
9. Brawley, S.; Mohan, R.; Nein, C.D. Localized Prostate Cancer: Treatment Options. *Am. Fam. Physician* **2018**, *97*, 798–805.
10. Wilkins, L.J.; Tosoian, J.J.; Sundi, D.; Ross, A.E.; Grimberg, D.; Klein, E.A.; Chapin, B.F.; Nyame, Y.A. Surgical management of high-risk, localized prostate cancer. *Nat. Rev. Urol.* **2020**, *17*, 679–690. [CrossRef]
11. Stattin, P.; Holmberg, E.; Johansson, J.-E.; Holmberg, L.; Adolfsson, J.; Hugosson, J.; National Prostate Cancer Register (NPCR) of Sweden. Outcomes in Localized Prostate Cancer: National Prostate Cancer Register of Sweden Follow-up Study. *JNCI J. Natl. Cancer Inst.* **2010**, *102*, 950–958. [CrossRef]
12. de Sá Moreira, E.; Robinson, D.; Hawthorne, S.; Zhao, L.; Hanson, M.; Kanas, G.; Turnure, M.; Davis, C.; Clark, O. Patterns of Care and Outcomes for Non-Metastatic Prostate Cancer in the United States: Results of the CancerMPact®Survey 2018. *Cancer Manag Res.* **2021**, *13*, 9127–9137. [CrossRef] [PubMed]
13. Attard, G.; Murphy, L.; Clarke, N.W.; Cross, W.; Jones, R.J.; Parker, C.C.; Gillessen, S.; Cook, A.; Brawley, C.; Amos, C.L.; et al. Abiraterone acetate and prednisolone with or without enzalutamide for high-risk non-metastatic prostate cancer: A meta-analysis of primary results from two randomised controlled phase 3 trials of the STAMPEDE platform protocol. *Lancet* **2021**, *399*, 447–460. [CrossRef]
14. Mateo, J.; Fizazi, K.; Gillessen, S.; Heidenreich, A.; Perez-Lopez, R.; Oyen, W.J.G.; Shore, N.; Smith, M.; Sweeney, C.; Tombal, B.; et al. Managing Nonmetastatic Castration-resistant Prostate Cancer. *Eur. Urol.* **2019**, *75*, 285–293. [CrossRef] [PubMed]
15. Smith, M.R. Progress in Nonmetastatic Prostate Cancer. *N. Engl. J. Med.* **2018**, *378*, 2531–2532. [CrossRef]
16. Nguyen, P.L.; Huang, H.C.; Davicioni, E.; Sandler, H.M.; Shipley, W.E.; Efstathiou, J.A.; Simko, J.; Pollack, A.; Dicker, A.; Roach, M.; et al. Validation of a 22-Gene Genomic Classifier in the NRG Oncology/RTOG 9202, 9413 and 9902 Phase III Randomized Trials: A Biopsy-Based Individual Patient Meta-Analysis in High-Risk Prostate Cancer. *Int. J. Radiat. Oncol. Biol. Phys.* **2021**, *111*, S50. [CrossRef]
17. Dawson, N.A.; Zibelman, M.; Lindsay, T.; Feldman, R.A.; Saul, M.; Gatalica, Z.; Korn, W.M.; Heath, E.I. An Emerging Landscape for Canonical and Actionable Molecular Alterations in Primary and Metastatic Prostate Cancer. *Mol. Cancer Ther.* **2020**, *19*, 1373–1382. [CrossRef]
18. Cancer Discovery-Research Watch, AR Enhancer Amplification Drives Castration-Resistant Prostate Cancer. *Cancer Discov.* **2018**, *8*, OF17. [CrossRef]

19. Takeda, D.Y.; Spisa´k, S.; Seo, J.H.; Bell, C.; O'Connor, E.; Korthauer, K.; Ribli, D.; Csabai, I.; Solymosi, N.; Szállási, Z.; et al. A Somatically Acquired Enhancer of the Androgen Receptor Is a Noncoding Driver in Advanced Prostate Cancer. *Cell* **2018**, *174*, 422–432. [CrossRef]
20. Viswanathan, S.; Ha, G.; Hoff, A.M.; Wala, J.A.; Carrot-Zhang, J.; Whelan, C.; Haradhvala, N.J.; Freeman, S.; Reed, S.; Rhoades, J.; et al. Structural Alterations Driving Castration-Resistant Prostate Cancer Revealed by Linked-Read Genome Sequencing. *Cell* **2018**, *174*, 433–447.e19. [CrossRef] [PubMed]
21. Quigley, D.A.; Dang, H.X.; Zhao, S.G.; Lloyd, P.; Aggarwal, R.; Alumkal, J.J.; Foye, A.; Kothari, V.; Perry, M.; Bailey, A.M.; et al. Genomic Hallmarks and Structural Variation in Metastatic Prostate Cancer. *Cell* **2018**, *174*, 758–769.e9, Erratum in *Cell* **2018**, *175*, 889. [CrossRef]
22. van Dessel, L.F.; van Riet, J.; Smits, M.; Zhu, Y.; Hamberg, P.; van der Heijden, M.S.; Bergman, A.M.; van Oort, I.M.; de Wit, R.; Voest, E.E.; et al. The genomic landscape of metastatic castration-resistant prostate cancers reveals multiple distinct genotypes with potential clinical impact. *Nat. Commun.* **2019**, *10*, 5251. [CrossRef]
23. de Bono, J.S.; Cook, N.; Yu, E.Y.; Lara, P.N.L.; Wang, J.S.; Yamasaki, Y.; Yamamiya, I.; Gao, P.; Calleja, E.M.; Rathkopf, D.E. First-in-Human Study of TAS3681, an Oral Androgen Receptor Antagonist with AR and AR Splice Variant Downregulation Activity, in Patients with mCRPC Refractory to Abiraterone And/or Enzalutamide and Chemotherapy. *J. Clin. Oncol.* **2021**, *39*, 5031. [CrossRef]
24. Fizazi, K.; Maldonado, X.; Foulon, S.; Roubaud, G.; McDermott, R.S.; Flechon, A.; Tombal, B.F.; Supiot, S.; Berthold, D.R.; Ronchin, P.; et al. A phase 3 trial with a 2x2 factorial design of abiraterone acetate plus prednisone and/or local radiotherapy in men with de novo metastatic castration-sensitive prostate cancer (mCSPC): First results of PEACE-1. *J. Clin. Oncol.* **2021**, *39*, 5000. [CrossRef]
25. James, N.D.; Sydes, M.R.; Clarke, N.W.; Mason, M.D.; Dearnaley, D.P.; Spears, M.R.; Ritchie, A.W.S.; Parker, C.C.; Russell, J.M.; Attard, G.; et al. Addition of docetaxel, zoledronic acid, or both to first-line long-term hormone therapy in prostate cancer (STAMPEDE): Survival results from an adaptive, multiarm, multistage, platform randomised controlled trial. *Lancet* **2016**, *387*, 1163–1177. [CrossRef]
26. Ravery, V.; Fizazi, K.; Oudard, S.; Drouet, L.; Eymard, J.-C.; Culine, S.; Gravis, G.; Hennequin, C.; Zerbib, M. The use of estramustine phosphate in the modern management of advanced prostate cancer. *Br. J. Urol.* **2011**, *108*, 1782–1786. [CrossRef] [PubMed]
27. Fizazi, K.; Le Maitre, A.; Hudes, G.; Berry, W.R.; Kelly, W.K.; Eymard, J.-C.; Logothetis, C.J.; Pignon, J.-P.; Michiels, S. Meta-analysis of Estramustine in Prostate Cancer (MECaP) Trialists' Collaborative Group. Addition of estramustine to chemotherapy and survival of patients with castration-refractory prostate cancer: A meta-analysis of individual patient data. *Lancet Oncol.* **2007**, *8*, 994–1000. [CrossRef]
28. Saylor, P.J.; Rumble, R.B.; Michalski, J.M. Bone Health and Bone-Targeted Therapies for Prostate Cancer: American Society of Clinical Oncology Endorsement Summary of a Cancer Care Ontario Guideline. *JCO Oncol. Pract.* **2020**, *16*, 389–393. [CrossRef]
29. Paller, C.J.; Carducci, M.A.; Philips, G.K. Management of bone metastases in refractory prostate cancer—Role of denosumab. *Clin. Interv. Aging* **2012**, *7*, 363–372. [CrossRef]
30. Hegemann, M.; Bedke, J.; Stenzl, A.; Todenhöfer, T. Denosumab treatment in the management of patients with advanced prostate cancer: Clinical evidence and experience. *Ther. Adv. Urol.* **2017**, *9*, 81–88. [CrossRef]
31. Pradhan, A.K.; Bhoopathi, P.; Talukdar, S.; Shen, X.-N.; Emdad, L.; Das, S.K.; Sarkar, D.; Fisher, P.B. Recombinant MDA-7/IL24 Suppresses Prostate Cancer Bone Metastasis through Downregulation of the Akt/Mcl-1 Pathway. *Mol. Cancer Ther.* **2018**, *17*, 1951–1960. [CrossRef]
32. Bolomsky, A.; Vogler, M.; Köse, M.C.; Heckman, C.A.; Ehx, G.; Ludwig, H.; Caers, J. MCL-1 inhibitors, fast-lane development of a new class of anti-cancer agents. *J. Hematol. Oncol.* **2020**, *13*, 1–19. [CrossRef]
33. Sweeney, C.; Bracarda, S.; Sternberg, C.N.; Chi, K.N.; Olmos, D.; Sandhu, S.; Massard, C.; Matsubara, N.; Alekseev, B.; Parnis, F.; et al. Ipatasertib plus abiraterone and prednisolone in metastatic castration-resistant prostate cancer (IPATential150): A multicentre, randomised, double-blind, phase 3 trial. *Lancet* **2021**, *398*, 131–142. [CrossRef]
34. Sartor, O.; de Bono, J.S.; Chi, K.N.; Fizazi, K.; Herrmann, K.; Rahbar, K.; Tagawa, S.T.; Nordquist, L.T.; Vaishampayan, N.; El-Haddad, G.; et al. Lutetium-177–PSMA-617 for Metastatic Castration-Resistant Prostate Cancer. *N. Engl. J. Med.* **2021**, *385*, 1091–1103. [CrossRef]
35. De Giorgi, U.; Sansovini, M.; Severi, S.; Nicolini, S.; Monti, M.; Gurioli, G.; Foca, F.; Casadei, C.; Conteduca, V.; Celli, M.; et al. Circulating androgen receptor gene amplification and resistance to 177Lu-PSMA-617 in metastatic castration-resistant prostate cancer: Results of a Phase 2 trial. *Br. J. Cancer* **2021**, *125*, 1226–1232. [CrossRef] [PubMed]
36. Tagawa, S.T.; Sun, M.; Sartor, A.O.; Thomas, C.; Singh, S.; Bissassar, M.; Fernandez, E.; Niaz, M.J.; Ho, B.; Vallabhajosula, S.; et al. Phase I study of 225Ac-J591 for men with metastatic castration-resistant prostate cancer (mCRPC). *J. Clin. Oncol.* **2021**, *39*, 5015. [CrossRef]
37. Sartor, O.; Coleman, R.; Nilsson, S.; Heinrich, D.; Helle, S.I.; O'Sullivan, J.M.; Fosså, S.D.; Chodacki, A.; Wiechno, P.; Logue, J.; et al. Effect of radium-223 dichloride on symptomatic skeletal events in patients with castration-resistant prostate cancer and bone metastases: Results from a phase 3, double-blind, randomised trial. *Lancet Oncol.* **2014**, *15*, 738–746. [CrossRef]
38. Robinson, D.; Van Allen, E.M.; Wu, Y.-M.; Schultz, N.; Lonigro, R.J.; Mosquera, J.-M.; Montgomery, B.; Taplin, M.-E.; Pritchard, C.C.; Attard, G.; et al. Integrative Clinical Genomics of Advanced Prostate Cancer. *Cell* **2015**, *161*, 1215–1228. [CrossRef] [PubMed]

39. Abida, W.; Armenia, J.; Gopalan, A.; Brennan, R.; Walsh, M.; Barron, D.; Danila, D.; Rathkopf, D.; Morris, M.; Slovin, S.; et al. Prospective Genomic Profiling of Prostate Cancer Across Disease States Reveals Germline and Somatic Alterations That May Affect Clinical Decision Making. *JCO Precis. Oncol.* **2017**, *2017*, PO.17.00029. [CrossRef]
40. Abeshouse, A.; Ahn, J.; Akbani, R.; Ally, A.; Amin, S.; Andry, C.D.; Annala, M.; Aprikian, A.; Armenia, J.; Arora, A.; et al. The Molecular Taxonomy of Primary Prostate Cancer. *Cell* **2015**, *163*, 1011–1025. [CrossRef]
41. Helleday, T. The underlying mechanism for the PARP and BRCA synthetic lethality: Clearing up the misunderstandings. *Mol. Oncol.* **2011**, *5*, 387–393. [CrossRef]
42. Hussain, M.; Mateo, J.; Fizazi, K.; Saad, F.; Shore, N.; Sandhu, S.; Chi, K.N.; Sartor, O.; Agarwal, N.; Olmos, D.; et al. Survival with Olaparib in Metastatic Castration-Resistant Prostate Cancer. *N. Engl. J. Med.* **2020**, *383*, 2345–2357. [CrossRef] [PubMed]
43. de Bono, J.S.; Mehra, N.; Scagliotti, G.V.; Castro, E.; Dorff, T.; Stirling, A.; Stenzl, A.; Fleming, M.T.; Higano, C.S.; Saad, F.; et al. Talazoparib monotherapy in metastatic castration-resistant prostate cancer with DNA repair alterations (TALAPRO-1): An open-label, phase 2 trial. *Lancet Oncol.* **2021**, *22*, 1250–1264. [CrossRef]
44. Schiewer, M.J.; Goodwin, J.F.; Han, S.; Brenner, J.C.; Augello, M.A.; Dean, J.L.; Liu, F.; Planck, J.L.; Ravindranathan, P.; Chinnaiyan, A.M.; et al. Dual Roles of PARP-1 Promote Cancer Growth and Progression. *Cancer Discov.* **2012**, *2*, 1134–1149. [CrossRef] [PubMed]
45. Gui, B.; Gui, F.; Takai, T.; Feng, C.; Bai, X.; Fazli, L.; Dong, X.; Liu, S.; Zhang, X.; Zhang, W.; et al. Selective targeting of PARP-2 inhibits androgen receptor signaling and prostate cancer growth through disruption of FOXA1 function. *Proc. Natl. Acad. Sci. USA* **2019**, *116*, 14573–14582. [CrossRef] [PubMed]
46. Asim, M.; Tarish, F.; Zecchini, H.I.; Sanjiv, K.; Gelali, E.; Massie, C.E.; Baridi, A.; Warren, A.Y.; Zhao, W.; Ogris, C.; et al. Synthetic lethality between androgen receptor signalling and the PARP pathway in prostate cancer. *Nat. Commun.* **2017**, *8*, 374. [CrossRef]
47. Pezaro, C. PARP inhibitor combinations in prostate cancer. *Ther. Adv. Med. Oncol.* **2020**, *12*, 1758835919897537. [CrossRef]
48. Li, L.; Karanika, S.; Yang, G.; Wang, J.; Park, S.; Broom, B.M.; Manyam, G.C.; Wu, W.; Luo, Y.; Basourakos, S.; et al. Androgen receptor inhibitor–induced "BRCAness" and PARP inhibition are synthetically lethal for castration-resistant prostate cancer. *Sci. Signal.* **2017**, *10*, eaam7479. [CrossRef]
49. Fay, E.K.; Graff, J.N. Immunotherapy in Prostate Cancer. *Cancers* **2020**, *12*, 1752. [CrossRef]
50. Abida, W.; Cheng, M.L.; Armenia, J.; Middha, S.; Autio, K.A.; Vargas, H.A.; Rathkopf, D.; Morris, M.J.; Danila, D.C.; Slovin, S.F.; et al. Analysis of the Prevalence of Microsatellite Instability in Prostate Cancer and Response to Immune Checkpoint Blockade. *JAMA Oncol.* **2019**, *5*, 471–478. [CrossRef]
51. Ravindranathan, D.; Russler, G.A.; Yantorni, L.; Drusbosky, L.M.; Bilen, M.A. Detection of Microsatellite Instability via Circulating Tumor DNA and Response to Immunotherapy in Metastatic Castration-Resistant Prostate Cancer: A Case Series. *Case Rep. Oncol.* **2021**, *14*, 190–196. [CrossRef]
52. De Almeida, D.V.P.; Fong, L.; Rettig, M.B.; Autio, K.A. Immune Checkpoint Blockade for Prostate Cancer: Niche Role or Next Breakthrough? *Am. Soc. Clin. Oncol. Educ. Book* **2020**, e89–e106. [CrossRef] [PubMed]
53. Cha, H.-R.; Lee, J.H.; Ponnazhagan, S. Revisiting Immunotherapy: A Focus on Prostate Cancer. *Cancer Res.* **2020**, *80*, 1615–1623. [CrossRef] [PubMed]
54. Graff, J.N.; Alumkal, J.J.; Thompson, R.; Moran, A.; Thomas, G.V.; Wood, M.A.; Drake, C.G.; Slottke, R.; Beer, T.M. Pembroluzimab (Pembro) plus enzalutamide (Enz) in metastatic castration resistant prostate cancer (mCRPC): Extended follow up. *J. Clin. Oncol.* **2018**, *36*, 5047. [CrossRef]
55. Yu, E.; Piulats, J.; Gravis, G.; Fong, P.; Todenhöfer, T.; Laguerre, B.; Arranz, J.; Oudard, S.; Massard, C.; Stoeckle, M.; et al. 73P Association between homologous recombination repair mutations and response to pembrolizumab (pembro) plus olaparib (ola) in metastatic castration-resistant prostate cancer (mCRPC): KEYNOTE-365 Cohort A biomarker analysis. *Ann. Oncol.* **2021**, *32*, S387. [CrossRef]
56. Rotte, A. Combination of CTLA-4 and PD-1 blockers for treatment of cancer. *J. Exp. Clin. Cancer Res.* **2019**, *38*, 255. [CrossRef]
57. Wu, Y.-M.; Cieślik, M.; Lonigro, R.J.; Vats, P.; Reimers, M.A.; Cao, X.; Ning, Y.; Wang, L.; Kunju, L.P.; de Sarkar, N.; et al. Inactivation of CDK12 Delineates a Distinct Immunogenic Class of Advanced Prostate Cancer. *Cell* **2018**, *173*, 1770–1782.e14. [CrossRef]
58. Powers, E.; Karachaliou, G.S.; Kao, C.; Harrison, M.R.; Hoimes, C.J.; George, D.J.; Armstrong, A.J.; Zhang, T. Novel therapies are changing treatment paradigms in metastatic prostate cancer. *J. Hematol. Oncol.* **2020**, *13*, 144. [CrossRef]
59. Dulos, J.; Carven, G.J.; van Boxtel, S.J.; Evers, S.; Driessen-Engels, L.J.A.; Hobo, W.; Gorecka, M.A.; de Haan, A.F.J.; Mulders, P.; Punt, C.J.A.; et al. PD-1 Blockade Augments Th1 and Th17 and Suppresses Th2 Responses in Peripheral Blood From Patients With Prostate and Advanced Melanoma Cancer. *J. Immunother.* **2012**, *35*, 169–178. [CrossRef]
60. Peng, W.; Liu, C.; Xu, C.; Lou, Y.; Chen, J.; Yang, Y.; Yagita, H.; Overwijk, W.W.; Lizée, G.; Radvanyi, L.; et al. PD-1 Blockade Enhances T-cell Migration to Tumors by Elevating IFN-γ Inducible Chemokines. *Cancer Res.* **2012**, *72*, 5209–5218. [CrossRef]
61. Ni, L.; Lu, J. Interferon gamma in cancer immunotherapy. *Cancer Med.* **2018**, *7*, 4509–4516. [CrossRef]
62. Manguso, R.T.; Pope, H.W.; Zimmer, M.D.; Brown, F.D.; Yates, K.B.; Miller, B.; Collins, N.B.; Bi, K.; LaFleur, M.W.; Juneja, V.R.; et al. In vivo CRISPR screening identifies Ptpn2 as a cancer immunotherapy target. *Nature* **2017**, *547*, 413–418. [CrossRef] [PubMed]
63. Krebs, M.; Solimando, A.G.; Kalogirou, C.; Marquardt, A.; Frank, T.; Sokolakis, I.; Hatzichristodoulou, G.; Kneitz, S.; Bargou, R.; Kübler, H.; et al. miR-221-3p Regulates VEGFR2 Expression in High-Risk Prostate Cancer and Represents an Escape Mechanism from Sunitinib In Vitro. *J. Clin. Med.* **2020**, *9*, 670. [CrossRef] [PubMed]

64. Michaelson, D.; Regan, M.M.; Oh, W.; Kaufman, D.S.; Olivier, K.; Michaelson, S.Z.; Spicer, B.; Gurski, C.; Kantoff, P.; Smith, M.R. Phase II study of sunitinib in men with advanced prostate cancer. *Ann. Oncol.* **2009**, *20*, 913–920. [CrossRef] [PubMed]
65. Armstrong, A.J.; Halabi, S.; Healy, P.; Lee, W.R.; Koontz, B.F.; Moul, J.W.; Mundy, K.; Creel, P.; Wood, S.; Davis, K.; et al. A phase 2 multimodality trial of docetaxel/prednisone with sunitinib followed by salvage radiation therapy in men with PSA recurrent prostate cancer after radical prostatectomy. *Prostate Cancer Prostatic Dis.* **2016**, *19*, 100–106. [CrossRef] [PubMed]
66. Kneitz, B.; Krebs, M.; Kalogirou, C.; Schubert, M.; Joniau, S.; Van Poppel, H.; Lerut, E.; Kneitz, S.; Scholz, C.-J.; Ströbel, P.; et al. Survival in Patients with High-Risk Prostate Cancer Is Predicted by miR-221, Which Regulates Proliferation, Apoptosis, and Invasion of Prostate Cancer Cells by Inhibiting IRF2 and SOCS3. *Cancer Res.* **2014**, *74*, 2591–2603. [CrossRef]
67. Perner, S.; Rupp, N.J.; Braun, M.; Rubin, M.; Moch, H.; Dietel, M.; Wernert, N.; Jung, K.; Stephan, C.; Kristiansen, G. Loss of SLC45A3 protein (prostein) expression in prostate cancer is associated withSLC45A3-ERGgene rearrangement and an unfavorable clinical course. *Int. J. Cancer* **2013**, *132*, 807–812. [CrossRef]
68. Mao, N.; Zhang, Z.; Lee, Y.S.; Choi, D.; Rivera, A.A.; Li, D.; Lee, C.; Haywood, S.; Chen, X.; Chang, Q.; et al. Defining the therapeutic selective dependencies for distinct subtypes of PI3K pathway-altered prostate cancers. *Nat. Commun.* **2021**, *12*, 5053. [CrossRef]
69. Wee, S.; Wiederschain, D.; Maira, S.-M.; Loo, A.; Miller, C.; Debeaumont, R.; Stegmeier, F.; Yao, Y.-M.; Lengauer, C. PTEN-deficient cancers depend on PIK3CB. *Proc. Natl. Acad. Sci. USA* **2008**, *105*, 13057–13062. [CrossRef]
70. Schwartz, S.; Wongvipat, J.; Trigwell, C.B.; Hancox, U.; Carver, B.S.; Rodrik-Outmezguine, V.; Will, M.; Yellen, P.; de Stanchina, E.; Baselga, J.; et al. Feedback Suppression of PI3Kα Signaling in PTEN-Mutated Tumors Is Relieved by Selective Inhibition of PI3Kβ. *Cancer Cell* **2014**, *27*, 109–122. [CrossRef]
71. Kaffenberger, S.D.; Barbieri, C. Molecular subtyping of prostate cancer. *Curr. Opin. Urol.* **2016**, *26*, 213–218. [CrossRef]
72. Geng, C.; He, B.; Xu, L.; Barbieri, C.; Eedunuri, V.K.; Chew, S.A.; Zimmermann, M.; Bond, R.; Shou, J.; Li, C.; et al. Prostate cancer-associated mutations in speckle-type POZ protein (SPOP) regulate steroid receptor coactivator 3 protein turnover. *Proc. Natl. Acad. Sci. USA* **2013**, *110*, 6997–7002. [CrossRef] [PubMed]
73. Boysen, G.; Barbieri, C.E.; Prandi, D.; Blattner, M.; Chae, S.-S.; Dahija, A.; Nataraj, S.; Huang, D.; Marotz, C.; Xu, L.; et al. SPOP mutation leads to genomic instability in prostate cancer. *eLife* **2015**, *4*, e09207. [CrossRef] [PubMed]
74. Tomlins, S.A.; Rhodes, D.R.; Yu, J.; Varambally, S.; Mehra, R.; Perner, S.; Demichelis, F.; Helgeson, B.E.; Laxman, B.; Morris, D.S.; et al. The role of SPINK1 in ETS rearrangement-negative prostate cancers. *Cancer Cell* **2008**, *13*, 519–528. [CrossRef]
75. Ku, S.-Y.; Gleave, M.; Beltran, H. Towards precision oncology in advanced prostate cancer. *Nat. Rev. Urol.* **2019**, *16*, 645–654. [CrossRef] [PubMed]
76. Manning, B.D.; Cantley, L.C. AKT/PKB Signaling: Navigating Downstream. *Cell* **2007**, *129*, 1261–1274. [CrossRef]
77. Taylor, B.S.; Schultz, N.; Hieronymus, H.; Gopalan, A.; Xiao, Y.; Carver, B.S.; Arora, V.K.; Kaushik, P.; Cerami, E.; Reva, B.; et al. Integrative Genomic Profiling of Human Prostate Cancer. *Cancer Cell* **2010**, *18*, 11–22. [CrossRef]
78. Sumanasuriya, S.; De Bono, J.S. Treatment of Advanced Prostate Cancer-A Review of Current Therapies and Future Promise. *Cold Spring Harb. Perspect. Med.* **2018**, *8*, a030635. [CrossRef]
79. Yoshimoto, M.; Ludkovski, O.; DeGrace, D.; Williams, J.L.; Evans, A.; Sircar, K.; Bismar, T.A.; Nuin, P.; Squire, J.A. PTEN genomic deletions that characterize aggressive prostate cancer originate close to segmental duplications. *Genes Chromosom. Cancer* **2011**, *51*, 149–160. [CrossRef]
80. Carver, B.S.; Chapinski, C.; Wongvipat, J.; Hieronymus, H.; Chen, Y.; Chandarlapaty, S.; Arora, V.K.; Le, C.; Koutcher, J.; Scher, H.; et al. Reciprocal Feedback Regulation of PI3K and Androgen Receptor Signaling in PTEN-Deficient Prostate Cancer. *Cancer Cell* **2011**, *19*, 575–586. [CrossRef]
81. Mirkheshti, N.; Park, S.; Jiang, S.; Cropper, J.; Werner, S.L.; Song, C.S.; Chatterjee, B. Dual targeting of androgen receptor and mTORC1 by salinomycin in prostate cancer. *Oncotarget* **2016**, *7*, 62240–62254. [CrossRef]
82. Kim, K.-Y.; Yu, S.-N.; Lee, S.-Y.; Chun, S.-S.; Choi, Y.-L.; Park, Y.-M.; Song, C.S.; Chatterjee, B.; Ahn, S.-C. Salinomycin-induced apoptosis of human prostate cancer cells due to accumulated reactive oxygen species and mitochondrial membrane depolarization. *Biochem. Biophys. Res. Commun.* **2011**, *413*, 80–86. [CrossRef] [PubMed]
83. Zafeiriou, Z.; Bianchini, D.; Chandler, R.; Rescigno, P.; Yuan, W.; Carreira, S.; Barrero, M.; Petremolo, A.; Miranda, S.; Riisnaes, R.; et al. Genomic Analysis of Three Metastatic Prostate Cancer Patients with Exceptional Responses to Carboplatin Indicating Different Types of DNA Repair Deficiency. *Eur. Urol.* **2018**, *75*, 184–192. [CrossRef]
84. Cheng, H.H.; Pritchard, C.C.; Boyd, T.; Nelson, P.S.; Montgomery, B. Biallelic Inactivation of BRCA2 in Platinum-sensitive Metastatic Castration-resistant Prostate Cancer. *Eur. Urol.* **2015**, *69*, 992–995. [CrossRef] [PubMed]
85. Kim, T.; Lee, Y.; Koo, K. Current Status and Future Perspectives of Androgen Receptor Inhibition Therapy for Prostate Cancer: A Comprehensive Review. *Biomolecules* **2021**, *11*, 492. [CrossRef] [PubMed]
86. Gustafson, J.L.; Neklesa, T.K.; Cox, C.S.; Roth, A.G.; Buckley, D.L.; Tae, H.S.; Sundberg, T.B.; Stagg, D.; Hines, J.; McDonnell, D.P.; et al. Small-Molecule-Mediated Degradation of the Androgen Receptor through Hydrophobic Tagging. *Angew. Chem. Int. Ed.* **2015**, *54*, 9659–9662. [CrossRef] [PubMed]
87. Luo, J.; Tian, J.; Chou, F.; Lin, C.; Xing, E.Z.; Zuo, L.; Niu, Y.; Yeh, S.; Chang, C. Targeting the androgen receptor (AR) with AR degradation enhancer ASC-J9®led to increase docetaxel sensitivity via suppressing the p21 expression. *Cancer Lett.* **2018**, *444*, 35–44. [CrossRef]

88. Wang, R.; Lin, W.; Lin, C.; Li, L.; Sun, Y.; Chang, C. ASC-J9®suppresses castration resistant prostate cancer progression via degrading the enzalutamide-induced androgen receptor mutant AR-F876L. *Cancer Lett.* **2016**, *379*, 154–160. [CrossRef]
89. Omlin, A.; Jones, R.; Van Der Noll, R.; Satoh, T.; Niwakawa, M.; Smith, S.A.; Graham, J.; Ong, M.; Finkelman, R.D.; Schellens, J.H.M.; et al. AZD3514, an oral selective androgen receptor down-regulator in patients with castration-resistant prostate cancer—Results of two parallel first-in-human phase I studies. *Investig. New Drugs* **2015**, *33*, 679–690. [CrossRef]
90. Salami, J.; Alabi, S.; Willard, R.R.; Vitale, N.J.; Wang, J.; Dong, H.; Jin, M.; McDonnell, D.P.; Crew, A.P.; Neklesa, T.K.; et al. Androgen receptor degradation by the proteolysis-targeting chimera ARCC-4 outperforms enzalutamide in cellular models of prostate cancer drug resistance. *Commun. Biol.* **2018**, *1*, 100. [CrossRef]
91. Kregel, S.; Wang, C.; Han, X.; Xiao, L.; Fernandez-Salas, E.; Bawa, P.; McCollum, B.L.; Wilder-Romans, K.; Apel, I.J.; Cao, X.; et al. Androgen receptor degraders overcome common resistance mechanisms developed during prostate cancer treatment. *Neoplasia* **2020**, *22*, 111–119. [CrossRef]
92. Petrylak, D.P.; Gao, X.; Vogelzang, N.J.; Garfield, M.H.; Taylor, I.; Moore, M.D.; Peck, R.A.; Burris, H.A. First-in-human phase I study of ARV-110, an androgen receptor (AR) PROTAC degrader in patients (pts) with metastatic castrate-resistant prostate cancer (mCRPC) following enzalutamide (ENZ) and/or abiraterone (ABI). *J. Clin. Oncol.* **2020**, *38*, 3500. [CrossRef]
93. Chen, L.; Han, L.; Mao, S.; Xu, P.; Xu, X.; Zhao, R.; Wu, Z.; Zhong, K.; Yu, G.; Wang, X. Discovery of A031 as effective proteolysis targeting chimera (PROTAC) androgen receptor (AR) degrader for the treatment of prostate cancer. *Eur. J. Med. Chem.* **2021**, *216*, 113307. [CrossRef] [PubMed]
94. Antonarakis, E.S.; Lu, C.; Wang, H.; Luber, B.; Nakazawa, M.; Roeser, J.C.; Chen, Y.; Mohammad, T.A.; Chen, Y.; Fedor, H.L.; et al. AR-V7 and resistance to enzalutamide and abiraterone in prostate cancer. *N. Engl. J. Med.* **2014**, *371*, 1028–1038. [CrossRef] [PubMed]
95. Scher, H.I.; Graf, R.P.; Schreiber, N.A.; McLaughlin, B.; Lu, D.; Louw, J.; Danila, D.C.; Dugan, L.; Johnson, A.; Heller, G.; et al. Nuclear-specific AR-V7 Protein Localization is Necessary to Guide Treatment Selection in Metastatic Castration-resistant Prostate Cancer. *Eur. Urol.* **2017**, *71*, 874–882. [CrossRef]
96. Ghildiyal, R.; Sawant, M.; Renganathan, A.; Mahajan, K.; Kim, E.H.; Luo, J.; Dang, H.X.; Maher, C.A.; Feng, F.Y.; Mahajan, N.P. Loss of Long Noncoding RNA NXTAR in Prostate Cancer Augments Androgen Receptor Expression and Enzalutamide Resistance. *Cancer Res.* **2021**, *82*, 155–168. [CrossRef]
97. Kelly, K.; Balk, S.P. Reprogramming to resist. *Science* **2017**, *355*, 29–30. [CrossRef]
98. Beltran, H.; Demichelis, F. Therapy considerations in neuroendocrine prostate cancer: What next? *Endocrine-Related Cancer* **2021**, *28*, T67–T78. [CrossRef]
99. Labrecque, M.; Coleman, I.M.; Brown, L.G.; True, L.D.; Kollath, L.; Lakely, B.; Nguyen, H.M.; Yang, Y.C.; Gil Da Costa, R.M.; Kaipainen, A.; et al. Molecular profiling stratifies diverse phenotypes of treatment-refractory metastatic castration-resistant prostate cancer. *J. Clin. Investig.* **2019**, *129*, 4492–4505. [CrossRef]
100. Xu, K.; Wu, Z.J.; Groner, A.C.; He, H.H.; Cai, C.; Lis, R.T.; Wu, X.; Stack, E.C.; Loda, M.; Liu, T.; et al. EZH2 Oncogenic Activity in Castration-Resistant Prostate Cancer Cells Is Polycomb-Independent. *Science* **2012**, *338*, 1465–1469. [CrossRef]
101. Ko, S.; Ahn, J.; Song, C.S.; Kim, S.; Knapczyk-Stwora, K.; Chatterjee, B. Lysine Methylation and Functional Modulation of Androgen Receptor by Set9 Methyltransferase. *Mol. Endocrinol.* **2011**, *25*, 433–444. [CrossRef]
102. Mu, P.; Zhang, Z.; Benelli, M.; Karthaus, W.R.; Hoover, E.; Chen, C.C.; Wongvipat, J.; Ku, S.; Gao, D.; Cao, Z.; et al. SOX2 promotes lineage plasticity and antiandrogen resistance in TP53- and RB1-deficient prostate cancer. *Science* **2017**, *355*, 84–88. [CrossRef] [PubMed]
103. Ku, S.Y.; Rosario, S.; Wang, Y.; Mu, P.; Seshadri, M.; Goodrich, Z.W.; Goodrich, M.M.; Labbé, D.P.; Gomez, E.C.; Wang, J.; et al. Rb1 and Trp53 cooperate to suppress prostate cancer lineage plasticity, metastasis, and antiandrogen resistance. *Science* **2017**, *355*, 78–83. [CrossRef] [PubMed]
104. Wu, C.; Peng, S.; Pilié, P.G.; Geng, C.; Park, S.; Manyam, G.C.; Lu, Y.; Yang, G.; Tang, Z.; Kondraganti, S.; et al. PARP and CDK4/6 Inhibitor Combination Therapy Induces Apoptosis and Suppresses Neuroendocrine Differentiation in Prostate Cancer. *Mol. Cancer Ther.* **2021**, *20*, 1680–1691. [CrossRef] [PubMed]
105. Pernigoni, N.; Zagato, E.; Calcinotto, A.; Troiani, M.; Mestre, R.P.; Calì, B.; Attanasio, G.; Troisi, J.; Minini, M.; Mosole, S.; et al. Commensal bacteria promote endocrine resistance in prostate cancer through androgen biosynthesis. *Science* **2021**, *374*, 216–224. [CrossRef] [PubMed]

cancers

MDPI

Article

Specialist versus Primary Care Prostate Cancer Follow-Up: A Process Evaluation of a Randomized Controlled Trial

Barbara M. Wollersheim [1,2], Kristel M. van Asselt [2], Floris J. Pos [3], Emine Akdemir [1], Shifra Crouse [1], Henk G. van der Poel [4], Neil K. Aaronson [1], Lonneke V. van de Poll-Franse [1,5,6] and Annelies H. Boekhout [1,*]

1 Division of Psychosocial Research and Epidemiology, The Netherlands Cancer Institute, Antoni van Leeuwenhoek Hospital, Plesmanlaan 121, 1066 CX Amsterdam, The Netherlands; b.wollersheim@nki.nl (B.M.W.); e.akdemir@nki.nl (E.A.); shifracrouse1996@gmail.com (S.C.); n.aaronson@nki.nl (N.K.A.); l.vd.poll@nki.nl (L.V.v.d.P.-F.)
2 Department of General Practice, Amsterdam UMC, Location AMC, University of Amsterdam, Meibergdreef 9, 1105 AZ Amsterdam, The Netherlands; k.m.vanasselt@amsterdamumc.nl
3 Department of Radiotherapy, The Netherlands Cancer Institute, Antoni van Leeuwenhoek Hospital, Plesmanlaan 121, 1066 CX Amsterdam, The Netherlands; f.pos@nki.nl
4 Department of Urology, The Netherlands Cancer Institute, Antoni van Leeuwenhoek Hospital, Plesmanlaan 121, 1066 CX Amsterdam, The Netherlands; h.vd.poel@nki.nl
5 Department of Research, Netherlands Comprehensive Cancer Organization (IKNL), Godebaldkwartier 419, 3511 DT Utrecht, The Netherlands
6 Department of Medical and Clinical Psychology, CoRPS—Center of Research on Psychology in Somatic Diseases, Tilburg University, Warandelaan 2, 5037 AB Tilburg, The Netherlands
* Correspondence: a.boekhout@nki.nl; Tel.: +31-20-512-9111

Citation: Wollersheim, B.M.; van Asselt, K.M.; Pos, F.J.; Akdemir, E.; Crouse, S.; van der Poel, H.G.; Aaronson, N.K.; van de Poll-Franse, L.V.; Boekhout, A.H. Specialist versus Primary Care Prostate Cancer Follow-Up: A Process Evaluation of a Randomized Controlled Trial. *Cancers* **2022**, *14*, 3166. https://doi.org/10.3390/cancers14133166

Academic Editors: José I. López and Claudia Manini

Received: 3 June 2022
Accepted: 24 June 2022
Published: 28 June 2022

Simple Summary: Worldwide, there is an increased focus on reorganizing prostate cancer survivorship care. In this study, we describe for the first time a process evaluation as part of a randomized controlled trial that is currently comparing the effectiveness of specialist- versus primary care-based prostate cancer follow-up. We found that within an RCT context, 67% patients and their GPs were willing to receive/provide primary care-based follow-up. Patients who received primary care-based follow-up care experienced this to be more personal, efficient, and sustainable. However, patients, GPs, and specialists also indicated several challenges that are described in this study and should be addressed to enable a smooth transition of prostate cancer follow-up to primary care.

Abstract: *Background:* A randomized controlled trial (RCT) is currently comparing the effectiveness of specialist- versus primary care-based prostate cancer follow-up. This process evaluation assesses the reach and identified constructs for the implementation of primary care-based follow-up. *Methods:* A mixed-methods approach is used to assess the reach and the implementation through the Consolidated Framework for Implementation Research. We use quantitative data to evaluate the reach of the RCT and qualitative data (interviews) to indicate the perspectives of patients (n = 15), general practitioners (GPs) (n = 10), and specialists (n = 8). Thematic analysis is used to analyze the interview transcripts. *Results:* In total, we reached 402 (67%) patients from 12 hospitals and randomized them to specialist- (n = 201) or to primary care-based (n = 201) follow-up. From the interviews, we identify several advantages of primary care- versus specialist-based follow-up: it is closer to home, more accessible, and the relationship is more personal. Nevertheless, participants also identified challenges: guidelines should be implemented, communication and collaboration between primary and secondary care should be improved, quality indicators should be collected, and GPs should be compensated. *Conclusion:* Within an RCT context, 402 (67%) patients and their GPs were willing to receive/provide primary care-based follow-up. If the RCT shows that primary care is equally as effective as specialist-based follow-up, the challenges identified in this study need to be addressed to enable a smooth transition of prostate cancer follow-up to primary care.

Keywords: prostate cancer survivors; follow-up care; primary health care; general practice; process evaluation; Consolidated Framework for Implementation Research

1. Introduction

Prostate cancer follow-up care is in need of a more sustainable follow-up care model because of the increasing number of prostate cancer survivors and the demands to improve its quality and efficiency [1,2]. Currently, in most Western countries, prostate cancer survivors are included in a hospital-based follow-up care program [3]. Different models of follow-up care for (prostate) cancer patients have been proposed [4]; this could be risk-based on clinical outcomes [5], including shared-care models and primary care-based follow-up care models. Currently, we are conducting a prospective, randomized, multicenter study (PROSPEC trial) to compare the (cost-)effectiveness of specialist- (usual care) versus primary care-based (intervention) follow-up of patients who have completed their primary treatment for localized prostate cancer [6].

When investigating the efficacy of such a potential shift in the organization of care, it is important to evaluate a number of relevant process outcomes [7,8]. Reorganizing this routine demands close collaboration between primary and secondary care providers, and it asks for change in the behavior of patients, clinicians, and in the organization of care [9]. To date, most studies have focused on cancer survivors' preferences for cancer follow-up care [10–13] and general practitioners' (GPs) willingness to provide cancer follow-up care [14–16]. Since most of these studies were conducted in a setting where specialist-based follow-up was current practice, it is important to investigate primary care-based follow-up in a setting where patients and clinicians actually have the opportunity to experience primary care-based care as well.

The aim of the current study was, in the context of an ongoing randomized clinical trial, to analyze the reach of the trial and to identify and evaluate constructs relevant to the implementation of primary care-based prostate cancer follow-up. We used the Consolidated Framework for Implementation Research (CFIR) to guide the evaluation of the implementation of primary care-based prostate cancer follow-up [17].

2. Materials and Methods

2.1. Design and Study Population

This mixed-methods process evaluation investigated a primary care-based follow-up program for prostate cancer survivors in the Netherlands in an RCT setting. A detailed description of the design of the RCT has been published previously [6]. Briefly, eligible prostate cancer survivors who had completed primary treatment (prostatectomy or radiotherapy) for localized prostate cancer and without evidence of recurrence were recruited between July 2018 and September 2021. In total, 402 consenting men were randomized to either specialist- (usual care) or primary care-based (intervention group) follow-up.

For the interviews, we recruited prostate cancer survivors (randomized to primary care-based follow-up), GPs (those who carried out the follow-up of at least one prostate cancer patient), and specialists (urologists, radiation oncologists, and physician assistants) who were participating in the PROSPEC trial for at least one year.

Trial Registration: Netherlands Trial Registry, Trial NL7068 (NTR7266). Prospectively registered on 11 June 2018.

Ethical approval was obtained from the institutional review board of a comprehensive cancer center in the Netherlands (IRBd19-251). All participants signed written informed consents before participating in the study.

2.2. Data Collection

We used a mix of quantitative and qualitative data to describe the reach and to identify and evaluate constructs relevant to the implementation of the trial (see Table 1). We used the CFIR determinant framework to guide the evaluation of the implementation of primary care-based prostate cancer follow-up [17].

To describe the reach, we collected data from the research logbook of the PROSPEC trial. The reach was calculated over patients who received the information letter of the trial. In addition, we conducted semi-structured interviews with patients, GPs, and specialists. We developed an interview guide using the CFIR domains (see Table S1). Data about prostate cancer survivors' socio-demographics (date of birth, marital status, and educational level) and clinical characteristics (date and type of treatment, and risk group) were collected as part of the RCT. GPs and specialists completed a brief questionnaire about socio-demographics (date of birth and sex) and their work situation (type of healthcare professional and type of GP practice/hospital).

Table 1. Components of the process evaluation, including CFIR domains [17].

Components	Definition	Source
Reach	The number and proportion of the target population participating in this intervention	Research logbook
CFIR Domains	Definition	Source
Intervention characteristics	The characteristics of the intervention when implemented in an organization	Qualitative interview questions
Outer setting characteristics	The political and social context within which an organization resides	Qualitative interview questions
Inner setting characteristics	The structural, political and cultural context through which the implementation process will proceed	Qualitative interview questions
Individual characteristics	The knowledge, beliefs, attitudes and expectations of the individuals involved in the intervention	Qualitative interview questions
Implementation process	Processes and change that are needed for a successful implementation	Qualitative interview questions

Abbreviations: CFIR = Consolidated Framework for Implementation Research.

2.3. Study Procedures

For the interviews, we used a purposive sampling strategy to include a maximum variation of patients, GPs, and specialists. Members of the research team personally invited participants. The research team informed the participants and provided information letters; written informed consent was obtained. All interviews were audio-recorded.

2.4. Data Analysis

For the interviews, descriptive statistics were used to characterize the study sample. The audio recordings of the interviews were transcribed verbatim. To safeguard the anonymity and confidentiality of the participants, names were removed from the transcripts. The interviews were analyzed in a systematic way by four researchers (BW, EA, SC, and AB). The transcripts were analyzed according to the procedure for thematic analysis as described by Braun and Clarke (see Table 2) [18]. Three researchers independently coded the transcripts of the interviews, using an inductive approach. We continued to recruit participants until we reached data saturation. A third author (AB), not involved in the initial theme development, was consulted to review the themes. Finally, consensus was reached and the final themes were developed (see Table S2).

Table 2. Overview transcript analysis.

Phase	Coding method	Performed by
1. Familiarizing yourself with your data		BW, EA, SC
2. Generating initial codes	Inductive approach	BW, EA, SC
Iterative process	Consensus-based codebook	BW, EA, SC
Review	Consensus-based codebook	BW, EA, SC, AB
Data saturation	Final codebook	BW, EA, SC, AB
3. Searching for themes	Using CFIR framework	BW, EA, SC
4. Reviewing themes	Using CFIR framework	BW, EA, SC
5. Defining and naming themes	Using CFIR framework	BW, EA, SC, AB
6. Producing the report		BW, AB

RQDA [19], R package [20] for computer-assisted qualitative data analysis, was used to perform the thematic analysis.

3. Results

In total, 597 patients with localized prostate cancer from 12 hospitals (1 academic hospital, 1 comprehensive cancer center, 6 top clinical hospitals, and 4 community hospitals) across different regions in the Netherlands were invited to participate in the RCT. The reach of the study is presented in Figure 1: 16 (3%) patients were not eligible, 157 (26%) patients declined (of whom most preferred follow-up in the hospital), and 22 (4%) GPs declined to participate. Ultimately, 402 (67%) patients were randomized to specialist- (n = 201) or to primary care-based (n = 201) follow-up.

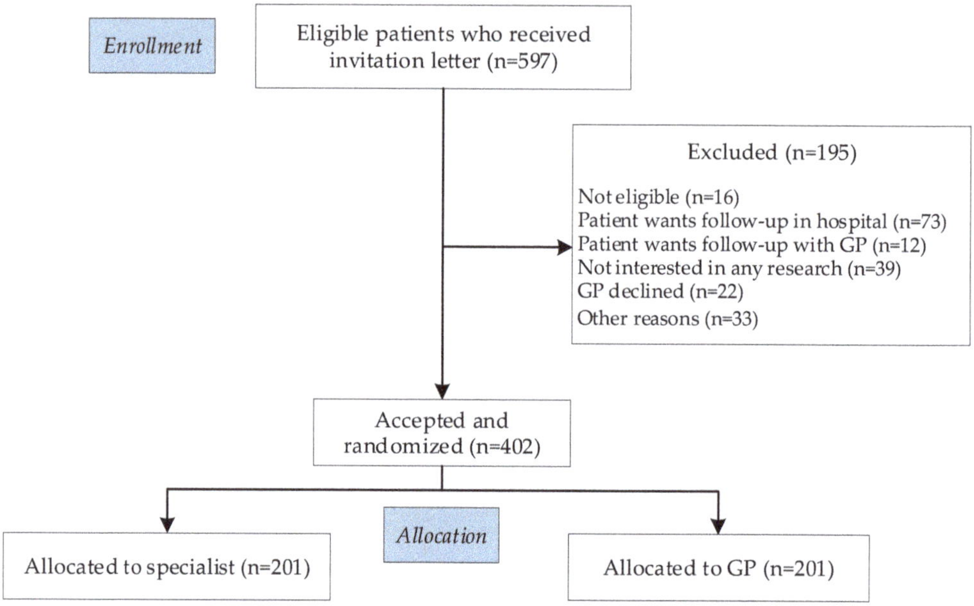

Figure 1. Consort flow diagram.

3.1. Interviews

In total, we interviewed 15 patients, 10 GPs, and 8 specialists (see Table 3). From the thematic analyses, we identified six overarching themes: structure of prostate cancer follow-up care, communication between primary and secondary care, clinical competencies, facilitators of and barriers to primary care-based follow-up, and organizational requirements for the implementation of primary care-based follow-up (see Table 4; including quotes to illustrate the data).

Table 3. Characteristics of interview participants.

Demographics	Patients, n = 15 (%)	GPs, n = 10 (%)	Specialists, n = 8 (%)
Age at interview in years, M (SD)	67 (6)	53 (10)	47 (7)
Sex	–		
Female	–	4 (40)	0 (0)
Male	15 (100)	6 (60)	8 (100)

Table 3. Cont.

Demographics	Patients, n = 15 (%)	GPs, n = 10 (%)	Specialists, n = 8 (%)
Marital status		–	–
Partner	14 (93)		
No partner	1 (7)		
Educational level [a]		–	–
Low	3 (20)		
Intermediate	1 (7)		
High	11 (73)		
Clinical characteristics			
Primary treatment		–	–
Radical prostatectomy	13 (87)		
Radiotherapy	2 (13)		
ADT	1 (7)		
Time since treatment in months, M (range)	20 (17–25)	–	–
LPC risk group [b]		–	–
Low	5 (33)		
Intermediate	5 (33)		
High	5 (33)		
Information healthcare professionals			
Type GP practice	–		–
Duo practice		5 (50)	
Group practice		5 (50)	
Type of healthcare professional	–	–	
Urologist			5 (62)
Radiation Oncologist			2 (25)
Physician Assistant			1 (13)
Type of hospital	–	–	
Academic hospital			1 (12)
Top clinical hospital			3 (38)
Comprehensive cancer center			3 (38)
Community hospital			1 (12)

Abbreviations: GP = general practitioner, ADT = androgen deprivation therapy, M = mean, SD = standard deviation, LPC = localized prostate cancer, – = not applicable. [a] Educational level was classified into low (no, lower (vocational) education), intermediate (secondary vocational education), and high (higher (vocational) education and university); [b] LPC risk group was classified according to the EAU guidelines [3].

Table 4. Quotes of patients, GPs, and specialists about primary care-based prostate cancer follow-up according to the themes from the transcript analysis.

Theme	Quotes (Examples)
Structure of prostate cancer follow-up care	P3: 'Once my wife was also very worried, and then I had my PSA checked because it does not help me if she gets nervous.'
	GP3: 'I see this person more often for all sorts of reasons, so sometimes it happened that I just, ehm, combined it (i.e., follow-up consult) with complaints of his respiratory system or something like that.'
	S8: 'We also offer people a psychologist or social worker, if there is a need. But the physical and oncological examination are the main aspects.'

Table 4. *Cont.*

Theme	Quotes (Examples)
Communication between primary and secondary care	P1: 'No, I did not experience any of that (i.e., communication).'
	GP9: 'In general, it is always difficult to reach a specialist, or you will be called back but not at the moment the patient is with you.'
	S6: 'No, I have never heard anything from the GPs. That shows how redundant we really are, at least for this part (i.e., follow-up).'
Clinical competencies of primary care-based follow-up	P4: 'What the GP did well, I must say, was covering everything ... like, how is it going physically, how is it going psychologically, do you have specific questions at a physical level, about urinary incontinence or erectile dysfunction, or are you tired, or do you still have ... ?'
	GP8: 'Especially information about prognosis, what are the chances that things can come back, I cannot of course, 1,2,3, I do not have those numbers ready of course, no.'
	S2: 'I have actually had no feedback from GPs who said, "Hey, I have a patient here with erectile problems and I am not sure what to do." Or you (i.e., study team) provide GPs with excellent information about this, or they do not have questions, or they do not call us. I am not quite sure.'
Facilitators of primary care-based follow-up	P5: 'And compared to the hospital, you know ... Emotionally that is better. Better to do this (i.e., follow-up) with your GP. And when that is an option, then that is very positive.'
	GP5: 'Well, I think it is very patient-friendly when he does not have to go to the hospital, it will save costs, the effort for me is little, and it is also pleasant for me that I can speak twice a year to someone who had prostate cancer.'
	S3: 'It really results in extra time in which you can take care of people with bigger problems, who really need the hospital setting.'
Barriers to primary care-based follow-up	P13: 'Yes, with the GP you have to undertake action yourself. That is, you know, a GP does not have a system to call people, so if you have complaints you have to go to the GP yourself.'
	GP1: 'The disadvantage is that we do not get one extra penny for it. But I do believe that, uh, primary care is capable of doing this. But then we kind of need ... or then we should receive compensation or extra staffing.'
	S1: 'That is my fear you know, that they (i.e., GPs) will not refer them (i.e., patients) back. Or that they are too late, or not frequently measure their PSA. And then we will lose the window of curability.'
Organizational requirement for the implementation	See text

Abbreviations: P = patient, GP = general practitioner, S = specialist.

3.2. Structure of Prostate Cancer Follow-Up Care

The structure of prostate cancer follow-up as described by the participants aligned with the provided guideline. In both primary and secondary care, follow-up visits focused on PSA measurements. All specialists indicated that they focus primarily on physical problems and that they only discuss psychosocial problems at the patient's request, and they usually refer patients with psychosocial problems to a nurse specialist, the GP, or a psychologist. The two radiation oncologists we interviewed indicated that they often speak with their patients who were treated with hormonal therapy about psychosocial problems (e.g., self-image and relationship issues). All GPs indicated that it was easy for them to discuss the psychosocial aspects of the cancer, but it was more difficult to discuss and manage prostate cancer-specific problems. Some of the GPs combined prostate cancer follow-up with other chronic disease care.

Almost all patients indicated having a good doctor–patient relationship. Some patients thought their follow-up appointments with the GP were less organized than what they had experienced in the hospital and some patients reported receiving more PSA measurements than described in the guideline. Two patients reported not being able to discuss prostate cancer-specific problems with their GP; one patient believed the GP could not help him and one patient struggled (in general) with requesting help. Within this theme, we observed that some patients were worried about the PSA value and therefore consulted the GP more often than the guideline prescribed. Some patients were proactive in consulting their GP, while others reported being more passive.

3.3. Communication between Primary and Secondary Care

None of the specialists experienced problems in the communication with primary care. All GPs indicated that the communication between primary and secondary care should be improved: not all GPs received clinical treatment information from the specialists and direct access to the specialist was perceived as difficult. GPs believed that the communication with specialists should be more accessible and transparent, preferably with one (nurse) specialist per cancer type per hospital. None of the patients was aware of any communication having taken place between primary and secondary care providers.

3.4. Clinical Competencies of Primary Care Follow-Up

Specialists agreed that GPs could take primary responsibility for follow-up care, but only when there are national guidelines in place, when the PSA value is normalized, and when there are no major complications after treatment or prostate cancer-specific problems that require active treatment. Some specialists indicated that they were not sure as to whether GPs are able to manage prostate cancer-specific problems because they had not had any contact or had not received any inquiries from them. In addition, some specialists specifically mentioned that they expect the GP to be more skilled in the area of psychosocial care.

All of the GPs believed that they are competent enough to provide prostate cancer follow-up. Some of the GPs mentioned that it was difficult to answer specific questions about their patients' cancer prognoses and to provide follow-up care to patients who had many prostate cancer-specific problems.

Most of the patients indicated that primary care is the appropriate place for their follow-up, especially when they have few complaints. Nevertheless, most patients mentioned they would like to be referred back to the hospital if they develop more prostate cancer-related symptoms.

3.5. Facilitators of Primary Care-Based Follow-Up

The participants mentioned several advantages of primary care-based follow-up: the GP is closer to home, it is more accessible, it is more efficient and less expensive, GPs might combine cancer follow-up with other chronic disease management, and the hospital can focus on patients who are undergoing active treatment. Patients also mentioned several advantages of primary care: it is easier to make an appointment, the GP spends more time with them, the GP knows the patient better, and there is less emotional burden at the GP's office, whereas the hospital environment can be emotionally upsetting.

3.6. Barriers to Primary Care-Based Follow-Up

Perceived barriers to primary care-based follow-up included limited knowledge and expertise among GPs. Some specialists were also concerned that patients would be lost to follow-up. Most of the GPs indicated having some issues with providing follow-up care, as it may result in an increased caseload of patients and a greater demand on capacity. The majority of the patients indicated no barriers to primary care-based follow-up, despite them having to be more proactive in consulting their GP.

3.7. Organizational Requirements for the Implementation of Primary Care-Based Follow-Up

Throughout the interviews, the participants mentioned several organizational requirements that have to be taken into account before implementing primary care-based follow-up. These requirements are presented in Figure 2.

- Collaboration between GPs, the Dutch College of General Practitioners, hospitals and health insurance companies
- Implementation of multidisciplinary evidence-based guidelines and transmural agreements, specifically including:
    - The first follow-up visit after treatment should take place at the hospital
    - Agreement on whether primary care-based follow-up becomes usual care or choice
    - Agreement on who is primary responsible for follow-up appointments
- Collection of quality indicators (i.e. PSA values, urinary incontinence, erectile function)
- Improvement of communication between primary and secondary care
- Improvement of information provision from secondary to primary care
- Compensation of GPs for the extra caseload
- Development and implementation of clinical pathways for the referral of patients back from primary to secondary care

Figure 2. Organizational requirements necessary for the implementation of primary care-based follow-up.

4. Discussion

We found that the PROSPEC trial reached 402 out of 597 prostate cancer survivors (67%). The reach of this trial was comparable to previously published RCTs investigating primary versus secondary follow-up care among patients with breast and colorectal cancer (accrual rates between 55–67%) [21–23]. This indicates that primary care-based follow-up is acceptable to the majority of cancer patients, and this seems to be similar across cancer types. Nevertheless, this number might be different outside of an RCT setting.

In line with previous studies, our results also indicated that GPs discussed the psychosocial context of cancer with patients, while not all specialists reported doing so [16,24]. Besides, patients mentioned that the GP-patient relationship is more personal. A previous study among prostate cancer survivors also reported that patients rate their GP higher than oncologists in terms of interpersonal relationship [25]. A good doctor–patient relationship is very important and associated with better health outcomes, especially in longitudinal care (seeing the same doctor) and with positive consultation experiences (patients' encounters with the doctor) [26].

Interestingly, we found that some patients were proactive in consulting their GPs for PSA measurements or problems. It is known from previous research that active participation in medical consultations increases commitment to treatment plans, results in better provision of information and support from physicians, and improves satisfaction with care [27]. If we want to implement primary care-based follow-up, it is important that future studies investigate how to support patients to ensure patients receive care according to the follow-up guideline.

As reported by studies before, GPs also argued that a transition of oncology care to primary care would result in a greater demand on capacity [15,28]. If we want primary care to be more involved in the cancer care continuum, it is critical to address this issue of GP workload and compensation.

Some GPs expressed a lack of confidence in managing prostate cancer-specific problems. This is in line with previous research, where GPs indicated having confidence in performing non-cancer tasks such as pain management and psychosocial support, but less confidence in surveillance testing and managing long-term effects of cancer [14,29]. Often, this uncertainty is associated with minimal training in follow-up care and a lack of evidence-based guidelines [14,28,30]. It is, therefore, important to implement (inter-)national multidisciplinary cancer follow-up care guidelines, including up-to-date disease management, for a successful implementation of primary care-based follow-up.

Another reason for a lack of confidence in relation to providing primary care-based follow-up can be the poor communication between primary and secondary care [14,28]. GPs reported that access to specialists and, consequently, access to information is difficult. Future studies should develop and implement intervention strategies aimed at improving the communication and information provision between primary and secondary care.

Limitations of this study include the fact that the results of the RCT on the effectiveness, adoption, and maintenance of primary care-based follow-up care are not yet available. In addition, we note that participants in the qualitative part of the current study had accepted the invitation to participate in our RCT. This, by definition, resulted in a selective sample of patients who most likely had good relationships with their GPs and were positive toward primary care-based follow-up. In addition, the patients we interviewed were relatively healthy, despite our attempts to recruit patients with a wide range of prostate cancer-related problems. Finally, we were unable to investigate attitudes toward primary care-based follow-up developed after sustained experience with it, as the patients, GPs, and specialists were experiencing it for the first time.

Strengths of this study include the use of the CFIR implementation research framework. This framework allows us to systematically evaluate a complex intervention and will contribute to a successful implementation of primary-based prostate cancer follow-up care [17]. In addition, the interviews were transcribed verbatim and analyzed by different researchers using a thematic analysis to assure the robustness of the findings. Furthermore, we aimed to include a heterogeneous population by including GPs and specialists working in different types of hospitals and practice sizes. Finally, we included both GPs who were in favor of and those who were critical toward primary care-based follow-up.

5. Conclusions

In conclusion, the results of this process evaluation indicate that patients, GPs, and specialists are positive toward primary care-based follow-up. Most of the participants indicated that primary care could make prostate cancer follow-up care more personal, efficient, and sustainable. If the RCT shows that primary care is equally as effective and safe as specialist-based follow-up, this process evaluation can be used to understand how the intervention is working, what challenges need to be overcome, and which requirements are necessary to enable successful implementation of prostate cancer follow-up in primary care.

Supplementary Materials: The following supporting information can be downloaded at https://www.mdpi.com/article/10.3390/cancers14133166/s1, Table S1: Interview guide based on CFIR domains (18); Table S2: Themes mapped to the CFIR framework.

Author Contributions: All authors contributed to the study's conception and design. Data collection and analysis: B.M.W., E.A., S.C., A.H.B. Interpretation of data: all authors. Drafting of the manuscript: B.M.W. Critical revision of the manuscript: B.M.W., K.M.v.A., F.J.P., E.A., S.C., H.G.v.d.P., N.K.A., L.V.v.d.P.-F. and A.H.B. All authors have read and agreed to the published version of the manuscript.

Funding: This work is funded by the Dutch Cancer Society (Delfandlaan 17, 1062 EA, Amsterdam, The Netherlands), grant number NKI 2015-7932.

Institutional Review Board Statement: All procedures performed in studies involving human participants were in accordance with the ethical standards of the institutional and/or national research committee and with the 1964 Helsinki Declaration and its later amendments or comparable ethical standards. Ethical approval was obtained from the institutional review board of the Antoni van Leeuwenhoek Hospital, a comprehensive cancer center located in Amsterdam, the Netherlands (IRBd19-251).

Informed Consent Statement: All participants signed written informed consents before participating in the study.

Data Availability Statement: The dataset used and analyzed during the current study will be available from the corresponding authors (stored in a data repository at the Netherlands Cancer Institute) on reasonable request.

Conflicts of Interest: The authors declare no conflict of interest.

References

1. Carioli, G.; Bertuccio, P.; Boffetta, P.; Levi, F.; La Vecchia, C.; Negri, E.; Malvezzi, M. European cancer mortality predictions for the year 2020 with a focus on prostate cancer. *Ann. Oncol.* **2020**, *31*, 650–658. [CrossRef] [PubMed]
2. Urquhart, R.; Cordoba, W.; Bender, J.; Cuthbert, C.; Easley, J.; Howell, D.; Kaal, J.; Kendell, C.; Radford, S.; Sussman, J. Risk Stratification and Cancer Follow-Up: Towards More Personalized Post-Treatment Care in Canada. *Curr. Oncol.* **2022**, *29*, 3215–3223. [CrossRef] [PubMed]
3. Mottet, N.; Cornford, P.; van den Bergh, R.C.N.; Briers, E.; De Santis, M.; Gillessen, S.; Grummet, J.; Henry, A.M.; van der Kwast, T.H.; Lam, T.B.; et al. Guidelines on Prostate Cancer. 2021. Available online: https://uroweb.org/guideline/prostate-cancer/ (accessed on 11 April 2022).
4. Nekhlyudov, L.; Mollica, M.A.; Jacobsen, P.B.; Mayer, D.K.; Shulman, L.N.; Geiger, A.M. Developing a Quality of Cancer Survivorship Care Framework: Implications for Clinical Care, Research, and Policy. *J. Natl. Cancer Inst.* **2019**, *111*, 1120–1130. [CrossRef] [PubMed]
5. Milonas, D.; Ruzgas, T.; Venclovas, Z.; Jievaltas, M.; Joniau, S. Impact of Grade Groups on Prostate Cancer-Specific and Other-Cause Mortality: Competing Risk Analysis from a Large Single Institution Series. *Cancers* **2021**, *13*, 1963. [CrossRef]
6. Wollersheim, B.M.; van Asselt, K.M.; van der Poel, H.G.; van Weert, H.; Hauptmann, M.; Retèl, V.P.; Aaronson, N.K.; van de Poll-Franse, L.V.; Boekhout, A.H. Design of the PROstate cancer follow-up care in Secondary and Primary hEalth Care study (PROSPEC): A randomized controlled trial to evaluate the effectiveness of primary care-based follow-up of localized prostate cancer survivors. *BMC Cancer* **2020**, *20*, 635. [CrossRef]
7. Hulscher, M.; Laurant, M.; Grol, R. Process evaluation of implementation strategies. In *Improving Patient Care: The Implementation of Change in Health Care*; Grol, R., Wensing, M., Eccles, M., Davis, D., Eds.; John Wiley & Sons: Hoboken, NJ, USA, 2013; pp. 333–349.
8. Oakley, A.; Strange, V.; Bonell, C.; Allen, E.; Stephenson, J. Process evaluation in randomised controlled trials of complex interventions. *BMJ* **2006**, *332*, 413–416. [CrossRef]
9. Grol, R.P.; Bosch, M.C.; Hulscher, M.E.; Eccles, M.P.; Wensing, M. Planning and studying improvement in patient care: The use of theoretical perspectives. *Milbank Q.* **2007**, *85*, 93–138. [CrossRef]
10. Huibertse, L.J.; van Eenbergen, M.; de Rooij, B.H.; Bastiaens, M.T.; Fossion, L.M.; de la Fuente, R.B.; Kil, P.J.; Koldewijn, E.L.; Meier, A.H.; Mommers, R.J.; et al. Cancer survivors' preference for follow-up care providers: A cross-sectional study from the population-based PROFILES-registry. *Acta Oncol.* **2017**, *56*, 278–287. [CrossRef]
11. Brandenbarg, D.; Roorda, C.; Stadlander, M.; de Bock, G.H.; Berger, M.Y.; Berendsen, A.J. Patients' views on general practitioners' role during treatment and follow-up of colorectal cancer: A qualitative study. *Fam. Pract.* **2017**, *34*, 234–238. [CrossRef]
12. Smith, T.G.; Strollo, S.; Hu, X.; Earle, C.C.; Leach, C.R.; Nekhlyudov, L. Understanding Long-Term Cancer Survivors' Preferences for Ongoing Medical Care. *J. Gen. Intern. Med.* **2019**, *34*, 2091–2097. [CrossRef]
13. Mayo, S.J.; Ajaj, R.; Drury, A. Survivors' preferences for the organization and delivery of supportive care after treatment: An integrative review. *Eur. J. Oncol. Nurs.* **2021**, *54*, 102040. [CrossRef] [PubMed]
14. Lawrence, R.A.; McLoone, J.K.; Wakefield, C.E.; Cohn, R.J. Primary Care Physicians' Perspectives of Their Role in Cancer Care: A Systematic Review. *J. Gen. Intern. Med.* **2016**, *31*, 1222–1236. [CrossRef] [PubMed]
15. Fidjeland, H.L.; Brekke, M.; Vistad, I. General practitioners' attitudes toward follow-up after cancer treatment: A cross-sectional questionnaire study. *Scand. J. Prim. Health Care* **2015**, *33*, 223–232. [CrossRef]
16. Wind, J.; Duineveld, L.A.; van der Heijden, R.P.; van Asselt, K.M.; Bemelman, W.A.; van Weert, H.C. Follow-up after colon cancer treatment in the Netherlands; a survey of patients, GPs, and colorectal surgeons. *Eur. J. Surg. Oncol. J. Eur. Soc. Surg. Oncol. Br. Assoc. Surg. Oncol.* **2013**, *39*, 837–843. [CrossRef] [PubMed]
17. Damschroder, L.J.; Aron, D.C.; Keith, R.E.; Kirsh, S.R.; Alexander, J.A.; Lowery, J.C. Fostering implementation of health services research findings into practice: A consolidated framework for advancing implementation science. *Implement. Sci. IS* **2009**, *4*, 50–65. [CrossRef]
18. Braun, V.; Clarke, V. Using thematic analysis in psychology. *Qual. Res. Psychol.* **2006**, *3*, 77–101. [CrossRef]
19. Huang, R. RQDA: R-Based Qualitative Data Analysis. Available online: http://rqda.r-forge.r-project.org (accessed on 24 February 2021).
20. R Core Team. R: A Language and Environment for Statistical Computing. Available online: https://www.r-project.org/ (accessed on 24 February 2021).
21. Grunfeld, E.; Levine, M.N.; Julian, J.A.; Coyle, D.; Szechtman, B.; Mirsky, D.; Verma, S.; Dent, S.; Sawka, C.; Pritchard, K.I.; et al. Randomized trial of long-term follow-up for early-stage breast cancer: A comparison of family physician versus specialist care. *J. Clin. Oncol.* **2006**, *24*, 848–855. [CrossRef]

22. Grunfeld, E.; Mant, D.; Yudkin, P.; Adewuyi-Dalton, R.; Cole, D.; Stewart, J.; Fitzpatrick, R.; Vessey, M. Routine follow up of breast cancer in primary care: Randomised trial. *BMJ (Clin. Res. Ed.)* **1996**, *313*, 665–669. [CrossRef]
23. Wattchow, D.A.; Weller, D.P.; Esterman, A.; Pilotto, L.S.; McGorm, K.; Hammett, Z.; Platell, C.; Silagy, C. General practice vs. surgical-based follow-up for patients with colon cancer: Randomised controlled trial. *Br. J. Cancer* **2006**, *94*, 1116–1121. [CrossRef]
24. Wollersheim, B.M.; Helweg, E.; Tillier, C.N.; van Muilekom, H.A.M.; de Blok, W.; van der Poel, H.G.; van Asselt, K.M.; Boekhout, A.H. The role of routine follow-up visits of prostate cancer survivors in addressing supportive care and information needs: A qualitative observational study. *Supportive Care Cancer* **2021**, *29*, 6449–6457. [CrossRef]
25. Hudson, S.V.; Ohman-Strickland, P.A.; Bator, A.; O'Malley, D.; Gundersen, D.; Lee, H.S.; Crabtree, B.F.; Miller, S.M. Breast and prostate cancer survivors' experiences of patient-centered cancer follow-up care from primary care physicians and oncologists. *J. Cancer Surviv. Res. Pract.* **2016**, *10*, 906–914. [CrossRef] [PubMed]
26. Ridd, M.; Shaw, A.; Lewis, G.; Salisbury, C. The patient-doctor relationship: A synthesis of the qualitative literature on patients' perspectives. *Br. J. Gen. Pract. J. R. Coll. Gen. Pract.* **2009**, *59*, e116–e133. [CrossRef] [PubMed]
27. Street, R.L., Jr.; Gordon, H.S.; Ward, M.M.; Krupat, E.; Kravitz, R.L. Patient participation in medical consultations: Why some patients are more involved than others. *Med. Care* **2005**, *43*, 960–969. [CrossRef]
28. Margariti, C.; Gannon, K.N.; Walsh, J.J.; Green, J.S.A. GP experience and understandings of providing follow-up care in prostate cancer survivors in England. *Health Soc. Care Community* **2020**, *28*, 1468–1478. [CrossRef] [PubMed]
29. Geramita, E.M.; Parker, I.R.; Brufsky, J.W.; Diergaarde, B.; van Londen, G.J. Primary Care Providers' Knowledge, Attitudes, Beliefs, and Practices Regarding Their Preparedness to Provide Cancer Survivorship Care. *J. Cancer Educ.* **2020**, *35*, 1219–1226. [CrossRef]
30. Taylor, S.; Johnson, H.; Peat, S.; Booker, J.; Yorke, J. Exploring the experiences of patients, general practitioners and oncologists of prostate cancer follow-up: A qualitative interview study. *Eur. J. Oncol. Nurs.* **2020**, *48*, 101820. [CrossRef]

cancers

MDPI

Article

Variables Associated with False-Positive PSA Results: A Cohort Study with Real-World Data

Blanca Lumbreras [1,2,*], Lucy Anne Parker [1,2], Juan Pablo Caballero-Romeu [3,4], Luis Gómez-Pérez [5,6], Marta Puig-García [1,2], Maite López-Garrigós [2,7], Nuria García [3] and Ildefonso Hernández-Aguado [1,2]

1 Department of Public Health, University Miguel Hernández de Elche, 03550 Alicante, Spain
2 CIBER de Epidemiología y Salud Pública (CIBERESP), 28029 Madrid, Spain
3 Department of Urology, University General Hospital of Alicante, 03010 Alicante, Spain
4 Alicante Institute for Health and Biomedical Research (ISABIAL), 03010 Alicante, Spain
5 Urology Department, General University Hospital of Elche, 03203 Elche, Spain
6 Pathology and Surgery Department, Miguel Hernández University of Elche, 03550 Alicante, Spain
7 Clinical Laboratory Department, University Hospital of San Juan de Alicante, Sant Joan d'Alacant, 03550 Alicante, Spain
* Correspondence: blumbreras@umh.es; Tel.: +34-965-919-510

Simple Summary: Controversy exists regarding prostate cancer (PC) screening using the prostate-specific antigen (PSA) test. It may reduce PC mortality risk but is associated with false-positive results. We aimed to evaluate the incidence of false-positive and false-negative results in a general clinical setting and the associated variables. We found a high rate of false-positive results (46.6%), resulting in a positive predictive value of 12.7%. Patients also showed a low rate of false-negative results (3.7%) with a negative predictive value of 99.5%. Age, alcohol intake, and having a urinary tract infection were associated with a higher probability of false-positive results; having diabetes mellitus type II was associated with a lower rate of false-positive results. This study showed a higher rate of false-positive results in clinical practice than in previous clinical trials, mainly in patients over 60 years.

Abstract: (1) Background: There are no real-world data evaluating the incidence of false-positive results. We analyzed the clinical and analytical factors associated with the presence of false-positive results in PSA determinations in practice. (2) Methods: A prospective cohort study of patients with a PSA test was performed in clinical practice. We followed the patients by reviewing their medical records for 2 years or until the diagnosis of PCa was reached, whichever came first. (3) Results: False-positive PSA rate was 46.8% (95% CI 44.2–49.2%) and false-negative PSA rate was 2.8% (95% CI 2–3.5%). Patients aged 61–70 years and those over 70 years were more likely to have a false-positive result than those under 45 years (aOR 2.83, 95% CI 1.06–7.55, $p = 0.038$, and aOR 4.62, 95% CI 1.75–12.22, $p = 0.002$, respectively). Patients with urinary tract infection were more likely to have a false-positive result (aOR 8.42, 95% CI 2.42–29.34, $p = 0.001$). Patients with diabetes mellitus were less likely to have a false-positive result (aOR 0.63, 95% CI 0.41–0.98, $p = 0.038$); (4) Conclusions: This study has generated relevant information that could be very useful for shared decision making in clinical practice.

Keywords: prostate-specific antigen; prostate cancer; false-positive results; real world-data

Citation: Lumbreras, B.; Parker, L.A.; Caballero-Romeu, J.P.; Gómez-Pérez, L.; Puig-García, M.; López-Garrigós, M.; García, N.; Hernández-Aguado, I. Variables Associated with False-Positive PSA Results: A Cohort Study with Real-World Data. *Cancers* **2023**, *15*, 261. https://doi.org/10.3390/cancers15010261

Academic Editor: José I. López

Received: 16 November 2022
Revised: 15 December 2022
Accepted: 23 December 2022
Published: 30 December 2022

1. Introduction

In recent years there has been a debate regarding the benefits and harms of PSA-based screening for prostate cancer (PCa) [1,2]. A systematic review of randomized controlled trials published in 2013 showed that PSA-based PCa screening did not significantly decrease PCa-specific mortality. In addition, harms related to the presence of overdiagnosis, overtreatment, and false-positive results were frequent [3]. Concerning the presence of false-positive results, previous evidence has also shown variations in PSA levels according to age [4], although there is no evidence of their implications in practice. A recent systematic review [5] showed that diabetes was significantly associated with lower PSA levels among asymptomatic men. Though, differences were small and unlikely to influence PCa detection. Similarly, treatments such as statins [6], metformin [7], or treatment for benign prostatic hypertrophy [8] can also affect PSA results leading to false-positive or -negative results. According to previous studies carried out in clinical trials [9], 10–12% of men undergoing regular PSA testing will experience a false-positive result. False-positive results can have a major impact on the clinical management and outcome of the patient, mainly due to possible adverse effects related to the diagnostic process (biopsy, surgery, and treatment). Biopsies, for instance, can cause infections as well as very serious complications such as urinary incontinence and sexual dysfunction [10].

These data led some international organizations, such as the US Preventive Services Task Force (UPSTF) in 2012, to establish recommendations against PSA-based screening for PCa [11]. In 2018, based on longer follow-up data from large screening trials, the UPSTF guideline was updated [12]. They stated that men aged 55–69 years should be informed about the benefits and harms of PCa screening with a grade C recommendation (the USPSTF recommends selectively offering or providing this service to individual patients based on professional judgment and patient preferences), while grade D (the USPSTF recommends against the service) was assigned to screening for men over 70 years. Aiming to reduce the harms related to false-positive results and overdiagnosis, the European Association of Urology guidelines have been recently updated [13] and recommend offering an individualized risk-adapted approach for the early detection of PCa to well-informed men older than 50 years old with at least 15 years of life expectancy.

However, most of these recommendations are based on randomized clinical trials, which include populations with different characteristics from those patients in clinical practice [14] (clinical trials tend to include healthier and younger patients and often represent a highly selected population of patients). Indeed, PSA levels in patients attending urology services are higher than those patients of the same age who participate in screening programs [15]. Therefore, both the baseline levels of the marker and the determinants that may affect a false-negative or false-positive result should be determined for each setting.

The risk-adapted approach for the early detection of PCa, as the European Association proposes, should be developed along with the evaluation of the probability of a patient having a false-positive or -negative result according to his individual characteristics. Nevertheless, at present, there are no real-world data evaluating the incidence of false-positive and -negative results in practice and the variables associated with them.

We aimed to evaluate the incidence of false-positive and false-negative results in a general clinical setting, including patients undergoing opportunistic screening or with symptoms suggestive of disease, and the variables associated.

2. Materials and Methods

This study was carried out and reported according to the Strengthening the Reporting of Observational Studies in Epidemiology (STROBE) Statement [16].

2.1. Study Design

A prospective observational cohort study of patients with a PSA determination for the early detection of PCa or in the presence of prostatic symptoms in general practice. The study protocol is registered at https://clinicaltrials.gov (ClinicalTrials.gov Identifier: NCT03978299, accessed on the 11 June 2019) and has been previously published [17].

2.2. Setting

The target population of the study were the residents of the catchment area of the two participating Health Departments 17 and 19, in the Valencian Community (these include 20 primary health centers and 2 hospitals: General University Hospital of Sant Joan d'Alacant and General University Hospital of Alicante, respectively). These are referral hospitals for all individuals living in their catchment areas and belong to the National Health System (NHS). Most of the Spanish population uses it as the main medical service (the publicly funded insurance scheme covers 98.5% of the Spanish population).

2.3. Study Population

We included men over 18 years with a PSA determination requested in a routine health examination from January to July 2018. Patients with a previous diagnosis of PCa or those who were being monitored for previous high PSA values were excluded.

We randomly selected a sample of patients with positive PSA results from within our cohort (total PSA value >10 mg/L or a total PSA between 4 and 10 mg/L if the value of the free PSA/total PSA fraction is <25% in at least two determinations) and a sample of patients with negative results (total PSA value is < 4 mg/L or a total PSA between 4 and 10 mg/L if the value of the free PSA/total PSA fraction is >25%) among a consecutive cohort of individuals undergoing PSA testing that have been described previously [18]. This previous study aimed to evaluate the potential non-compliance of PSA testing with current guidelines in general practice. PSA determinations are performed centrally in the laboratories of these two hospitals using the same protocol. A blood sample is extracted, and after centrifugation (15 min), the PSA level is determined in the serum using the chemiluminescent assay technique (the analyte is stable for 4 days at 2–8 °C). The detection limit of the technique is 0.05 ng/mL.

2.4. Study Size and Recruitment Procedure

The predictive positive value of PSA is estimated at 21%, and the predictive negative value at 91%, according to previous evidence [19]. Given that we included symptomatic and asymptomatic patients, we estimated a prevalence of PCa in our study higher than 5%. We calculated that we needed to include at least 665 patients with a negative PSA result and 690 PSA-positive patients (for a precision of 2% with 95% CIs). Considering the potential loss during follow-up, we increased this sample by 20%.

2.5. Data Collection

Both hospitals have digital medical records from which data were extracted.

We recorded the following data from medical records: demographic characteristics (age) and clinical characteristics (patient who had the PSA test as part of opportunistic screening or due to the presence of symptoms suggestive of disease), toxic habits (smoking habit and alcohol consumption), family history of PCa, any pharmacological treatment prescribed and specifically treatment with statins, metformin, treatment for BPH, diuretics and ASA, PSA values, anthropometric measures, and other comorbidities.

2.6. Cohort Follow-Up

We followed both cohorts (positive and negative PSA results) for 2 years by reviewing their medical records (every 3 months for patients with a positive PSA result and annually for patients with a negative result) until the diagnosis of PCa or end of follow-up, whichever came first. We also collected results from digital rectal exams and biopsies. For those patients with a diagnosis of PCa, data regarding the Gleason score and the International Society of Urological Pathology grade were also collected.

2.7. Outcomes

A false-positive result was defined if the determination of serum PSA was positive and the result of digital rectal examination(s) and/or subsequent biopsy or biopsies were negative, according to the recommendations of the European Association of Urology [20]. A false-negative result was defined if the determination of serum PSA was negative and the patient was diagnosed with PCa within 2 years.

2.8. Statistical Analysis

We used descriptive statistics to summarize the population and calculated the proportion of false-positive and false-negative results for the diagnosis of PCa and the associated variables (Chi-squared test). To analyze clinical and analytical factors associated with the presence of false-positive results in PSA determinations, we calculated odds ratios and their 95% confidence interval with logistic regression. Independent variables included were urinary tract infection, diabetes mellitus type II, hyperlipidemia, age, alcohol consumption, family history of prostate cancer, and prescription of any pharmacological treatment.

The analysis was performed using the Stata IC 15 program. All p values and CIs were two-sided, 95%.

3. Results

3.1. Description of The Patients Included in Both Cohorts and Accuracy of the PSA Test

Of the 1664 patients included in the study, 833 (51%) had a positive PSA result, and 831 (49%) had a negative PSA result. Out of 833 patients with a positive PSA result, 106 (97.2%) had a diagnosis of PCa; of the 831 patients with a negative PSA result, 3 (2.8%) had a diagnosis of PCa. The positive predictive value (PPV) of the PSA determination was 12.7% (95% CI 10.4–15%), and the negative predictive value (NPV) was 99.6% (95% CI 99.2–100%). The false-positive PSA rate was 46.8% (95% CI 44.2–49.2%), and the false-negative PSA rate was 2.8% (95% CI 2–3.5%).

Median follow-up was 19.51 months (IQR 14.39–22.24), and median follow-up until the patient was diagnosed with PCa was 5 months (IQR 11–16).

Of the 1664 patients included in the study, 33 (1.9%) had a family history of PCa and 2 of them (6.1%) developed PCa, 169 (10.1%) had no family history of prostate cancer, and 19 (11.2%) developed cancer; however, no data on family history of PCa were available for 1462 patients (87.9%).

3.2. False-Positive Rate and Associated Sociodemographic and Clinical Characteristics

Table 1 shows the distribution of the number of patients with PCa and the false-positive PSA rates, according to the different sociodemographic and clinical variables, for symptomatic and asymptomatic patients.

Overall, there were differences in the false-positive rate according to patients' age: the rate increased from 26.3% in patients younger than 45 years to 55.2% in patients over 70 years ($p < 0.001$). In asymptomatic patients, a difference in the false-positive rate according to patients' age was also found ($p < 0.001$).

Most patients had a previous PSA (1347/1665; 80.9%), and they were less likely to have a false-positive result than patients who had a PSA test for the first time (45.1% vs. 53.8%, $p = 0.008$). In asymptomatic patients, a difference in the false-positive rate according to having a previous PSA was also found ($p = 0.023$).

Patients who never had drunk alcohol (386/1665; 23.2%) were less likely to have a false-positive result (46.2%) than those who were current (56.6%) or ex-drinkers (58.5%) ($p = 0.014$). In asymptomatic patients, a difference in the false-positive rate according to alcohol intake was also found ($p = 0.004$).

3.3. False-Positive Rate Associated with The Presence of Comorbidities

Table 2 shows the distribution of the number of patients with PCa and the false-positive PSA rates, according to the different patients' comorbidities, for symptomatic and asymptomatic patients.

Table 1. Description of the sociodemographic and clinical characteristics associated with false-positive rates for both symptomatic and asymptomatic patients.

	Total			Symptomatic			Asymptomatic		
Variables	**Total Cancer/Patients (109/1664; 6.5%)**	**False-Positive PSA Rate [1] (728/1555; 46.8%)**	***p*-Value [2]**	**Total Cancer/Patients (17/303; 5.6)**	**False-Positive PSA Rate [1] (137/286; 47.9)**	***p*-Value [2]**	**Total Cancer/Patients (92/1361; 6.8)**	**False-Positive PSA Rate [1] (591/1269; 46.6)**	***p*-Value [2]**
Age			<0.001			0.070			<0.001
<45	0/57; 0	15/57; 26.3		0/19; 0	5/19; 26.3		0/38; 0	10/38; 26.3	
45–50	1/81; 1.2	22/80; 27.5		1/14; 7.1	6/13; 46.2		0/67; 0	16/67; 23.9	
51–60	12/369; 3.3	139/357; 38.9		3/59; 5.1	21/56; 37.5		9/310; 2.9	118/301; 39.2	
61–70	51/527; 9.7	229/476; 48.1		5/98; 5.1	52/93; 55.9		46/429; 10.7	177/383; 46.2	
>70	45/630; 7.1	323/585; 55.2		8/113; 7.1	53/105; 50.5		37/517; 7.2	270/480; 56.3	
PSA previous			0.008			0.147			0.023
No	29/315; 9.2	154/286; 53.8		6/55; 10.9	28/49; 57.1		23/260; 8.8	126/237; 53.2	
Yes	80/1347; 5.9	572/1267; 45.1		11/247; 4.5	108/236; 45.8		69/1100; 6.3	464/1031; 45	
Family history of PCa			0.003			0.899			0.002
No	19/169; 11.2	89/150; 59.3		3/34; 8.8	16/31; 51.6		16/135; 11.9	73/119; 61.3	
Yes	2/33; 6.1	17/31; 54.8		1/9; 11.1	4/8; 50		1/24; 4.2	13/23; 56.5	
Unknown	88/1462; 6	622/1374; 45.3		13/260; 5	117/247; 47.4		75/1202; 6.2	505/1127; 44.8	
Tobacco			0.274			0.056			0.677
Never smoker	21/377; 5.6	184/356; 51.7		4/69; 5.8	39/65; 60		17/308; 5.5	145/291; 49.8	
Current smoker	27/362; 7.5	154/335; 46		4/66; 6.1	24/62; 38.7		23/296; 7.8	130/273; 47.6	
Ex-smoker	40/546; 7.3	239/506; 47.2		6/103; 5.8	49/97; 50.5		34/44	190/409; 46.5	
Alcohol			0.014			0.836			0.004
Never drinker	22/386; 5.7	168/364; 46.2		5/75; 6.7	38/70; 54.3		17/311; 5.5	130/294; 44.2	
Current drinker	26/346; 7.5	181/320; 56.6		3/69; 4.3	36/66; 54.5		23/277; 8.3	145/254; 57.1	
Ex-drinker	3/56; 5.4	31/53; 58.5		0/7; 0	3/7; 42.9		3/49/6.1	28/46; 60.9	

[1] False-positive rate: (False-positive results/patients without cancer) * 100; [2] *p*-value: Differences in the false-positive PSA rate.

Table 2. Description of the comorbidities associated with false-positive rates for both symptomatic and asymptomatic patients.

Variables	Total			With Symptoms			Without Symptoms		
	Total Cancer/Patients (109/1664; 6.5%)	**False-Positive PSA Rate [1] (728/1555; 46.8%)**	***p*-Value [2]**	**Total Cancer/Patients (17/303; 5.6)**	**False-Positive PSA Rate [1] (137/286; 47.9)**	***p*-Value [2]**	**Total Cancer/Patients (92/1361; 6.8)**	**False-Positive PSA Rate [1] (591/1269; 46.6)**	***p*-Value [2]**
Cardiovascular Disease			0.702			0.846			0.637
No	103/1508; 6.8	660/1405; 47		15/281; 5.3	127/266; 47.7		88/1227; 7.2	533/1139; 46.8	
Yes	6/156; 3.8	68/150; 45.3		2/22; 9.1	10/20; 50		4/134; 3	58/130; 44.6	
Hyperlipidemia			0.016			0.814			0.012
No	96/1391; 6.9	624/1295; 48.2		16/263; 6.1	119/247; 48.2		80/1128; 7.1	505/1048; 48.2	
Yes	13/273; 4.8	104/260; 40		1/40; 2.5	18/39; 46.2		12/233; 5.2	86/221;38.9	
Type II Diabetes Mellitus			0.002			0.008			0.019
No	98/1450; 6.8	654/1352; 48.4		15/275; 5.5	131/260; 50.4		83/1175; 7.1	523/1092; 47.9	
Yes	11/214; 5.1	74/203; 36.5		2/28; 7.1	6/26; 23.1		9/186; 4.8	68/177: 38.4	
Urinary tract infection			<0.001			0.007			<0.001
No	108/1617; 6.7	691/1509; 45.8		17/293; 5.8	128/46.4; 46.4		91/1324; 6.9	563/1233; 45.7	
Yes	1/47; 2.1	37/46; 80.4		0	9/10; 90		1/37; 2.7	28/36; 77.8	
IMC > 30			0.310			0.927			0.348
No	48/667; 7.2	291/619; 47		7/127; 5.5	52/120; 43.3		41/540; 7.6	239/499; 47.9	
Yes	20/333; 6	137/313; 43.8		2/56; 3.6	23/54; 42.6		18/277; 6.5	114/259; 44%	
HTA			0.175			0.590			0.212
No	18/352; 5.1	164/334; 49.1		2/70; 2.9	33/68; 48.5		16/282; 5.7	131/266; 49.2	
Yes	82/1154; 7.1	481/1072; 44.9		14/204; 6.9	85/190; 44.7		68/950; 7.2	396/882; 44.9	

[1] False-positive rate: (False-positive results/patients without cancer) * 100; [2] *p*-value: Differences in the false-positive PSA rate.

Patients with hyperlipidemia were less likely to have a false-positive result (40% vs. 48.2%, $p = 0.016$). In asymptomatic patients, a difference in the false-positive rate according to the presence of hyperlipidemia was also found ($p = 0.012$). The presence of diabetes mellitus type II was also associated with a lower probability of having a false-positive result (36.5% vs. 48.4%, $p = 0.002$). Both symptomatic and asymptomatic patients with diabetes mellitus type II had a lower probability of having a false positive than those without diabetes ($p = 0.008$ and $p = 0.019$, respectively). In contrast, the presence of urinary tract infection was associated with a higher probability of false-positive results (80.4% vs. 45.8%, $p < 0.001$) for both symptomatic ($p < 0.001$) and asymptomatic ($p < 0.001$) patients.

3.4. False-Positive Rate Associated with the Prescription of Medication

Patients with a prescribed pharmacological treatment were less likely to have a false-positive result (44.8% vs. 54.1%, $p = 0.003$). In asymptomatic patients, a difference in the false-positive rate according to the prescription of pharmacological treatment was also found ($p = 0.012$) (Table 3). However, there were no differences in the false-positive rate according to the specific pharmacological treatments.

In multivariable analysis, patients with urinary tract infections were more likely to have a false-positive result (aOR 8.42, 95% CI 2.42–29.34, $p = 0.001$). This was also the case for patients who were current drinkers in comparison with never drinkers (aOR 1.48, 95% CI 1.10–2.02, $p = 0.011$). Patients with diabetes mellitus were less likely to have a false-positive result (aOR 0.63, 95% CI 0.41–0.98, $p = 0.038$). Moreover, patients aged 61–70 years and those over 70 years were more likely to have a false-positive result than those under 45 years (aOR 2.83, 95% CI 1.06–7.55, $p = 0.038$, and aOR 4.62, 95% CI 1.75–12.22, $p = 0.002$, respectively).

3.5. Description of the PCa Cases and their Main Characteristics

Of the 109 patients with PCa, 94 (86.2%) were able to be classified according to the Gleason score: 31 (33%) were classified as Gleason 6, 36 (38.3%) as Gleason 7, and 27 (28.7%) as Gleason 8–10. There were statistically significant differences in the median PSA value for each Gleason category ($p = 0.006$) (Table 4).

According to the International Society of Urological Pathology grade, 43 (39.4%) patients were classified as: grade 1, 22 (51.2%); grade 2, 7 (16.3%); grade 3, 10 (23.3%); grade 4, 5 (11.6%), and grade 5, 1 (2.3%). There were also differences in the median PSA values according to this classification: from a PSA median of 4 (IQR 1.67–6.04) in patients with grade 1 to a PSA median of 10.54 in patients with grade 5 ($p < 0.001$).

Of the 1665 patients included in the study, 303 (18.2%) had symptoms suggestive of prostate disease, and 17 (5.6%) developed PCa; of the 1361 (81.8%) patients without symptoms, 92 (6.8%) developed PCa. Cancer patients who had been symptomatic at the time of the PSA test had a Gleason score of 6–7 (7, 46.7%) and a Gleason score of 8–10 (8, 53.3%). At the same time, those who had been asymptomatic had a Gleason score < 6 (6, 7.7%), a Gleason score of 6–7 (54, 68.4%), and a Gleason score of 8–10 (19, 24.4%), $p = 0.059$.

In patients who did not have a PCa diagnosis, there were differences in PSA level according to age ($p = 0.001$); these differences were not found in patients with PCa ($p = 0.072$) (Figure 1).

In patients without PCa, having hyperlipidemia, diabetes, or a prescribed pharmacological treatment was associated with lower PSA levels ($p = 0.009$, $p = 0.001$, and $p < 0.001$, respectively); having an infection of the urinary tract was associated with higher PSA levels (<0.001). These differences were not observed in patients with PCa (Table 5).

Table 3. Description of the pharmacological treatments associated with false-positive rate for both symptomatic and asymptomatic patients.

	Total			Symptomatic			Asymptomatic		
Variables	**Total Cancer/Patients (109/1664; 6.5%)**	**False-Positive PSA Rate [1] (728/1555; 46.8%)**	***p*-value [2]**	**Total Cancer/Patients (17/303; 5.6)**	**False-Positive PSA Rate [1] (137/286; 47.9)**	***p*-value [2]**	**Total Cancer/Patients (92/1361; 6.8)**	**False-Positive PSA Rate [1] (591/1269; 46.6)**	***p*-Value [2]**
Treatment			0.003			0.611			0.003
No	1/334; 0.3	180/333; 54.1		0/59; 0	30/59; 50.8		1/275; 0.4	150/274; 54.7	
Yes	108/1330; 8.1	548/1222; 44.8		17/244; 7	107/227; 47.1		91/1086; 8.4	441/995; 44.3	
Statins			0.138			0.122			0.363
No	82/1201; 6.8	537/1119; 48		11/218; 5	105/207; 50.7		71/983; 7.2	432/912; 47.4	
Yes	27/463; 5.8	191/436; 43.8		6/85; 7.1	32/79; 40.5		21/378; 5.6	159/357; 44.5	
Metformin			0.415			0.059			0.979
No	101/1456; 6.1	629/1355; 46.4		16/267; 6	115/251; 45.8		85/1189; 7.1	514/1104; 46.6	
Yes	8/208; 3.8	99/200; 49.5		1/36; 2.8	22/35; 62.9		7/172; 4.1	77/165; 46.7	
Treatment for BPH			0.082			0.497			0.103
No	70/1305; 5.4	592/1235; 47.9		12/226; 5.3	105/214; 49.1		58/1079; 5.4	487/1021; 47.7	
Yes	39/359; 10.9	136/320; 42.5		5/77; 6.5	32/72; 44.4		34/282; 12.1	104/248; 41.9	
Diuretic			0.245			0.801			0.157
No	96/1482; 6.5	656/47.3		16/270; 5.9	121/254; 47.6		80/1212; 6.6	535/1132; 47.3	
Yes	13/182; 7.1	72/169; 42.6		1/33/3	16/32; 50		12/149; 8.1	56/137; 40.9	
ASA			0.234			0.481			0.329
No	95/1412; 6.7	625/1317; 47.5		11/266; 4.1	124/255; 48.6		84/1146; 7.3	501/1062; 47.2	
Yes	14/252; 5.6	103/238; 43.3		6/37; 16.2	13/31; 41.9		8/215; 3.7	90/207; 43.5	

[1] False-positive rate: (False-positive results/patients without cancer) * 100; [2] *p*-value: Differences in the false-positive PSA rate.

Table 4. Description of the PSA levels (ng/mL) in those patients with PCa according to Gleason classification.

	Gleason			*p*-Value
PSA Levels	**6**	**7**	**8–10**	
PSA (ng/mL) (median, IQR)	6.66 (4.91–10.13)	6.32 (5.15–8.89)	13.37 (6.44–31.24)	0.004

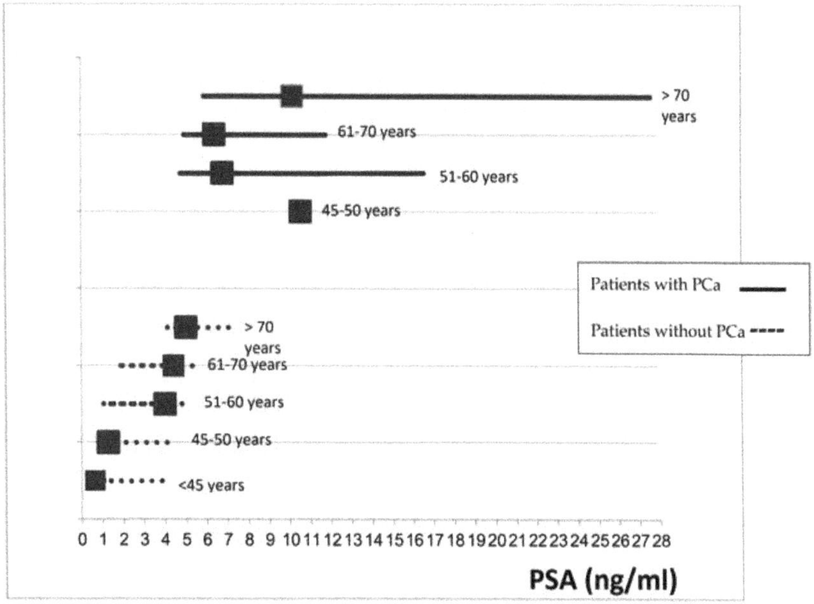

Figure 1. Comparison of PSA levels (median, IQR) between patients with and without PCa by age groups.

Table 5. Differences in the PSA levels (ng/mL) according to the presence of some comorbidities and prescribed pharmacological treatment for patients without and with PCa.

	Total	**Without PCa**		**With PCa**	
Variables	**PSA Levels (ng/mL) (Median, IQR)**	**PSA Levels (ng/mL) (Median, IQR)**	***p*-Value**	**PSA Levels (ng/mL) (Median, IQR)**	***p*-Value**
Hyperlipidemia			0.009		0.082
• No	4.53 (1.89–6.36)	4.44 (1.67–6.01)		8.20 (5.18–16.84)	
• Yes	4.27 (1.24–5.15)	4.24 (1.18–5.13)		5.87 (4.44–6.61)	
Diabetes			0.001		0.974
• No	4.53 (1.90–6.26)	4.45 (1.73–5.91)		6.82 (5.08–16.80)	
• Yes	4.19 (1–5.39)	4.04 (0.92–5.08)		7.98 (5.93–14.45)	
UTI			<0.001		0.993
• No	4.45 (1.69–6.07)	4.36 (1.15–5.77)		6.86 (5.09–16.54)	
• Yes	5.71 (4.64–10.33)	5.70 (4.64–9.51)		10.33	
Pharmacological treatment			<0.001		0.484
• No	4.70 (2.29–6.35)	4.70 (2.29–6.34)		5.68 (5.08–6.28)	
• Yes	4.39 (1.64–6.07)	4.32 (1.41–5.74)		7.17 (5.11–16.77)	

4. Discussion

We found a high rate of false-positive results both in symptomatic (47.9%) and asymptomatic patients (46.6%), resulting in a PPV of 12.7% in clinical practice. Patients in this study also showed a low rate of false-negative results (3.7%) with an NPV of 99.5%.

This is the first study based on real-world data that includes both symptomatic and asymptomatic patients, and thus, it is difficult to compare our results with similar studies. However, it was possible to compare our results with previous evidence based on clinical trials, and we found a greater rate of false-positive results. Results from randomized clinical trials showed the PPV of an elevated PSA in the 20–30% range [21], higher than the PPV shown in our study. Furthermore, according to a recent meta-analysis of clinical trials, 4–19% of all screening participants had a false-positive screening result. In our study, PSA values higher than 4.0 ng/mL were used as an indication for biopsy among men of all ages. Given that most of the available clinical trials used a 3.0 ng/mL threshold, we consider that false-positive results in these clinical trials should be higher. These differences in the false-positive rates could reflect the different populations included in clinical practice (older patients and with several comorbidities such as UTI) in contrast with clinical trials.

The level of PSA in the blood increases with age by about 3.2% per year [22], leading to an increase in the false-positive rate in the PSA test. Therefore, according to previous results, the use of this threshold for recommending prostate biopsy is not adequate for men of all ages [23]. Nevertheless, the usefulness of age-adjusted PSA thresholds is controversial due to the risk of missing a high proportion of clinically significant cancers in older men. In our study, there were no differences in the PSA values in those patients with PCa according to age, while there were differences in patients without PCa. These results suggest that an age-adjusted PSA threshold for biopsy may be of limited use despite the increase in the false-positive results in the PSA test with patients' age. In addition, given the low incidence of false-negative results and the PSA levels of the patients with PCa, the value of 4 ng/mL seems a reasonable threshold. At the same time, prostate cancer overdiagnosis (cases that would not have caused clinical consequences during a man's lifetime if left untreated) has a strong relationship to age [24]. A relevant percentage of the prostate cancer cases detected in asymptomatic patients included in the study could be overdiagnosed cases. Hence, restricting screening in patients older than 70 years old, as most available guidelines recommend, could importantly reduce overdiagnosis.

We have shown that several factors are associated with a higher probability of false-positive results: age, alcohol consumption, and having a urinary tract infection. In contrast, having diabetes mellitus type II was associated with a lower rate of false-positive results. Differences related to the presence of diabetes mellitus, urinary tract infection, and age were only observed in men who had no urinary symptomatology when the PSA test was ordered. This could be explained by the lack of precision in patients with symptoms or that the presence of symptoms acted as an effect modifier. Patients with hyperlipidemia and diabetes showed lower PSA levels than those without these pathologies. A recent systematic review showed that diabetes was significantly associated with lower PSA levels among asymptomatic men older than 60 years, leading to a lower probability of having a false-positive PSA result [5]. The implication is that the predictive value of a positive test is higher among patients with diabetes or hyperlipidemia. One could consider that the lower range of PSA values identified in diabetics compared to non-diabetics and in patients with hyperlipidemia compared to those without hyperlipidemia justifies a different cut-off point when considering a positive PSA result in this population. However, given that there were no differences in PSA values among patients with PCa, it is unlikely that using a specific cut-off would improve the accuracy of PSA screening.

In any case, this could be a useful observation for clinicians when interpreting PSA results in routine practice.

Several studies have developed risk prediction models to avoid false-positive results based on biomarkers and sociodemographic and clinical variables [25]. However, none of these models have been validated in clinical practice with positive results. The STHLM3 study [26] developed and validated a model combining plasma protein biomarkers, genetic polymorphisms, and clinical variables. It demonstrated a significant improvement in the specificity of PCa screening with the same sensitivity as PSA testing. This study included men in a specific age range, 50–69 years, whereas those undergoing PSA testing in clinical practice came from a wider age range. Given the differences between the populations included in the clinical trials and clinical practice, we consider that further research should be conducted in clinical practice to evaluate the addition of other clinical and genetic factors. In contrast with previous studies, we did not show a relationship between BMI and PSA levels. A previous study of Korean men showed a significant inverse relationship with BMI in overweight and obese men aged 40–59 years. However, there was no relationship between serum PSA and BMI in men older than 60 years [27]. In previous studies aimed at developing a predictive model for PCa diagnosis based on marching learning, the inclusion of PSA level and patient age increased the accuracy of the model, while BMI had only a minimal effect [28].

In our study, there were no differences in false-positive and false-negative PSA results according to whether the patient had a pharmacological prescription in univariate analysis but not in multivariable analysis. There were also no differences in PSA levels according to the type of treatment. In a previous longitudinal study [6], PSA levels declined to a statistically significant extent after the initiation of statin treatment, which can complicate cancer detection. In another study [7], metformin was found to have a dose-dependent inverse relationship with serum PSA levels. A previous clinical trial found that finasteride decreased PSA levels [29] and concluded that new prostate biomarkers should be interpreted with caution in patients receiving these treatments. However, another research [30] concluded that although both dutasteride and finasteride reduced PSA levels, dutasteride should not be considered equivalent to finasteride in the reduction rate of PSA. In our study, the absence of statistical significance when analyzing treatment type could perhaps be explained by the limited statistical power. Moreover, nearly 80% of patients had a current prescription for a pharmacological treatment (and most of them had two or more concomitant treatments). Hence, it was difficult to analyze the impact of an individual treatment on the false-positive rate in the univariate analysis.

There were also no differences in the probability of having a PCa according to the presence of symptoms (5.6% vs. 6.8%), and the false-positive rate was similar for both symptomatic and asymptomatic patients. A recent systematic review of randomized clinical trials [1] showed that in men without symptoms benefits of PSA-based PCa screening did not outweigh the harms, but our study showed that the presence of symptoms does not seem to have a relevant impact on the PCa diagnosis.

A population study in Norway showed how opportunistic PSA testing substantially increased the incidence of localized and regional prostate cancers among men aged 50–74 years [31]. In our study, there were no statistically significant differences according to the Gleason score distribution between symptomatic and asymptomatic patients, but symptomatic patients were more likely to have a Gleason score of 8–10 than asymptomatic patients. However, given the follow-up period of the study, we were unable to assess the impact on long-term mortality.

Recently, the European Association of Urology, the European Association of Nuclear Medicine-European Society for Radiotherapy, the Oncology-European Society of Urogenital Radiology-International, and the Society of Geriatric Oncology have published the guidelines on screening, diagnosis, and local treatment of clinically localized PCa [20]. As a screening strategy, they advise a risk-adapted screening to identify men who may develop PCa, from age 50 based on individualized life expectancy. However, this risk-adapted screening should be offered to men at increased risk from the age of 45 and to carriers of breast cancer susceptibility gene (BRCA) mutations. The use of multiparametric MRI is also recommended to avoid unnecessary biopsies. A retrospective study found that bi-parametric prostate MRI was a powerful tool in the detection of clinically significant PCa, but PSA density did not appear to significantly improve its diagnostic performance [32]. In addition, a recent clinical trial found that MRI-directed targeted biopsy for screening and early detection in persons with elevated PSA levels reduced the risk of overdiagnosis [33].

International guidelines strongly recommend incorporating shared decision making into PSA-based prostate cancer screening [20]. Patients' knowledge and attitudes toward PCa screening are decisive factors in the adoption of opportunistic screening [34], and there are variations across the world [35]. In Italy, for example, knowledge about PCa screening amongst male subjects is quite high, although knowledge of PCa risk factors, mainly both genetic/hereditary, is low, and the practice of DRE is underutilized [36]. Considering the new recommendations published by European Commission and Urologist Associations [20], education interventions should be implemented to allow patients to participate in shared decision-making regarding PSA opportunistic screening.

This study has several limitations. Firstly, we followed our patients until they were diagnosed with PCa or for 2 years (whichever came first), and therefore, this period was too short to assess the effect of PSA determination on mortality. In addition, we aimed to evaluate the PSA false-positive and -negative rate, and we did not consider the accuracy of PSA together with the determination of digital rectal examination or other genetic biomarkers (such as carriers of the Breast Cancer type 2 mutation). Given the high rate of false-positive results in the PSA determination, these factors should be considered for a final biopsy decision. Lastly, we collected the data from medical records, and thus, there are some missing data such as the Gleason classification, smoking habit, or the family history of PCa, among others.

5. Conclusions

In conclusion, this study showed a high rate of false-positive results in clinical practice for both symptomatic and asymptomatic patients, mainly in patients over 50 years. This study has generated relevant information related to the frequency and variables associated with the presence of false-positive results, which could be very useful for shared decision making. Although randomized control trials contribute to meaningful results, studies carried out in clinical practice are also needed to better support approaches to shared decision making.

Author Contributions: Conceptualization, B.L. and I.H.-A.; methodology, B.L., I.H.-A., L.A.P., J.P.C.-R., L.G.-P. and M.L.-G.; software, M.P.-G. and N.G.; validation, B.L., L.A.P., J.P.C.-R., L.G.-P., M.P.-G., M.L.-G., N.G. and I.H.-A.; formal analysis, B.L. and M.P.-G.; investigation, B.L., L.A.P., J.P.C.-R., L.G.-P., M.P.-G., M.L.-G., N.G. and I.H.-A.; resources, B.L., J.P.C.-R., L.G.-P., M.L.-G. and N.G.; data curation, B.L., J.P.C.-R., L.G.-P., M.L.-G., M.P.-G. and N.G.; writing—original draft preparation, B.L.; writing—review and editing, B.L., L.A.P., J.P.C.-R., L.G.-P., M.P.-G., M.L.-G., N.G. and I.H.-A.; visualization, B.L., I.H.-A., L.A.P., J.P.C.-R. and L.G.-P.; supervision, B.L. and I.H.-A.; project administration, B.L.; funding acquisition, B.L. All authors have read and agreed to the published version of the manuscript.

Funding: This research was funded by the Institute of Health Carlos III (Ministry of Economy and Competitiveness, MINECO) and by the European Regional Development Fund (ERDF) "A Way to Make Europe", grant number PI17/01883.

Institutional Review Board Statement: The study was conducted according to the guidelines of the Declaration of Helsinki and approved by the Ethics Committee of Clinical Research Ethics Committee (CEIC) of the Hospital Sant Joan d'Alacant (protocol code 17/324 and date of approval 27 July 2017).

Informed Consent Statement: Patient consent was waived since the data collection was only obtained from the clinical history of patients (with a large sample size of approximately 1300 subjects), and we considered it unfeasible to obtain the informed consent of patients without large losses of cases and significant selection biases. For this reason, and due to the absence of any significant risk to the patients from their records being accessed, the CEIC approved a waiver of the informed consent requirement. In the research database, patients will be anonymized using dissociated codes, unidentifiable and meaningless to any other information system and which will not allow the identification of individual patients or their crossing with other databases. Since the project database will not contain any data that would allow the identification of patients, no declaration to the Data Protection Agency is required.

Data Availability Statement: The data presented in this study are available on request from the corresponding author.

Acknowledgments: We acknowledge Jessica Gorlin for the English editing of the manuscript.

Conflicts of Interest: The authors declare no conflict of interest. The funders had no role in the design of the study; in the collection, analyses, or interpretation of data; in the writing of the manuscript, or in the decision to publish the results.

References

1. Paschen, U.; Sturtz, S.; Fleer, D.; Lampert, U.; Skoetz, N.; Dahm, P. Assessment of prostate-specific antigen screening: An evidence-based report by the German Institute for Quality and Efficiency in Health Care. *BJU Int.* **2022**, *129*, 280–289. [CrossRef] [PubMed]
2. Heidegger, I. PSA screening—A matter of debate? *Memo* **2019**, *12*, 244–248. [CrossRef]
3. Ilic, D.; Neuberger, M.M.; Djulbegovic, M.; Dahm, P. Screening for prostate cancer. *Cochrane Database Syst. Rev.* **2013**, *1*, CD004720. [CrossRef] [PubMed]
4. Mansourian, A.R.; Ghaemi, E.O.; Ahmadi, A.R.; Marjani, A.; Moradi, A.; Saifi, A. Age related prostate specific antigen reference range among men in south-east caspian sea. *Pakistan J. Biol. Sci.* **2007**, *10*, 1496–1500. [CrossRef]
5. Bernal-Soriano, M.C.; Lumbreras, B.; Hernández-Aguado, I.; Pastor-Valero, M.; López-Garrigos, M.; Parker, L.A. Untangling the association between prostate-specific antigen and diabetes: A systematic review and meta-analysis. *Clin. Chem. Lab. Med.* **2020**, *59*, 11–26. [CrossRef]
6. Hamilton, R.J.; Goldberg, K.C.; Platz, E.A.; Freedland, S.J. The influence of statin medications on prostate-specific antigen levels. *J. Natl. Cancer Inst.* **2008**, *100*, 1511–1518. [CrossRef]
7. Jayalath, V.H.; Ireland, C.; Fleshner, N.E.; Hamilton, R.J.; Jenkins, D.J. The relationship between metformin and serum prostate-specific antigen levels. *Prostate* **2016**, *76*, 1445–1453. [CrossRef]
8. Marberger, M.; Freedland, S.J.; Andriole, G.L.; Emberton, M.; Pettaway, C.; Montorsi, F.; Teloken, C.; Rittmaster, R.S.; Somerville, M.C.; Castro, R. Usefulness of prostate specific antigen (PSA) rise as a marker of prostate cancer in men treated with dutasteride: Lessons from the REDUCE study. *BJU Int.* **2012**, *109*, 1162–1169. [CrossRef]
9. Fenton, J.J.; Weyrich, M.S.; Durbin, S.; Liu, Y.; Bang, H.; Melnikow, J. *Prostate-Specific Antigen-Based Screening for Prostate Cancer: A Systematic Evidence Review for the U.S. Preventive Services Task Force*; Agency for Healthcare Research and Quality (US): Rockville, MD, USA, 2018.
10. Loeb, S.; van den Heuvel, S.; Zhu, X.; Bangma, C.H.; Schröder, F.H.; Roobol, M.J. Infectious complications and hospital admissions after prostate biopsy in a European randomized trial. *Eur. Urol.* **2012**, *61*, 1110. [CrossRef]
11. Moyer, V.A. U.S. Preventive Services Task Force. Screening for prostate cancer: U.S. Preventive Services Task Force recommendation statement. *Ann. Intern. Med.* **2012**, *157*, 120–134. [CrossRef]
12. U.S. Preventive Services Task Force; Grossman, D.C.; Curry, S.J.; Owens, D.K.; Bibbins-Domingo, K.; Caughey, A.B.; Davidson, K.W.; Doubeni, C.A.; Ebell, M.; Epling, J.W., Jr.; et al. Screening for Prostate Cancer: U.S. Preventive Services Task Force Recommendation Statement. *JAMA* **2018**, *319*, 1901–1913. [CrossRef] [PubMed]
13. Van Poppel, H.; Roobol, M.J.; Chapple, C.R.; Catto, J.W.F.; N'Dow, J.; Sønksen, J.; Stenzl, A.; Wirth, M. Prostate-specific Antigen Testing as Part of a Risk-Adapted Early Detection Strategy for Prostate Cancer: European Association of Urology Position and Recommendations for 2021. *Eur. Urol.* **2021**, *80*, 703–711. [CrossRef] [PubMed]
14. Lewis, J.H.; Kilgore, M.L.; Goldman, D.P.; Trimble, E.L.; Kaplan, R.; Montello, M.J.; Housman, M.G.; Escarce, J.J. Participation of patients 65 years of age or older in cancer clinical trials. *J. Clin. Oncol.* **2003**, *21*, 1383–1389. [CrossRef] [PubMed]
15. Kobayashi, T.; Kinoshita, H.; Nishizawa, K.; Mitsumori, K.; Ogawa, O.; Kamoto, T. Age-associated increase of prostate-specific antigen in a high level of men visiting urological clinics. *Int. J. Urol.* **2005**, *12*, 733–738. [CrossRef] [PubMed]
16. von Elm, E.; Altman, D.G.; Egger, M.; Pocock, S.J.; Gøtzsche, P.C.; Vandenbroucke, J.P.; STROBE Initiative. The Strengthening the Reporting of Observational Studies in Epidemiology (STROBE) statement: Guidelines for reporting observational studies. *J. Clin. Epidemiol.* **2008**, *61*, 344–349.
17. Bernal-Soriano, M.C.; Parker, L.A.; López-Garrigos, M.; Hernández-Aguado, I.; Caballero-Romeu, J.P.; Gómez-Pérez, L.; Alfayate-Guerra, R.; Pastor-Valero, M.; García, N.; Lumbreras, B. Factors associated with false negative and false positive results of prostate-specific antigen (PSA) and the impact on patient health: Cohort study protocol. *Medicine* **2019**, *98*, e17451. [CrossRef]
18. Bernal-Soriano, M.C.; Parker, L.A.; López-Garrigós, M.; Hernández-Aguado, I.; Gómez-Pérez, L.; Caballero-Romeu, J.P.; Pastor-Valero, M.; García, N.; Alfayate-Guerra, R.; Lumbreras, B. Do the Prostate-Specific Antigen (PSA) Tests That Are Ordered in Clinical Practice Adhere to the Pertinent Guidelines? *J. Clin. Med.* **2021**, *10*, 2650. [CrossRef] [PubMed]
19. Wolf, A.M.; Wender, R.C.; Etzioni, R.B.; Thompson, I.M.; D'Amico, A.V.; Volk, R.J.; Brooks, D.D.; Dash, C.; Guessous, I.; Andrews, K.; et al. American Cancer Society guideline for the early detection of prostate cancer: Update 2010. *CA Cancer J. Clin.* **2010**, *60*, 70–98. [CrossRef]

20. Mottet, N.; Bellmunt, J.; Bolla, M.; Briers, E.; Cumberbatch, M.G.; De Santis, M.; Fossati, N.; Gross, T.; Henry, A.M.; Joniau, S.; et al. EAU-ESTRO-SIOG Guidelines on Prostate Cancer. Part 1: Screening, Diagnosis, and Local Treatment with Curative Intent. *Eur. Urol.* **2017**, *71*, 618–629. [CrossRef]
21. Lilja, H.; Ulmert, D.; Vickers, A.J. Prostate-specific antigen and prostate cancer: Prediction, detection and monitoring. *Nat. Rev. Cancer* **2008**, *8*, 268–278. [CrossRef]
22. Oesterling, J.E.; Jacobsen, S.J.; Chute, C.G.; Guess, H.A.; Girman, C.J.; Panser, L.A.; Lieber, M.M. Serum prostate-specific antigen in a community-based population of healthy men: Establishment of age-specific reference ranges. *JAMA J. Am. Med. Assoc.* **1993**, *270*, 860–864. [CrossRef]
23. Luboldt, H.J.; Schindler, J.F.; RÜbben, H. Age-Specific Reference Ranges for Prostate-Specific Antigen as Marker for Prostate Cancer. *EAU-EBU Update Ser.* **2006**, *5*, 38–48. [CrossRef]
24. Vickers, A.J.; Sjoberg, D.D.; Ulmert, D.; Vertosick, E.; Roobol, M.J.; Thompson, I.; Heijnsdijk, E.A.; De Koning, H.; Atoria-Swartz, C.; Scardino, P.T.; et al. Empirical estimates of prostate cancer overdiagnosis by age and prostate-specific antigen. *BMC Med.* **2014**, *12*, 26. [CrossRef] [PubMed]
25. Vickers, A.J.; Cronin, A.M.; Aus, G.; Pihl, C.G.; Becker, C.; Pettersson, K.; Scardino, P.T.; Hugosson, J.; Lilja, H. Impact of recent screening on predicting the outcome of prostate cancer biopsy in men with elevated prostate-specific antigen: Data from the European Randomized Study of Prostate Cancer Screening in Gothenburg, Sweden. *Cancer* **2010**, *116*, 2612–2620. [CrossRef] [PubMed]
26. Grönberg, H.; Adolfsson, J.; Aly, M.; Nordström, T.; Wiklund, P.; Brandberg, Y.; Thompson, J.; Wiklund, F.; Lindberg, J.; Clements, M.; et al. Prostate cancer screening in men aged 50-69 years (STHLM3): A prospective population-based diagnostic study. *Lancet Oncol.* **2015**, *16*, 1667–1676. [CrossRef]
27. Yang, W.J.; Cheon, S.H.; Kim, Y.S.; Kim, D.J.; Seong, H.D.; Song, Y.S.; Cho, S.Y.; Cho, I.R.; Kim, S.I.; Kim, S.J.; et al. Relationship between prostate-specific antigen and body mass index according to age: Lower prostate-specific antigen in middle-aged overweight and obese Korean men. *Urol. Int.* **2010**, *85*, 143–146. [CrossRef]
28. Barlow, H.; Mao, S.; Khushi, M. Predicting High-Risk Prostate Cancer Using Machine Learning Methods. *Data* **2019**, *4*, 129. [CrossRef]
29. Hernandez, J.; Gelfond, J.; Goros, M.; Liss, M.A.; Liang, Y.; Ankerst, D.; Thompson, I.M., Jr.; Leach, R.J. The effect of 3-month finasteride challenge on biomarkers for predicting cancer outcome on biopsy: Results of a randomized trial. *PLoS ONE* **2018**, *9*, e0204823. [CrossRef]
30. Choi, Y.H.; Cho, S.Y.; Cho, I.R. The different reduction rate of prostate-specific antigen in dutasteride and finasteride. *Korean J. Urol.* **2010**, *51*, 704–708. [CrossRef]
31. Møller, M.H.; Kristiansen, I.S.; Beisland, C.; Rørvik, J.; Støvring, H. Trends in stage-specific incidence of prostate cancer in Norway, 1980–2010: A population-based study. *BJU Int.* **2016**, *118*, 547–555. [CrossRef]
32. Cuocolo, R.; Stanzione, A.; Rusconi, G.; Petretta, M.; Ponsiglione, A.; Fusco, F.; Longo, N.; Persico, F.; Cocozza, S.; Brunetti, A.; et al. PSA-density does not improve bi-parametric prostate MR detection of prostate cancer in a biopsy naïve patient population. *Eur. J. Radiol.* **2018**, *104*, 64–70. [CrossRef] [PubMed]
33. Hugosson, J.; Månsson, M.; Wallström, J.; Axcrona, U.; Carlsson, S.V.; Egevad, L.; Geterud, K.; Khatami, A.; Kohestani, K.; Pihl, C.G.; et al. Prostate Cancer Screening with PSA and MRI Followed by Targeted Biopsy Only. *N. Engl. J. Med.* **2022**, *387*, 2126–2137. [CrossRef] [PubMed]
34. Arafa, M.A.; Rabah, D.M.; Wahdan, I.H. Awareness of general public towards cancer prostate and screening practice in Arabic communities: A comparative multi-center study. *Asian Pac. J. Cancer Prev.* **2012**, *13*, 4321–4326. [CrossRef]
35. Cullati, S.; Charvet-Bérard, A.I.; Perneger, T.V. Cancer screening in a middle-aged general population: Factors associated with practices and attitudes. *BMC Public Health.* **2009**, *29*, 118. [CrossRef] [PubMed]
36. Mirone, V.; Imbimbo, C.; Arcaniolo, D.; Franco, M.; La Rocca, R.; Venturino, L.; Spirito, L.; Creta, M.; Verze, P. Knowledge, attitudes, and practices towards prostate cancer screening amongst men living in the southern Italian peninsula: The Prevention and Research in Oncology (PRO) non-profit Foundation experience. *World J. Urol.* **2017**, *35*, 1857–1862. [CrossRef] [PubMed]

cancers

MDPI

Systematic Review

Comparison of Multiparametric Magnetic Resonance Imaging with Prostate-Specific Membrane Antigen Positron-Emission Tomography Imaging in Primary Prostate Cancer Diagnosis: A Systematic Review and Meta-Analysis

Yi Zhao [1,*], Benjamin S. Simpson [2], Naomi Morka [3], Alex Freeman [4], Alex Kirkham [5], Daniel Kelly [6], Hayley C. Whitaker [7], Mark Emberton [7,8] and Joseph M. Norris [7,8]

1 School of Medicine, Imperial College London, London SW7 2BX, UK
2 UCL Cancer Institute, University College London, London WC1E 6BT, UK; b.simpson@ucl.ac.uk
3 UCL Medical School, University College London, London WC1E 6BT, UK; naomi.morka.17@ucl.ac.uk
4 Department of Pathology, University College London Hospitals NHS Foundation Trust, London NW1 2PG, UK; alex.freeman2@nhs.net
5 Department of Radiology, University College London Hospitals NHS Foundation Trust, London NW1 2PG, UK; alexkirkham@nhs.net
6 School of Healthcare Sciences, Cardiff University, Cardiff CF10 3AT, UK; kellydm@cardiff.ac.uk
7 UCL Division of Surgery & Interventional Science, University College London, London WC1E 6BT, UK; hayley.whitaker@ucl.ac.uk (H.C.W.); m.emberton@ucl.ac.uk (M.E.); joseph.norris@ucl.ac.uk (J.M.N.)
8 Department of Urology, University College London Hospitals NHS Foundation Trust, London NW1 2PG, UK
* Correspondence: yi.zhao18@imperial.ac.uk

Citation: Zhao, Y.; Simpson, B.S.; Morka, N.; Freeman, A.; Kirkham, A.; Kelly, D.; Whitaker, H.C.; Emberton, M.; Norris, J.M. Comparison of Multiparametric Magnetic Resonance Imaging with Prostate-Specific Membrane Antigen Positron-Emission Tomography Imaging in Primary Prostate Cancer Diagnosis: A Systematic Review and Meta-Analysis. *Cancers* **2022**, *14*, 3497. https://doi.org/10.3390/cancers14143497

Academic Editor: Claudia Manini

Received: 16 June 2022
Accepted: 12 July 2022
Published: 19 July 2022

Simple Summary: Multiparametric magnetic-resonance imaging (mpMRI) is a routinely used imaging modality for diagnosing prostate cancer but misses 10–20% of prostate tumours. Recently, prostate-specific membrane antigen positron-emission tomography (PSMA PET) has been proposed as an alternative to mpMRI for diagnosis. Our systematic review and meta-analysis aimed to compare the diagnostic performance between mpMRI and PSMA PET modalities prior to biopsy. Ten articles directly comparing the performance of both modalities in the same patient cohort were investigated. PSMA PET/CT was superior in diagnosing patients with prostate cancer over mpMRI, but not in defining the location of the cancer. Early evidence suggests that the addition of PSMA PET within the diagnostic pathway may enhance the detection of clinically significant prostate cancer.

Abstract: Multiparametric magnetic-resonance imaging (mpMRI) has proven utility in diagnosing primary prostate cancer. However, the diagnostic potential of prostate-specific membrane antigen positron-emission tomography (PSMA PET) has yet to be established. This study aims to systematically review the current literature comparing the diagnostic performance of mpMRI and PSMA PET imaging to diagnose primary prostate cancer. A systematic literature search was performed up to December 2021. Quality analyses were conducted using the QUADAS-2 tool. The reference standard was whole-mount prostatectomy or prostate biopsy. Statistical analysis involved the pooling of the reported diagnostic performances of each modality, and differences in per-patient and per-lesion analysis were compared using a Fisher's exact test. Ten articles were included in the meta-analysis. At a per-patient level, the pooled values of sensitivity, specificity, and area under the curve (AUC) for mpMRI and PSMA PET/CT were 0.87 (95% CI: 0.83–0.91) vs. 0.93 (95% CI: 0.90–0.96, $p < 0.01$); 0.47 (95% CI: 0.23–0.71) vs. 0.54 (95% CI: 0.23–0.84, $p > 0.05$); and 0.84 vs. 0.91, respectively. At a per-lesion level, the pooled sensitivity, specificity, and AUC value for mpMRI and PSMA PET/CT were lower, at 0.63 (95% CI: 0.52–0.74) vs. 0.79 (95% CI: 0.62–0.92, $p < 0.001$); 0.88 (95% CI: 0.81–0.95) vs. 0.71 (95% CI: 0.47–0.90, $p < 0.05$); and 0.83 vs. 0.84, respectively. High heterogeneity was observed between studies. PSMA PET/CT may better confirm the presence of prostate cancer than mpMRI. However, both modalities appear comparable in determining the localisation of the lesions.

Keywords: prostate-specific membrane antigen positron-emission tomography; multiparametric magnetic-resonance imaging; primary diagnosis; prostate cancer; systematic review

1. Introduction

The introduction of multiparametric magnetic-resonance imaging (mpMRI) has improved the diagnostic pathway for suspected prostate cancer (PCa) [1]. Recently, a novel imaging modality, prostate-specific membrane antigen emission tomography (PSMA PET), has demonstrated potential as an adjunctive or alternative imaging technique for primary prostate cancer diagnosis [2]. PSMA, a type 2 transmembrane glycoprotein, is known to be overexpressed in prostate tumours [3,4], and its level of expression correlates with high serum levels of prostate-specific antigen (PSA) and a higher Gleason score. As such, PSMA may provide greater utility as a more targeted and specific marker of prostate cancer [5].

Although mpMRI is now commonly used in the diagnosis process, previous studies showed that around 10–20% of missed diagnoses are clinically significant prostate tumours [6,7]. In addition, PSMA PET has demonstrated efficacy as a useful staging tool for prostate cancer and for detecting metastases [5,8].

It is unclear whether PSMA PET may offer an improved ability to diagnose primary prostate cancer over mpMRI and whether it has sufficient sensitivity to pinpoint tumour location. In this study, we systematically reviewed the evidence comparing the diagnostic accuracies of mpMRI and PSMA-PET for detecting clinically significant diseases. The comparators tested between the two imaging modalities included sensitivity, specificity, and overall AUC values for both the presence of PCa (patient-level) and location of the lesion (lesion-level).

2. Evidence Acquisition

2.1. Study Design

This review was prospectively registered with the PROSPERO International Registry (CRD42021239296). The protocol for this systematic review and meta-analysis has been published previously and was based on the Preferred Reporting Items for Systematic Review and Meta-Analysis Protocols (PRISMA-P) statement [9].

2.2. Literature Search

A systematic literature search was conducted across four databases—MEDLINE, PubMed, EMBASE, and Cochrane—to retrieve all relevant studies. Controlled Medical Subject Heading (MeSH) terms were selected to refine the relevance of studies and reduce the number of unrelated studies. Multiple synonyms of the term "mpMRI" and "PSMA PET" were used in the search strategy to account for variations in terminology. The final search strategy contained 17 components linked by AND/OR operator terms: Prostat* AND (Cancer OR Tumo* OR malignan* OR adenocarcinoma OR lesion* OR Disease) AND (PSMA OR "prostate-specific membrane antigen positron emission tomography") AND (MR OR magnetic resonance imaging OR MP-MRI OR multiparametric MRI OR multiparametric magnetic resonance imaging OR multiparametric MRI OR "multiparametric magnetic resonance imaging") AND Diagnosis.

2.3. Study Selection

All retrieved studies published between July 1977 to December 2021 were uploaded to Rayyan, a semi-automated tool to assist in the further selection of articles efficiently and accurately [10]. Figure 1 illustrates an overview of the study selection process. In order to be included, studies had to compare the diagnostic accuracies of PSMA PET and mpMRI for the primary diagnosis of prostate cancer. Studies of interest were those comparing the sensitivity and specificity of both modalities separately. The reference standard for histopathology was whole-mount prostatectomy or prostate biopsy. Expert opinions, correspondence articles, conference abstracts, review articles, and case reports were excluded. Any studies that were not written in the English language were also excluded. Included studies made a direct comparison between PSMA PET and mpMRI. Articles that focused on investigating the combined accuracy of both modalities or solely on the diagnostic accuracy of PSMA PET or mpMRI alone were also removed.

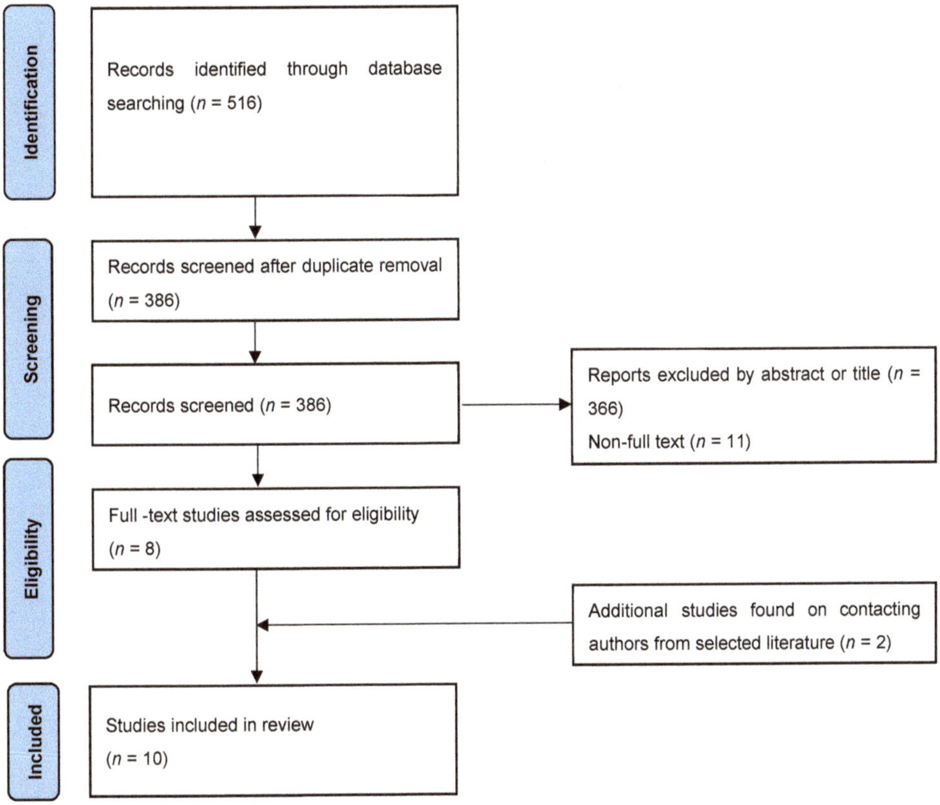

Figure 1. PRISMA flow diagram of evidence acquisition. PRISMA—Preferred Reporting Items for Systematic Reviews and Meta-Analysis.

2.4. Data Collection

All extracted data were collected using a standardised form and were checked independently by each reviewer. The data collection types included the following: the year of publication and the study authors, the study design and the patient demographics, the specification concerning the methodology, and the reported number of true/false positives and true/false negatives [9]. Three investigators (Y.Z., J.M.N. and B.S.S.) independently screened all eligible studies, assessing both the titles and abstracts for relevance. Reference sections of included articles were also manually searched to identify missed studies and additional data. Full-text articles were then retrieved for further review of eligibility.

2.5. Quality Assessment

Risk-of-bias assessment was conducted using the QUADAS-2 score [11]. The description of this method has also been described in previous systematic review articles [12]. Scoring of the QUADAS-2 score is split into four main domains: patient selection, index test, reference standard, flow, and timing. This bias assessment was conducted to assess the applicability and reliability of the data produced. Studies with low quality or suggesting a high level of bias were excluded or included with appropriate commentary [13].

2.6. Data Synthesis

Primarily, our endpoint was statistically significant differences in quantitative measurements such as sensitivity, specificity, PPV, and NPV in determining diagnostic accuracies between PSMA PET and mpMRI. An additional focus was to derive critical themes within the retrieved literature, such as the utility of different mpMRI scoring systems, including the Prostate Imaging-Reporting and Data System (PI-RADS), Likert score, and other radiogenomic features, as well as the criteria used to define clinically significant prostate cancer (csPCa), including PI-RADS or Likert score thresholds.

2.7. Meta-Analysis

The individual study's true positives (TPs), false negatives (FNs), true negatives (TNs), and false positives (FPs) were extracted to build a 2 × 2 contingency table based on the detection of csPCa via mpMRI and PSMA PET/CT. Pooled quantitative sensitivities and specificities were compared using bivariate analysis, with 95% confidence intervals (CIs) presented. The summary receiver operating characteristic (SROC) curves were then generated using the area-under-the-curve (AUC) values presented. Normality was assessed using density plots for the distribution of untransformed, logit, and double-arcsine-transformed proportions and confirmed using a Shapiro–Wilk test. The set of values most resembling a normal distribution was used in the combined analysis. Heterogeneity and inter-study variation were quantified through I^2, and a random-effects model was applied for estimation with partial pooling. Leave-one-out analysis was performed to detect potential outliers, and studies with a statistically significant influence on the fitted model were removed and the model re-fitted. Summary comparisons between PSMA PET and mpMRI were estimated once heterogeneity had been minimised through outlier removal. A Fisher's exact test was conducted to assess statistically significant differences between the two diagnostic tests, with $p < 0.05$ considered statistically significant. All data analysis and visualisation were performed in the R statistical environment (version 4.1.1, 10 August 2021) using the "mada" and "meta" packages.

3. Evidence Synthesis

3.1. Study Characteristics

Overall, 516 articles were retrieved: 135 from EMBASE, 71 from Medline, 373 from PubMed, and none from Cochrane. From these studies, ten articles were eligible for further analysis (Table 1) [14–23]. The included studies were published between 2016 and 2021. A total of 918 patients and 540 lesions were included for intra-individual comparison between mpMRI and PSMA PET/CT imaging. All studies used 3.0 Tesla for MRI imaging, and two studies used both 1.5 Tesla power and 3.0 Tesla power in mpMRI imaging [15,19]. One study used the PI-RADS scoring system version 1.0 [22], while eight studies adopted PI-RADS v2 [14–21], and one study adopted the newest PI-RADS v2.1 [23]. For mpMRI, a lesion with a PI-RADS score > 3 was considered highly indicative of clinically significant prostate cancer in nine studies [14,15,17–23]. One study used a PI-RADS score > 4 as the threshold for clinically significant prostate cancer [16]. All but two studies used a 68 Ga-PSMA-11 tracer (HBED-CC), with one study using an 18 F-PSMA-1007 tracer and one using 68 Ga-PSMA-617 [21,23]. The range of the PSMA tracer injected was between 131.7 and 310 MBq in all studies. PSMA PET images were interpreted visually where regions of interest were compared with background uptake in all studies. A high suspicion of clinically significant cancer was defined using a 3- or 4-point Likert scale [17–19], based on the SUVmax value [14] or higher update to the background activity [15,21–23]. A score of equivocal and above, or probably positive and above, was considered a clinically significant cancer [18,19]. The histopathological definition of csPCa was based on a Gleason score ≥ 7 (3 + 4 or 4 + 3) [16–18,21–23] or the International Society of Urological Pathology

(ISUP) grading [14,19,20], with three studies incorporating the tumour size in their csPCa definition [15,16,22]. The age of the included patients ranged from 62–69 years old. The range of mean PSA values in the included studies was 5.6–17.4 ng/dL (Table 1). Four articles were identified for PSMA PET/MRI analysis, but were not considered for meta-analysis due to the different definitions of PSMA PET/MRI [16,24–26].

3.2. Meta-Analysis

Sensitivity and specificity for both mpMRI and PSMA PET/CT were reported separately at per-patient and per-lesion levels. In the per-patient-level analysis, each case was regarded as an individual patient receiving both imaging modalities. In the per-lesion-level analysis, each case was regarded as an individual lesion identified in each histopathological sample.

3.2.1. Per-Patient Analysis

Four studies were included in the paired analysis between mpMRI and PSMA PET/CT (Figure 2) [18–21]. A total of 707 patients were included, with 464 patients having proven csPCa. The pooled sensitivity for mpMRI and PSMA PET/CT was 0.87 (95% CI: 0.83–0.91) vs. 0.93 (95% CI: 0.90–0.96, p = 0.001657), and the pooled specificity was 0.47 (95% CI: 0.23–0.71) vs. 0.54 (95% CI: 0.23–0.84, p = 0.5225), respectively (Figure 2). The AUC values were 0.84 vs. 0.91, respectively (Figure 2). The heterogeneity between studies was large for specificity analysis for mpMRI (I^2 = 0.94) and PSMA PET/CT (I^2 = 0.97), with statically significant Cochrane Q statistics $p < 0.01$.

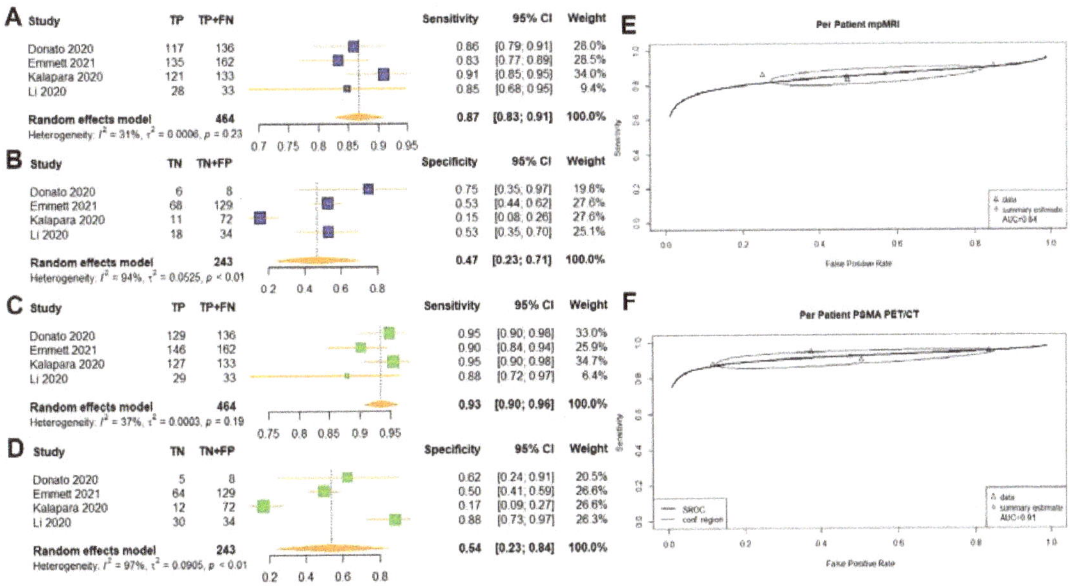

Figure 2. (**A–F**) Reported sensitivity and specificity values for both mpMRI and PSMA PET/CT with AUC values on SROC curves for per-patient analysis. Forest plots for pooled sensitivities and specificities are displayed in bold and as diamonds in the graphs for mpMRI (**A**,**B**) and PSMA PET/CT (**C**,**D**). The SROC curves indicate the summary estimates in circles (**E** for mpMRI; **F**, PSMA PET/CT). Triangles represent included study, with dotted lines representing the confidence interval and solid lines for the SROCs. AUC values are displayed in the legend. mpMRI = multiparametric magnetic-resonance imaging; PSMA PET/CT—prostate-specific membrane antigen positron-emission tomography/computed tomography; AUC—area under the curve, SROC—summary receiver operating characteristic.

Table 1. Overview of studies included for both per-patient and per-lesion analysis.

Authors	Year	Ref	No. of Patients	No. of Lesions	MRI Imaging Power	PI-RADS Version	mpMRI Positivity Criteria	PSMA PET Scoring System	PSMA Tracer	PSMA Tracer Injected (MBq)	PSMA PET/CT Positivity Criteria	Mean Age (yrs)	Mean PSA Value (ng/dL)	Reference Standard	Clinically Significant Definition
Berger	2018	[14]	50	84	3T	v2.0	PI-RAD ≥ 3	SUVmax	68Ga-PSMA-11	-	SUVmax > 2.5	64.9 (59.3–70.5)	10.6 (2.5–18.7)	WMP	ISUP ≥ 1
Bettermann	2019	[15]	17	193 (772 quadrants)	3T/1.5T	v2.0	PI-RADS ≥ 3	Uptake against background	68Ga-PSMA-11	172 (138–206)	Uptake superior to the background activity in >1 slice	67 (48–76)	17.4 (6.01–218.0)	WMP	Lesions extending > 3 mm into another quadrant
Chen	2019	[16]	54	90	3T	v2.0	PI-RADS ≥ 4	MI-ES Score	68Ga-PSMA-11	131.7 (130.6–177.6)	MI-ES ≥ 2	69 (55–84)	13.53 (4.04–110.00)	WMP	Cancer volume ≥ 0.5 cm³/GS ≥ 3 + 4 /Stage ≥ pT3
Donato 2019	2019	[17]	58	88	3T	v2.0	PI-RADS ≥ 3	3-point Likert Scale [a]	68Ga-PSMA-11	150.0 (142.5–157.5)	SUVmax > 5 (Equivocal)	65.5 (60–68)	7.35 (5.6–12)	WMP	GS ≥ 3 + 4
Donato 2020	2020	[18]	144	-	3T	v2.0	PI-RADS ≥ 3	3-point Likert Scale [a]	68Ga-PSMA-11	150.0 (142.5–157.5)	>Equivocal	66.5 (61.7–71.25)	8.6 (6–12.25)	Ultrasound-guided transperineal targeted biopsies	GS ≥ 3 + 4
Emmett	2021	[19]	291	-	3T/1.5T	v2.0	PI-RADS ≥ 3	4-point Certainty Scale [b]	68Ga-PSMA-11	1.8–2.2 MBq/kg	Positive (Probably/Definite)	64.0 (58.7–69.9)	5.6 (4.2–7.5)	Systematic transperineal biopsies	ISUP ≥ 2
Kalapara	2020	[20]	205	-	3T	v2.0	PI-RAD ≥ 3	Binary Scale	68Ga-PSMA-11	1.8–2.2 MBq/kg	Lesion with the highest avidity by SUVmax	67 (61–72)	7.18 (4.90–10.20)	WMP	ISUP ≥ 3
Li	2020	[21]	67	-	3T	v2.0	PI-RAD ≥ 3	Uptake against background	68Ga-PSMA-617	111–185	Uptake superior to the background activity	68 (42–85)	10.48 (3.15–19.76)	Transrectal ultrasound biopsy	GS ≥ 7
Rhee	2016	[22]	22	71 (540 segments)	3T	v1.0	PI-RAD ≥ 3	Uptake against background	68Ga-PSMA-11	150	Uptake superior to the background activity	62 (55–69)	6.1 (0–14.6)	WMP	GS ≥ 4 + 3 +/− tumour size ≥ 6 mm
Zamboglou	2021	[23]	10	14 (601 segments) *	3T	v2.1	PI-RAD ≥ 3	Uptake against background	[18F]PSMA-1007	310 (249–370)	Uptake superior to the background activity	-	-	WMP	GS ≥ 7 *

[a]—likely, equivocal, unlikely; [b]—definitely negative, probably negative, probably positive, definitely positive; MI-ES—Molecular Imaging PMSA Expression; GS—Gleason score; ISUP—International Society of Urological Pathology; WMP—Whole-mount prostatectomy. * Obtained by contacting the author directly.

3.2.2. Per-Lesion Analysis

Six studies investigated the diagnostic accuracy of mpMRI and PSMA PET/CT at a lesion level (Figure 3) [14–17,22,23]. A "lesion" was defined as individual tissue slices analysed by both imaging modalities with histopathological confirmation of the lesion location in the included studies. Of the 2175 lesions included in the analysis from 211 patients, 1325 were considered csPCa (Table 1). The pooled sensitivity values of mpMRI and PSMA PET/CT were lower at 0.63 (95% CI: 0.52–0.74) vs. 0.79 (95% CI: 0.62–0.92, $p = 1.848 \times 10^{-12}$), and the pooled specificity values were 0.88 (95% CI: 0.81–0.95) vs. 0.71 (95% CI: 0.47–0.90, $p = 0.0226$), respectively (Figure 3). The AUC values were 0.83 vs. 0.84, respectively (Figure 3). Heterogeneity remained large for the pooled sensitivity of mpMRI and for both the sensitivity and specificity of PSMA PET/CT, with statically significant Cochrane Q statistics $p < 0.01$.

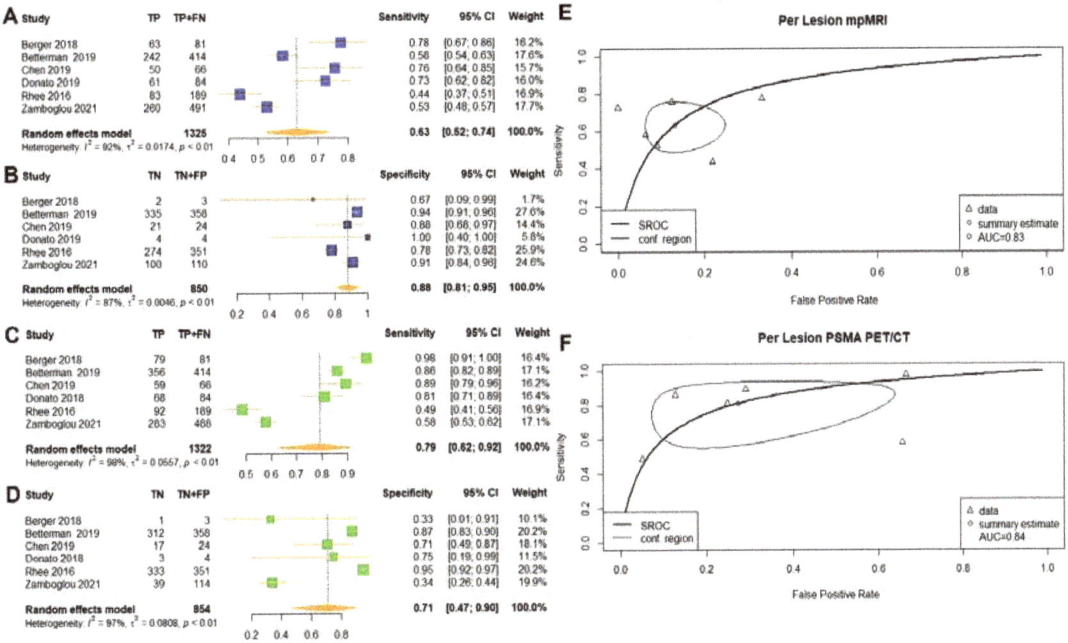

Figure 3. (**A–F**) Reported sensitivity and specificity values for both mpMRI and PSMA PET/CT with AUC values on SROC curves for per-lesion analysis. Forest plots for pooled sensitivities and specificities are displayed in bold and as diamonds in the graphs for mpMRI (**A**,**B**) and PSMA PET/CT (**C**,**D**). The SROC curves indicate the summary estimates in circles (**E** for mpMRI; **F**, PSMA PET/CT). Triangles represent included study, with dotted lines representing the confidence interval and solid lines for the SROCs. AUC values are displayed in the legend. mpMRI—multiparametric magnetic-resonance imaging; PSMA PET/CT—prostate-specific membrane antigen positron-emission tomography/computed tomography; AUC—area under the curve; SROC—summary receiver operating characteristic.

3.3. *Risk of Bias*

The risk of bias analysis was assessed using the QUADAS-2 tool [11]. The overall risk of bias was high for the included studies (Figure 4). Although most studies recruited patients prospectively, eight patients were recruited with known prostate cancer retrospectively [14,15,17,18,20,22,27].

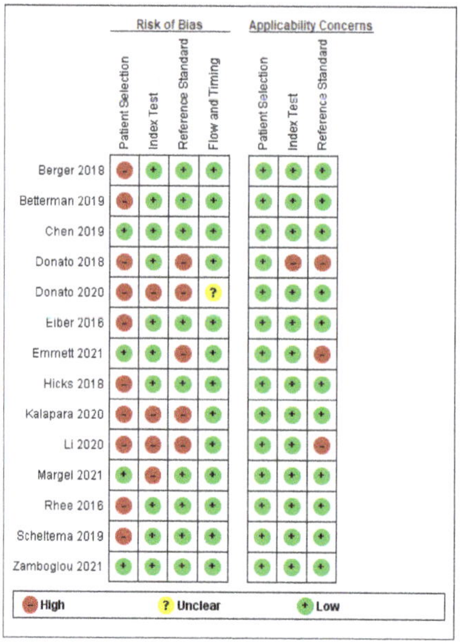

Figure 4. QUADAS-2 score indicates the risk of bias analysis in assessing the low, high, or unclear risk for patient selection, index test, reference standard, flow, and timing for individual studies. An add-on analysis on applicability concerns is also included.

Three studies investigated the index tests without the knowledge of histopathology [18,20,21], and five studies investigated the reference test unblinded [17–21]. All but one study stated the time interval between mpMRI and PSMA PET/CT [18]. For applicability concerns of this meta-analysis, all studies had low concern for patient selection, while one study had high concern for the index test [17] and three studies had high concern for the reference standard [17,19,21].

4. Discussion

We reported the first meta-analysis comparing the diagnostic accuracy of mpMRI and PSMA for detecting clinically significant cancer in matched-patient cohorts. The meta-analysis showed that PSMA PET/CT might be favourable in identifying patients with csPCa (Figure 2). However, we were unable to confirm if this modality is superior in identifying suspected csPCa lesions (Figure 3). The results should be interpreted with the large heterogeneity observed in our study.

The performance of mpMRI and PSMA PET/CT in the per-patient analysis was comparable with the existing literature investigating individual imaging modalities in diagnosing PCa. Zhen et al. investigated the pooled diagnostic accuracy for mpMRI from 29 studies, reporting a good sensitivity of 0.87 (95% CI: 0.81–0.91) and a moderate specificity of 0.68 (95% CI: 0.56–0.79) [28]. Satapathy et al. reported the pooled diagnostic accuracy of PSMA PET/CT from seven studies, with a favourable sensitivity of 0.97 (95% CI: 0.90–0.99) and a moderate specificity of 0.66 (95% CI: 0.52–0.78) [2]. Our meta-analysis revealed a comparable sensitivity of mpMRI (0.87) and slightly reduced sensitivity in PSMA PET/CT (0.93) (Figure 2). The difference may be because our meta-analysis focused on the diagnosis of csPCa as opposed to the diagnosis of PCa in both Zhen et al. and Satapathy et al. [2,28]. However, the specificity for both modalities was remarkably reduced in our analysis compared with the existing literature (mpMRI: 0.47 vs. 0.68; PSMA PET/CT: 0.54 vs. 0.66). The PRECISION trial showed an inverse association of negative MRI-targeted biopsies with lesion conspicuity reported by the PI-RADS v2.0 criteria [29]. Starvrinides and colleagues attempted to capture the characteristics of false-positive MRI lesions, which are distinct from clinically significant diseases [30]. The authors highlighted the use of MRI-calculated PSA density (PSAD) and apparent diffusion coefficient (ADC) as potential predictors of significant disease, a finding verified by the literature [30–32]. Although ADC was incorporated in the PI-RADS v2.0 criteria, benign diseases such as prostatitis and prostatic atrophy are known to decrease the signal on the ADC map. Moreover, PSMA expression in non-cancerous prostatic conditions such as inflammation and benign tumours may explain the false-positive cases of PSMA PET/CT as observed in the per-patient analysis [21,33]. Comprehensive interpretations of imaging findings with patient-specific variables, including PSAD and clinical histories, may aid in the distinction of lesions that are likely to be PCa [33]. In our review, PSAD was reported in three studies, while the clinical histories of the included patients were absent from the included studies [17,18,20]. Future studies should include the PSAD and the diagnosis of other prostatic diseases to reduce the diagnosis of false-positive cases.

The purpose of conducting per-lesion analysis was to assess the multifocality of patients with PCa; however, the pooled sensitivity in per-lesion analysis for both the mpMRI and PSMA-PET/CT were unsatisfactory in our review (Figure 3). Previous literature has shown the drawback of segmenting the prostate into sextants. This may lead to the inappropriate assignment of tumour foci on boundaries between sextants, thereby reducing the diagnostic performance compared to only examining targeted histopathological findings [25,34]. In our review, three studies in the per-lesion analysis segmented the specimen into different segments, which may have contributed to the limited diagnostic performance [15,22,23]. In addition, Rhee et al. and Berger et al. conducted their studies when the use of PSMA PET/CT in the diagnosis of prostate cancer was relatively new and lacked the reporting guidelines to make accurate diagnoses [14,22]. Both studies investigated the diagnostic performance from a small patient sample, thereby limiting the generalisability of the results [14,22]. Heterogeneity remained high for both the sensitivity and specificity results, which limits the ability of our study to detect differences between these techniques. Multifocality is a common feature of prostate cancer, as more than one distinct tumour nodule may be present within a prostate gland [22,35]. Although secondary lesions may present with a smaller volume than the index lesions, recent studies suggest that the volume of a tumour may not indicate the biological significance, and smaller tumours may be of greater clinical significance in their impact on prognosis [35]. The per-lesion diagnostic accuracy of PSMA PET/CT has been widely discussed in the context of the staging of prostate cancers [36,37]. However, in the context of primary prostate cancer diagnosis, the existing literature focused on the diagnostic accuracy of per-patient analysis [2]. The accurate localisation of prostate cancer lesions is critical to accurate biopsy and treatment planning [29,38,39]. The low sensitivity for both mpMRI and PSMA PET/CT reported in our study may highlight the need for methods that improve the ability of these techniques to define a lesion's specific location.

It is noteworthy that the histological criteria for csPCa varied between studies, which may affect the reporting of clinically significant cases and, therefore, the sensitivities and specificities of both modalities. This may also contribute to the high inter-study heterogeneities observed at both the patient and lesion levels. Although GS 7 remains the most common histopathological definition for csPCa [16–18,21–23], the existing literature has highlighted the clinical significance of the GS 3 + 4 and GS 4 + 3 groups. The GS 4 + 3 groups have a worse prognosis than their counterparts in terms of risk of progression, metastasis, and survival [40,41]. The ISUP grading system has attempted to address the clinical discrepancy between GS 3 + 4 (ISUP 2) and GS 4 + 3 (ISUP 3) by differentiating the two groups in their scoring [42]. However, given the low representation of pattern 4 in <5% of overall tumour volume and the recent introduction of ISUP, it is recommended to report both ISUP grading and GS in the current reporting of prostate cancer [42,43]. The use of a stricter diagnostic criterion for clinically significant prostate tumours may lead to larger estimates of false positives for both mpMRI and PSMA PET/CT. Kalapara et al. showed a lower specificity than other studies in the per-patient analysis [20]. The authors defined clinically significant disease as having ISUP grades of 3–5, which was higher than other studies adopting the ISUP grading system [19]. Future studies, therefore, may be improved through a more standardised classification of csPCa.

Factors affecting the interpretation of mpMRI and PSMA PET/CT images may warrant further investigation. For mpMRI, most included studies commonly used a PI-RADS score of >3 as being potentially indicative of csPCa, but one study in our analysis highlighted that 11 out of 15 patients with PI-RADS 2 scores were proven to have csPCa following histopathological assessment [18]. This may prompt further insight into the more modern PI-RADS v2.0 and v2.1 scoring systems in reducing false-negative reports. It is known that inter-reader variability and reader experience in assigning PI-RADS scores and the subsequent detection of csPCa has been shown to affect the detection rates of csPCa [44–46]. However, image quality has also been shown to significantly affect the diagnostic performance of mpMRI [47,48]. Future studies may provide information on image quality using metrics such as the Prostate Imaging Quality (PI-QUAL) score [49].

In PSMA PET/CT analysis, the discrepancies in scoring criteria were more apparent in the clinical suspicion of csPCa. It was not apparent, from the retrieved studies, what the effect is of different volumes of tracer being used and how this may affect imaging interpretation or scores, such as the measurement of SUVmax and MI-ES scores [14,16]. These scoring systems may require further studies concerning a suitable threshold for detecting clinically significant prostate cancer at primary diagnosis, and inter-observer variability within this context [50]. Differences in the PSMA PET/CT scoring criteria may also have contributed to the high observed inter-study heterogeneity within our study, which warrants further analysis in the future. Similar to mpMRI, imaging quality and reader experience have been reported to impact interpretation and diagnostic performance [51,52]. A recent guideline proposed the E-PSMA criteria, which showed lower inter-observer variability and may contribute to the future standardisation of PSMA PET/CT scoring [51].

Studies comparing mpMRI and PSMA PET/MRI were excluded from this meta-analysis owing to differences in PET/MRI methodologies observed in the literature. For example, two studies conducted PET/MRI scans using a hybrid imaging system whereby both MRI and PET imaging acquisition were acquired, simultaneously, in one setting [24,25]. Another two studies defined PSMA PET/MRI as a combination of imaging analyses for mpMRI and PET/CT imaging whereby the images were acquired separately [16,26]. Due to the intrinsic differences in PSMA PET/MRI methods, the number of articles for each PSMA PET/MRI method was insufficient to conduct a meta-analysis. The performance of hybrid PSMA PET/MRI in primary diagnosis has been reported previously [37]. The pooled sensitivity and specificity for the per-patient and per-lesion analyses were 61.5% and 90.9%, and 94.9% and 62.5%, respectively [37]. However, the study did not investigate the performance of combined mpMRI and PSMA PET/CT imaging analysis, which may warrant further systematic review. Moreover, as the study aimed to investigate the performance of PSMA PET/MRI solely. Further diagnostic test accuracy analysis comparing mpMRI alone, PSMA PET alone, and PSMA PET/MRI may be useful to evaluate the optimum imaging technique for the diagnosis of primary csPCa. The aforementioned study was particularly interesting as mpMRI showed high sensitivity in the per-patient analysis and high specificity in the per-lesion analysis. This result was contrary to the reported strengths of PSMA PET/MRI [37].

Genetic factors contributing to mpMRI- or PSMA PET/CT-visible and -invisible prostate cancers may explain the results in our study [12,33]. Visible mpMRI tumours are associated with increased Decipher and Oncotype scores and a greater frequency of phosphatase and tensin homologue (PTEN) loss; no comparable genetic evidence of increased aggression in mpMRI-invisible tumours has been reported [12]. However, genes involving cell structure (such as the actin filament-based process and cytoskeleton organisation) were downregulated in mpMRI-invisible tumours and associated with lower tissue density [12,27,53]. The findings may explain the misdiagnosis of low-cellularity prostate cancer, as mpMRI primarily investigates water molecule content and movement in cancerous prostate cells with high cellularity and contributed to limited sensitivity in our meta-analysis (Figures 2 and 3) [54]. Further validation trials such as the ReIMAGINE Trail (NCT04063566) may depict the role of genetic biomarkers in the use of mpMRI for prostate cancer diagnosis. In contrast to mpMRI, PSMA PET/CT imaging is not dependent on the degree of cellularity, but instead, on the expression of PSMA ligands [33]. It remains to be seen whether PSMA PET/CT also identifies the most high-risk lesions.

Furthermore, the apical expression of PSMA is markedly increased in PCa cells compared with non-cancerous cells [33]. This may require an alternative imaging modality to detect mpMRI-invisible tumours, as PSMA ligand expression is associated with the FOLH1 gene, which is a separate genetic pathway to cell structure expression [12,27,53,55,56]. PSMA induces the activation of phosphoinositide 3-kinase (PI3K) independently of PTEN loss, which contributes to the proliferation of prostate cancer [56]. However, the association between PSMA expression and cellularity remains unknown, currently. Therefore, the combination of PSMA PET/CT and mpMRI imaging for mpMRI-invisible tumours, identified via pre-imaging genetic risk stratification, may be appropriate regarding overall diagnostic accuracy and cost-effectiveness. Previous studies have demonstrated high sensitivity and specificity in combination imaging approaches [16,26]. However, the studies did not categorise patients according to mpMRI-visible and mpMRI-invisible PCa, as all patients were examined using uniform imaging methodologies. Therefore, the additional diagnostic value of PSMA PET when a lesion is mpMRI-invisible may warrant further investigation.

There were several limitations to this systematic review and meta-analysis. First, the limited number of studies investigating the paired analysis between mpMRI and PSMA PET/CT may hinder the robustness of our results. Second, many studies used different definitions of csPCa, and it was impossible to conduct subgroup analysis based on these csPCa definitions to determine their effect. This may explain the high heterogeneity amongst included studies. Third, the insignificant statistical result in the per-patient analysis may require further investigation to determine if any differences exist in the specificity of mpMRI and PSMA PET/CT. Our meta-analysis was unable to be conducted for PSMA PET/MRI fusion techniques owing to the different definitions for PET/MRI fusion.

Nevertheless, our study represents the first meta-analysis of the diagnostic accuracy of mpMRI and PSMA PET/CT on primary prostate cancer. It is based on a robust research methodology with strict criteria for study selection. Given the current limitations, further research should continue to contribute to the evidence base and address the heterogeneity observed in the study. Future research should standardise the interpretation of PSMA PET/CT images and histopathology scoring systems to address the methodological discrepancies.

5. Conclusions

This meta-analysis shows that, at the per-patient level, PSMA PET/CT may perform better than mpMRI in detecting primary prostate cancer. In contrast, both modalities were comparable in locating specific lesions in patients. PSMA PET/CT is a whole-body procedure and may add intrinsic value compared to pelvic mpMRI. However, considerable heterogeneity was observed in our study. Therefore, there is a need to standardise imaging interpretation and histopathology scoring systems to reduce variation between studies. Further analyses should focus on the diagnostic performance of combined mpMRI and PSMA PET/CT imaging modalities.

Author Contributions: N.M., B.S.S. and J.M.N. drafted the manuscript and created the study concept; A.K., A.F., D.K., H.C.W. and M.E. provided supervision and guidance during the study; Y.Z. is the guarantor of this work. All authors have read and agreed to the published version of the manuscript.

Funding: Norris is funded by the Medical Research Council (MRC) (MR/S00680X/1). Simpson is funded by the Royal Marsden Cancer Charity. Emberton receives research support from the United Kingdom's National Institute of Health Research (NIHR) UCLH/UCL Biochemical Research Centre.

Conflicts of Interest: Norris receives funding from the MRC. Simpson receives funding from the Royal Marsden Cancer Charity. Whitaker receives funding from Prostate Cancer UK, the Urology Foundation, and Rosetrees Trust. Kirkham, Freeman, and Emberton have stock interest in Nuada Medical Ltd. Emberton acts as a consultant, trainer, and proctor to Sonatherm Inc., Angiodynamics Inc and Exact Imaging Inc.

References

1. Ahmed, H.U.; Bosaily, A.E.-S.; Brown, L.C.; Gabe, R.; Kaplan, R.; Parmar, M.K.; Collaco-Moraes, Y.; Ward, K.; Hindley, R.G.; Freeman, A.; et al. Diagnostic Accuracy of Multi-Parametric MRI and TRUS Biopsy in Prostate Cancer (PROMIS): A Paired Validating Confirmatory Study. *Lancet* **2017**, *389*, 815–822. [CrossRef]
2. Satapathy, S.; Singh, H.; Kumar, R.; Mittal, B.R. Diagnostic Accuracy of 68Ga-PSMA PET/CT for Initial Detection in Patients with Suspected Prostate Cancer: A Systematic Review and Meta-Analysis. *Am. J. Roentgenol.* **2021**, *216*, 599–607. [CrossRef]
3. Bouchelouche, K.; Choyke, P.L. Advances in PSMA Positron Emission Tomography (PET) of Prostate Cancer. *Curr. Opin. Oncol.* **2018**, *30*, 189–196. [CrossRef]
4. Wright, G.L.; Haley, C.; Beckett, M.L.; Schellhammer, P.F. Expression of Prostate-Specific Membrane Antigen in Normal, Benign, and Malignant Prostate Tissues. *Urol. Oncol. Semin. Orig. Investig.* **1995**, *1*, 18–28. [CrossRef]
5. Hoffmann, M.A.; Wieler, H.J.; Baues, C.; Kuntz, N.J.; Richardsen, I.; Schreckenberger, M. The Impact of 68Ga-PSMA PET/CT and PET/MRI on the Management of Prostate Cancer. *Urology* **2019**, *130*, 1–12. [CrossRef]
6. Bass, E.J.; Pantovic, A.; Connor, M.; Gabe, R.; Padhani, A.R.; Rockall, A.; Sokhi, H.; Tam, H.; Winkler, M.; Ahmed, H.U. A Systematic Review and Meta-Analysis of the Diagnostic Accuracy of Biparametric Prostate MRI for Prostate Cancer in Men at Risk. *Prostate Cancer Prostatic Dis.* **2020**, 1–16. [CrossRef]
7. Radtke, J.P.; Kuru, T.H.; Boxler, S.; Alt, C.D.; Popeneciu, I.V.; Huettenbrink, C.; Klein, T.; Steinemann, S.; Bergstraesser, C.; Roethke, M.; et al. Comparative Analysis of Transperineal Template Saturation Prostate Biopsy versus Magnetic Resonance Imaging Targeted Biopsy with Magnetic Resonance Imaging-Ultrasound Fusion Guidance. *J. Urol.* **2015**, *193*, 87–94. [CrossRef]
8. Hofman, M.S.; Lawrentschuk, N.; Francis, R.J.; Tang, C.; Vela, I.; Thomas, P.; Rutherford, N.; Martin, J.M.; Frydenberg, M.; Shakher, R.; et al. Prostate-Specific Membrane Antigen PET-CT in Patients with High-Risk Prostate Cancer before Curative-Intent Surgery or Radiotherapy (ProPSMA): A Prospective, Randomised, Multicentre Study. *Lancet* **2020**, *395*, 1208–1216. [CrossRef]
9. Zhao, Y.; Morka, N.; Simpson, B.S.S.; Freeman, A.; Kirkham, A.; Kelly, D.; Whitaker, H.C.; Emberton, M.; Norris, J.M. Prostate-Specific Membrane Antigen Positron Emission Tomography Compared to Multiparametric MRI for Prostate Cancer Diagnosis: A Protocol for a Systematic Review and Meta-Analysis. *BMJ Open* **2021**, *11*, e052277. [CrossRef]
10. Ouzzani, M.; Hammady, H.; Fedorowicz, Z.; Elmagarmid, A. Rayyan-a Web and Mobile App for Systematic Reviews. *Syst. Rev.* **2016**, *5*, 210. [CrossRef]
11. Whiting, P.F.; Rutjes, A.W.S.; Westwood, M.E.; Mallett, S.; Deeks, J.J.; Reitsma, J.B.; Leeflang, M.M.G.; Sterne, J.A.C.; Bossuyt, P.M.M. QUADAS-2: A Revised Tool for the Quality Assessment of Diagnostic Accuracy Studies. *Ann. Intern. Med.* **2011**, *155*, 529–536. [CrossRef] [PubMed]
12. Norris, J.M.; Simpson, B.S.; Parry, M.A.; Allen, C.; Ball, R.; Freeman, A.; Kelly, D.; Kim, H.L.; Kirkham, A.; You, S.; et al. Genetic Landscape of Prostate Cancer Conspicuity on Multiparametric Magnetic Resonance Imaging: A Systematic Review and Bioinformatic Analysis. *Eur. Urol. Open Sci.* **2020**, *20*, 37–47. [CrossRef]
13. Genetic Landscape of Prostate Cancer Conspicuity on Multiparametric MRI: A Protocol for a Systematic Review and Bioinformatic Analysis. *BMJ Open* **2020**, *10*, e034611. Available online: https://bmjopen.bmj.com/content/10/1/e034611.abstract (accessed on 1 March 2021). [CrossRef] [PubMed]
14. Berger, I.; Annabattula, C.; Lewis, J.; Shetty, D.V.; Kam, J.; Maclean, F.; Arianayagam, M.; Canagasingham, B.; Ferguson, R.; Khadra, M.; et al. 68 Ga-PSMA PET/CT vs. MpMRI for Locoregional Prostate Cancer Staging: Correlation with Final Histopathology. *Prostate Cancer Prostatic Dis.* **2018**, *21*, 204–211. [CrossRef] [PubMed]
15. Bettermann, A.S.; Zamboglou, C.; Kiefer, S.; Jilg, C.A.; Spohn, S.; Kranz-Rudolph, J.; Fassbender, T.F.; Bronsert, P.; Nicolay, N.H.; Gratzke, C.; et al. [68Ga-]PSMA-11 PET/CT and Multiparametric MRI for Gross Tumor Volume Delineation in a Slice by Slice Analysis with Whole Mount Histopathology as a Reference Standard—Implications for Focal Radiotherapy Planning in Primary Prostate Cancer. *Radiother. Oncol.* **2019**, *141*, 214–219. [CrossRef]
16. Chen, M.; Zhang, Q.; Zhang, C.; Zhao, X.; Marra, G.; Gao, J.; Lv, X.; Zhang, B.; Fu, Y.; Wang, F.; et al. Combination of 68Ga-PSMA PET/CT and Multiparametric MRI Improves the Detection of Clinically Significant Prostate Cancer: A Lesion-by-Lesion Analysis. *J. Nucl. Med.* **2019**, *60*, 944–949. [CrossRef]
17. Donato, P.; Roberts, M.J.; Morton, A.; Kyle, S.; Coughlin, G.; Esler, R.; Dunglison, N.; Gardiner, R.A.; Yaxley, J. Improved Specificity with 68Ga PSMA PET/CT to Detect Clinically Significant Lesions "Invisible" on Multiparametric MRI of the Prostate: A Single Institution Comparative Analysis with Radical Prostatectomy Histology. *Eur. J. Nucl. Med. Mol. Imaging* **2019**, *46*, 20–30. [CrossRef]
18. Donato, P.; Morton, A.; Yaxley, J.; Ranasinghe, S.; Teloken, P.E.; Kyle, S.; Coughlin, G.; Esler, R.; Dunglison, N.; Gardiner, R.A.; et al. 68Ga-PSMA PET/CT Better Characterises Localised Prostate Cancer after MRI and Transperineal Prostate Biopsy: Is 68Ga-PSMA PET/CT Guided Biopsy the Future? *Eur. J. Nucl. Med. Mol. Imaging* **2020**, *47*, 1843–1851. [CrossRef]

19. Emmett, L.; Buteau, J.; Papa, N.; Moon, D.; Thompson, J.; Roberts, M.J.; Rasiah, K.; Pattison, D.A.; Yaxley, J.; Thomas, P.; et al. The Additive Diagnostic Value of Prostate-Specific Membrane Antigen Positron Emission Tomography Computed Tomography to Multiparametric Magnetic Resonance Imaging Triage in the Diagnosis of Prostate Cancer (PRIMARY): A Prospective Multicentre Study. *Eur. Urol.* **2021**, *80*, 682–689. [CrossRef]
20. Kalapara, A.A.; Nzenza, T.; Pan, H.Y.C.; Ballok, Z.; Ramdave, S.; O'Sullivan, R.; Ryan, A.; Cherk, M.; Hofman, M.S.; Konety, B.R.; et al. Detection and Localisation of Primary Prostate Cancer Using 68 Gallium Prostate-Specific Membrane Antigen Positron Emission Tomography/Computed Tomography Compared with Multiparametric Magnetic Resonance Imaging and Radical Prostatectomy Specimen Pathology. *Br. J. Urol.* **2020**, *126*, 83–90. [CrossRef]
21. Li, Y.; Han, D.; Wu, P.; Ren, J.; Ma, S.; Zhang, J.; Song, W.; Lin, X.; Jiao, D.; Shi, S.; et al. Comparison of 68Ga-PSMA-617 PET/CT with MpMRI for the Detection of PCa in Patients with a PSA Level of 4–20 Ng/Ml before the Initial Biopsy. *Sci. Rep.* **2020**, *10*, 10963. [CrossRef] [PubMed]
22. Rhee, H.; Thomas, P.; Shepherd, B.; Gustafson, S.; Vela, I.; Russell, P.J.; Nelson, C.; Chung, E.; Wood, G.; Malone, G.; et al. Prostate Specific Membrane Antigen Positron Emission Tomography May Improve the Diagnostic Accuracy of Multiparametric Magnetic Resonance Imaging in Localized Prostate Cancer. *J. Urol.* **2016**, *196*, 1261–1267. [CrossRef] [PubMed]
23. Zamboglou, C.; Kramer, M.; Kiefer, S.; Bronsert, P.; Ceci, L.; Sigle, A.; Schultze-Seemann, W.; Jilg, C.A.; Sprave, T.; Fassbender, T.F.; et al. The Impact of the Co-Registration Technique and Analysis Methodology in Comparison Studies between Advanced Imaging Modalities and Whole-Mount-Histology Reference in Primary Prostate Cancer. *Sci. Rep.* **2021**, *11*, 5836. [CrossRef] [PubMed]
24. Hicks, R.M.; Simko, J.P.; Westphalen, A.C.; Nguyen, H.G.; Greene, K.L.; Zhang, L.; Carroll, P.R.; Hope, T.A. Diagnostic Accuracy of 68Ga-PSMA-11 PET/MRI Compared with Multiparametric MRI in the Detection of Prostate Cancer. *Radiology* **2018**, *289*, 730–737. [CrossRef] [PubMed]
25. Eiber, M.; Weirich, G.; Holzapfel, K.; Souvatzoglou, M.; Haller, B.; Rauscher, I.; Beer, A.J.; Wester, H.-J.; Gschwend, J.; Schwaiger, M.; et al. Simultaneous 68Ga-PSMA HBED-CC PET/MRI Improves the Localization of Primary Prostate Cancer. *Eur. Urol.* **2016**, *70*, 829–836. [CrossRef] [PubMed]
26. Scheltema, M.J.; Chang, J.I.; Stricker, P.D.; van Leeuwen, P.J.; Nguyen, Q.A.; Ho, B.; Delprado, W.; Lee, J.; Thompson, J.E.; Cusick, T.; et al. Diagnostic Accuracy of 68Ga-Prostate-Specific Membrane Antigen (PSMA) Positron-Emission Tomography (PET) and Multiparametric (Mp)MRI to Detect Intermediate-Grade Intra-Prostatic Prostate Cancer Using Whole-Mount Pathology: Impact of the Addition of 68Ga-PSMA PET to MpMRI. *BJU Int.* **2019**, *124*, 42–49. [CrossRef]
27. Li, P.; You, S.; Nguyen, C.; Wang, Y.; Kim, J.; Sirohi, D.; Ziembiec, A.; Luthringer, D.; Lin, S.-C.; Daskivich, T.; et al. Genes Involved in Prostate Cancer Progression Determine MRI Visibility. *Theranostics* **2018**, *8*, 1752–1765. [CrossRef]
28. Zhen, L.; Liu, X.; Yegang, C.; Yongjiao, Y.; Yawei, X.; Jiaqi, K.; Xianhao, W.; Yuxuan, S.; Rui, H.; Wei, Z.; et al. Accuracy of Multiparametric Magnetic Resonance Imaging for Diagnosing Prostate Cancer: A Systematic Review and Meta-Analysis. *BMC Cancer* **2019**, *19*, 1244. [CrossRef]
29. Kasivisvanathan, V.; Rannikko, A.S.; Borghi, M.; Panebianco, V.; Mynderse, L.A.; Vaarala, M.H.; Briganti, A.; Budäus, L.; Hellawell, G.; Hindley, R.G.; et al. MRI-Targeted or Standard Biopsy for Prostate-Cancer Diagnosis. *N. Engl. J. Med.* **2018**, *378*, 1767–1777. [CrossRef]
30. Stavrinides, V.; Syer, T.; Hu, Y.; Giganti, F.; Freeman, A.; Karapanagiotis, S.; Bott, S.R.J.; Brown, L.C.; Burns-Cox, N.; Dudderidge, T.J.; et al. False Positive Multiparametric Magnetic Resonance Imaging Phenotypes in the Biopsy-Naïve Prostate: Are They Distinct from Significant Cancer-Associated Lesions? Lessons from PROMIS. *Eur. Urol.* **2021**, *79*, 20–29. [CrossRef]
31. Litjens, G.J.S.; Elliott, R.; Shih, N.N.C.; Feldman, M.D.; Kobus, T.; Hulsbergen-van de Kaa, C.; Barentsz, J.O.; Huisman, H.J.; Madabhushi, A. Computer-Extracted Features Can Distinguish Non-cancerous Confounding Disease from Prostatic Adenocarcinoma at Multiparametric MR Imaging. *Radiology* **2016**, *278*, 135–145. [CrossRef] [PubMed]
32. Tamada, T.; Prabhu, V.; Li, J.; Babb, J.S.; Taneja, S.S.; Rosenkrantz, A.B. Prostate Cancer: Diffusion-Weighted MR Imaging for Detection and Assessment of Aggressiveness-Comparison between Conventional and Kurtosis Models. *Radiology* **2017**, *284*, 100–108. [CrossRef] [PubMed]
33. de Galiza Barbosa, F.; Queiroz, M.A.; Nunes, R.F.; Costa, L.B.; Zaniboni, E.C.; Marin, J.F.G.; Cerri, G.G.; Buchpiguel, C.A. Nonprostatic Diseases on PSMA PET Imaging: A Spectrum of Benign and Malignant Findings. *Cancer Imaging* **2020**, *20*, 23. [CrossRef] [PubMed]
34. Li, M.; Huang, Z.; Yu, H.; Wang, Y.; Zhang, Y.; Song, B. Comparison of PET/MRI with Multiparametric MRI in Diagnosis of Primary Prostate Cancer: A Meta-Analysis. *Eur. J. Radiol.* **2019**, *113*, 225–231. [CrossRef]
35. Matsumoto, K.; Omura, M.; Takeda, T.; Kosaka, T.; Hashiguchi, A.; Takamatsu, K.; Yasumizu, Y.; Tanaka, N.; Morita, S.; Mizuno, R.; et al. Grading of Multifocal Prostate Cancer Cases in Which the Largest Volume and the Highest Grade Do not Coincide within One Lesion. *J. Urol.* **2021**, *206*, 338–345. [CrossRef]
36. Perera, M.; Papa, N.; Roberts, M.; Williams, M.; Udovicich, C.; Vela, I.; Christidis, D.; Bolton, D.; Hofman, M.S.; Lawrentschuk, N.; et al. Gallium-68 Prostate-Specific Membrane Antigen Positron Emission Tomography in Advanced Prostate Cancer—Updated Diagnostic Utility, Sensitivity, Specificity, and Distribution of Prostate-Specific Membrane Antigen-Avid Lesions: A Systematic Review and Meta-Analysis. *Eur. Urol.* **2020**, *77*, 403–417. [CrossRef]
37. Evangelista, L.; Zattoni, F.; Cassarino, G.; Artioli, P.; Cecchin, D.; Dal Moro, F.; Zucchetta, P. PET/MRI in Prostate Cancer: A Systematic Review and Meta-Analysis. *Eur. J. Nucl. Med. Mol. Imaging* **2021**, *48*, 859–873. [CrossRef]

38. Rebello, R.J.; Oing, C.; Knudsen, K.E.; Loeb, S.; Johnson, D.C.; Reiter, R.E.; Gillessen, S.; Van der Kwast, T.; Bristow, R.G. Prostate Cancer. *Nat. Rev. Dis. Primers* **2021**, *7*, 1–27. [CrossRef]
39. Zamboglou, C.; Sachpazidis, I.; Koubar, K.; Drendel, V.; Wiehle, R.; Kirste, S.; Mix, M.; Schiller, F.; Mavroidis, P.; Meyer, P.T.; et al. Evaluation of Intensity Modulated Radiation Therapy Dose Painting for Localized Prostate Cancer Using 68Ga-HBED-CC PSMA-PET/CT: A Planning Study Based on Histopathology Reference. *Radiother. Oncol.* **2017**, *123*, 472–477. [CrossRef]
40. Chan, T.Y.; Partin, A.W.; Walsh, P.C.; Epstein, J.I. Prognostic Significance of Gleason Score 3+4 versus Gleason Score 4+3 Tumor at Radical Prostatectomy. *Urology* **2000**, *56*, 823–827. [CrossRef]
41. van den Bergh, R.C.N.; Roemeling, S.; Roobol, M.J.; Aus, G.; Hugosson, J.; Rannikko, A.S.; Tammela, T.L.; Bangma, C.H.; Schröder, F.H. Gleason Score 7 Screen-Detected Prostate Cancers Initially Managed Expectantly: Outcomes in 50 Men. *BJU Int.* **2009**, *103*, 1472–1477. [CrossRef] [PubMed]
42. Egevad, L.; Delahunt, B.; Srigley, J.R.; Samaratunga, H. International Society of Urological Pathology (ISUP) Grading of Prostate Cancer—An ISUP Consensus on Contemporary Grading. *APMIS* **2016**, *124*, 433–435. [CrossRef] [PubMed]
43. van Leenders, G.J.L.H.; van der Kwast, T.H.; Grignon, D.J.; Evans, A.J.; Kristiansen, G.; Kweldam, C.F.; Litjens, G.; McKenney, J.K.; Melamed, J.; Mottet, N.; et al. The 2019 International Society of Urological Pathology (ISUP) Consensus Conference on Grading of Prostatic Carcinoma. *Am. J. Surg. Pathol.* **2020**, *44*, e87. [CrossRef] [PubMed]
44. Kang, H.C.; Jo, N.; Bamashmos, A.S.; Ahmed, M.; Sun, J.; Ward, J.F.; Choi, H. Accuracy of Prostate Magnetic Resonance Imaging: Reader Experience Matters. *Eur. Urol. Open Sci.* **2021**, *27*, 53–60. [CrossRef]
45. Muller, B.G.; Shih, J.H.; Sankineni, S.; Marko, J.; Rais-Bahrami, S.; George, A.K.; de la Rosette, J.J.M.C.H.; Merino, M.J.; Wood, B.J.; Pinto, P.; et al. Prostate Cancer: Interobserver Agreement and Accuracy with the Revised Prostate Imaging Reporting and Data System at Multiparametric MR Imaging. *Radiology* **2015**, *277*, 741–750. [CrossRef]
46. Sonn, G.A.; Fan, R.E.; Ghanouni, P.; Wang, N.N.; Brooks, J.D.; Loening, A.M.; Daniel, B.L.; To'o, K.J.; Thong, A.E.; Leppert, J.T. Prostate Magnetic Resonance Imaging Interpretation Varies Substantially Across Radiologists. *Eur. Urol. Focus* **2019**, *5*, 592–599. [CrossRef]
47. Giannarini, G.; Valotto, C.; Girometti, R.; Dal Moro, F.; Briganti, A.; Padhani, A.R. Measuring the Quality of Diagnostic Prostate Magnetic Resonance Imaging: A Urologist's Perspective. *Eur. Urol.* **2021**, *79*, 440–441. [CrossRef]
48. Giganti, F.; Kasivisvanathan, V.; Kirkham, A.; Punwani, S.; Emberton, M.; Moore, C.M.; Allen, C. Prostate MRI Quality: A Critical Review of the Last 5 Years and the Role of the PI-QUAL Score. *BJR* **2022**, *95*, 20210415. [CrossRef]
49. Giganti, F.; Lindner, S.; Piper, J.W.; Kasivisvanathan, V.; Emberton, M.; Moore, C.M.; Allen, C. Multiparametric Prostate MRI Quality Assessment Using a Semi-Automated PI-QUAL Software Program. *Eur. Radiol. Exp.* **2021**, *5*, 48. [CrossRef]
50. Toriihara, A.; Nobashi, T.; Baratto, L.; Duan, H.; Moradi, F.; Park, S.; Hatami, N.; Aparici, C.M.; Davidzon, G.; Iagaru, A. Comparison of 3 Interpretation Criteria for 68Ga-PSMA11 PET Based on Inter- and Intrareader Agreement. *J. Nucl. Med.* **2020**, *61*, 533–539. [CrossRef]
51. Ceci, F.; Oprea-Lager, D.E.; Emmett, L.; Adam, J.A.; Bomanji, J.; Czernin, J.; Eiber, M.; Haberkorn, U.; Hofman, M.S.; Hope, T.A.; et al. E-PSMA: The EANM Standardized Reporting Guidelines v1.0 for PSMA-PET. *Eur. J. Nucl. Med. Mol. Imaging* **2021**, *48*, 1626–1638. [CrossRef]
52. Rowe, S.P.; Pienta, K.J.; Pomper, M.G.; Gorin, M.A. Proposal for a Structured Reporting System for Prostate-Specific Membrane Antigen–Targeted PET Imaging: PSMA-RADS Version 1.0. *J. Nucl. Med.* **2018**, *59*, 479–485. [CrossRef]
53. Salami, S.S.; Kaplan, J.B.; Nallandhighal, S.; Takhar, M.; Tosoian, J.J.; Lee, M.; Yoon, J.; Hovelson, D.H.; Plouffe, K.R.; Kaffenberger, S.D.; et al. Biologic Significance of Magnetic Resonance Imaging Invisibility in Localized Prostate Cancer. *JCO Precis. Oncol.* **2019**, *3*, PO.19.00054. [CrossRef]
54. Stabile, A.; Giganti, F.; Rosenkrantz, A.B.; Taneja, S.S.; Villeirs, G.; Gill, I.S.; Allen, C.; Emberton, M.; Moore, C.M.; Kasivisvanathan, V. Multiparametric MRI for Prostate Cancer Diagnosis: Current Status and Future Directions. *Nat. Rev. Urol.* **2020**, *17*, 41–61. [CrossRef]
55. Will, L.; Sonni, I.; Kopka, K.; Kratochwil, C.; Giesel, F.L.; Haberkorn, U. Radiolabeled Prostate-Specific Membrane Antigen Small-Molecule Inhibitors. *Q. J. Nucl. Med. Mol. Imaging* **2017**, *61*, 168–180. [CrossRef]
56. Kaittanis, C.; Andreou, C.; Hieronymus, H.; Mao, N.; Foss, C.A.; Eiber, M.; Weirich, G.; Panchal, P.; Gopalan, A.; Zurita, J.; et al. Prostate-Specific Membrane Antigen Cleavage of Vitamin B9 Stimulates Oncogenic Signaling through Metabotropic Glutamate Receptors. *J. Exp. Med.* **2018**, *215*, 159–175. [CrossRef]

cancers

Review

Clinical Applications of PSMA PET Examination in Patients with Prostate Cancer

Sazan Rasul [1] and Alexander R. Haug [1,2,*]

[1] Department of Biomedical Imaging and Image-Guided Therapy, Division of Nuclear Medicine, Medical University of Vienna, 1090 Vienna, Austria; sazan.rasul@meduniwien.ac.at
[2] Christian Doppler Laboratory for Applied Metabolomics (CDL AM), Medical University of Vienna, 1090 Vienna, Austria
* Correspondence: alexander.haug@meduniwien.ac.at; Tel.: +43-1-40400-39360; Fax: +43-1-40400-55520

Simple Summary: The prostate specific membrane antigens, abbreviated as PSMAs, are type II membrane proteins that are highly ex-pressed on the surface of malignant prostate tissue in prostate cancer (PCa), particularly in aggressive, andro-gen-deprived, metastatic, and hormone-refractory PCa. Today, radionuclides that bind to these PSMA peptides are widely available for diagnostic and therapeutic purposes to specifically image and target prostate tumor cells at molec-ular level. In this descriptive review, we aimed to emphasize the usefulness of PSMA positron emission tomography (PET) examination in the management of patients with various stages of PCa. In addition, we outlined the main pitfalls and limitations of this scan to avoid misinterpretation of the results and to improve the decision making process in rela-tion to the patient's further treatment. We concluded that PSMA PET examination in primary PCa patients has an es-sential role in the high-risk group. It is the new imaging standard in patients with in biochemical recurrence PCa and plays an important role in treatment decision. Furthermore, PSMA PET scan is a gold standard for the evaluation of PSMA targeted therapies in patients having progress of the disease. Future prospective studies, particularly on the im-pact of PSMA PET on therapy stratification, may further strengthen the role of PSMA in the treatment of PCa patients.

Abstract: With the progressive aging of the population in industrially developed countries, as well as advances in diagnostic and biopsy techniques and improvements in patient awareness, the incidence of prostate cancer (PCa) is continuously increasing worldwide. Therefore, PCa is currently considered as the second leading cause of tumor-related death. Early detection of the tumor and its metastasis is essential, as the rate of disease recurrence is high and occurs in 27% to 53% of all patients who underwent curative therapy with radical prostatectomy or local radiotherapy. In this regard, the prostate specific membrane antigens, abbreviated as PSMAs, are type II membrane proteins that are highly expressed on the surface of malignant prostate tissue in PCa, particularly in aggressive, androgen-deprived, metastatic, and hormone-refractory PCa, and they are inversely associated with the androgen level. Up to 95% of adenocarcinomas of the prostate express PSMA receptors on their surface. Today, radionuclides that bind to these PSMA peptides are widely accepted for diagnostic and therapeutic purposes to specifically image and target prostate tumor cells at the molecular level, a process referred to as targeted theranostics. Numerous studies have demonstrated that the integration of these peptides into diagnostic and therapeutic procedures plays a critical role in the primary staging and treatment decisions of especially high-risk PCa, expands therapeutic options for patients with advanced stage of prostate tumor, and prolongs patients' survival rate. In this review article, we intend to briefly spotlight the latest clinical utilization of the PSMA-targeted radioligand PET imaging modality in patients with different stages of PCa. Furthermore, limitations and pitfalls of this diagnostic technique are presented.

Keywords: prostate cancer; diagnosis; PSMA; PSA; tumor; PET scan

Citation: Rasul, S.; Haug, A.R. Clinical Applications of PSMA PET Examination in Patients with Prostate Cancer. *Cancers* **2022**, *14*, 3768. https://doi.org/10.3390/cancers14153768

Academic Editors: José I. López and Claudia Manini

Received: 23 June 2022
Accepted: 31 July 2022
Published: 2 August 2022

1. Introduction

Prostate cancer (PCa) remains one of the primary causes of tumor-related deaths in men. It is estimated that, after lung cancer, PCa is the second leading cause of cancer death, with the average age of men at time of diagnosis being about 66 years. Since the 5 year survival rate of PCa is highly dependent on tumor stage (ranging from 100% in early-stage tumors to as low as 30% in patients with advanced cancer), early detection and accurate staging of the disease are essential [1,2]. Prostate-specific antigen (PSA) that is produced by the epithelial cells of the prostate predicts prostate cancer and provides an important indicator of tumor recurrence after primary therapy for PCa [3]. Although surgical radical prostatectomy (RP) is the curative therapy of PCa, up to 50% of all patients might develop an increase in PSA in terms of biochemical recurrence (BCR) after 5 years. At this clinical stage, systemic androgen deprivation therapies (ADT) may be considered the treatment of choice in most patients. In addition to the serious side-effects of this therapy, such as hot flashes, impotence and sexual dysfunction, metabolic syndrome, bone loss, and increased risk of cardiovascular disease, which negatively affect the quality of life of these patients, patients develop castration resistance after a certain period of therapy.

Prostate-specific membrane antigen, which is abbreviated as PSMA and called glutamate carboxypeptidase type II, is a type II membrane protein, composed of collectively 750 amino acids and located on chromosome 11. Its level increasingly expressed of the surface of the inflammatory, benign, and malignant prostatic tissues. However, the level of this antigen increases up to thousand times its normal level in the case of prostate cancer (PCa), especially in the case of aggressive types with high Gleason scores (GS), as well as in androgen-deprived, metastatic, and hormone-refractory PCa [4].

PSMA was discovered in late 1980s and. since then, it has become an attractive target for the diagnosis and therapy of prostate cancer and its associated metastasis. Therefore, it has become the focus of numerous therapeutic approaches such as PSMA-based immunotherapies, including anti-PSMA antibodies and PSMA-targeted prodrug treatments. In the early 2000s, antibodies directed against the cytoplasmic part of the PSMA antigen were labeled with [^{111}indium] and used for the first time for the scintigraphical imaging [5]. However, to improve image quality and reduce radiation exposure for the patients, the researchers at Heidelberg University were able to develop in 2012 for the first time a [^{68}Ga]Gallium-labeled PSMA ligand suitable for positron emission tomography (PET) scanning that targets the extracellular component of PSMA [6]. In the General Hospital of Vienna, Medical University of Vienna, Austria, we could perform the first [^{68}Ga]Gallium PSMA PET examination for patients with PCa in May 2013. Since then, we have conducted more than 5000 PSMA PET examinations for patients with primary PCa, as well for patients with biochemical recurrent PCa and patients with metastatic hormone-sensitive (mHSPC) and castration-resistant PCa (mCRPC) [7–10].

Meanwhile, there are many available PSMA ligands labeled either with [^{68}Ga]Gallium or [^{18}F]Fluoride such as [^{68}Ga]Ga-PSMA-11, [^{68}Ga]Ga-PSMA-I&T, [^{68}Ga]Ga-PSMA-11 gozetotide, [^{18}F]F-PSMA-1007, [^{18}F]F-DCFPyL, [^{18}F]rh-PSMA7, [^{18}F]JK-PSMA-7, and [^{68}Ga]Ga-PSMAR2. Until now, the United States FDA has only approved [^{18}F]F-DCFPyL and [^{68}Ga]Ga-PSMA-11 gozetotide for clinical use, and they are, therefore, the only PSMA PET tracers that are currently commercially available. The approval of these tracers via FDA was based on clinical outcomes of the phase II/III multicenter Osprey study [11,12], as well as Condor [13] and Vision trials [14], which could demonstrate the safety and the diagnostic performance of these tracers in patients with locally advanced PCa, as well as in patients with metastatic and biochemical recurrent tumor.

In this brief review, we aimed to focus on the current clinically relevant indications of PSMA PET examinations in patients with various stages of PCa. In addition, we aimed to outline the main pitfalls and limitations of this scan to improve the diagnostic and therapeutic management of patients with PCa.

2. Clinical Report of the PSMA PET Scan

Before writing the clinical report and interpreting the PSMA PET scan results, it is important to know that PSMA is not specific to prostate tissue and is physiologically expressed on the cellular surface of other organs such as the lacrimal, parotid, and salivary glands as well as brain, proximal tubule of kidneys, small intestine, liver, and spleen [15]. This physiological expression of PSMA in these organs varies from organ to organ, with the highest PSMA expression in the kidneys and salivary glands and the lowest in brain tissue. However, the intense upregulation of this receptor beyond its normal level results in lesions and metastases associated with PCa having visually and quantitatively much higher PSMA tracer avidity on PET scan; this greatly facilitates their differentiation from background and neighboring organs physiologically expressing PSMA. Furthermore, the presence of computed tomography (CT) and magnetic resonance imaging (MRI) as a part of the integrated PET examination might sometimes play a critical role in the morphological confirmation of these lesions. Nevertheless, PSMA PET scans in patients with PCa should be carefully assessed to avoid pitfalls and misinterpretation of results and findings. After all, this has a direct impact on further steps in the therapy and management of these patients. Moreover, the biodistribution of all abovementioned PSMA PET tracers are almost similar except for [^{18}F]F-PSMA-1007, as this tracer is almost exclusively excreted via the hepatobiliary duct [16].

For lesion evaluation, the European Association of Nuclear Medicine recommends visual and quantitative evaluation of the lesions expressing PSMA. Visual evaluation involves determining the PSMA expression level of the lesion compared to background, also called PSMA expression V. Quantitative evaluation involves measuring the standard uptake values (SUV) of the focal tracer uptake and is referred to as PSMA expression Q [16]. This recommendation was based on the results of the PROMISE study that divided PSMA expression into four scores. Score 0 is when there is no PSMA expression or expression below the blood pool. Score 1 is when expression is low or equal to or above the blood pool and lower than the liver. Score 2 is intermediate with expression equal to or above the liver and lower than the parotid gland. Score 3 is high with expression equal to or above the parotid gland [16]. Lesions with scores of 2 and 3 are considered very typical for prostate cancer-related lesions, and they are favorable for PSMA-directed radioligand therapy.

3. Clinical Indications of PSMA PET Examinations in Prostate Cancer

The most clinically relevant indications for PSMA PET examinations in patients with PCa are in the primary staging of the tumor, in BCR and castration-resistant PCa, and in the evaluation for PSMA targeted therapies. However, it is important to know that up to 5% of prostate adenocarcinomas do not express PSMA [17,18].

3.1. Roles of PSMA PET in Primary Staging of Prostate Cancer

In this setting, although studies showed that the primary tumor is nearly always detected by PSMA PET reaching a sensitivity up to 99.3%, and that the PET metrics correlate with histology grades and International Society of Urological Pathology (ISUP) classifications and GS, high variation in sensitivity (between 33% to 92%) with a very good specificity of 82% to 100% was found in the detection of lymph node metastases [7,19]. Current guidelines of European Association of urology recommend this examination only for high-risk PCa patients, i.e., patients who have an initial PSA level $\geq$ 20 ng/mL, a pathologically ISUP grade of 3 to 5, or a clinical stage of T3 or greater at the time of PCa diagnosis, which means the tumor has exceeded the prostate capsule. In this context and in relation to local grading of the tumor, studies have compared integrated PSMA PET-MRI with multiparametric (mp)MRI in patients with primary PCa. These studies showed that integrated PET-MRI has superior diagnostic accuracy compared with mpMRI and is particularly useful in tumors with equivocal Prostate Imaging Reporting and Data System (PIRADS) classification of 3 [20–22]. Recently, we were able to confirm these results in a meta-analysis of five prospective studies with collectively 497 men and could highlight the favorable diagnostic precision of PSMA PET targeted biopsy to detect clinically significant PCa [23]. Furthermore, local characteristics of the tumor such as tumor infiltration into the seminal vesicle, as well as into the other neighbor organs like the urinary balder, rectum, and neurovascular bundle, can exactly be determined using combined PET-MR examination [7]. Muehlematter et al. could illustrate in a retrospective assessment of 40 consecutive men who performed mpMRI and PSMA PET-MRI followed by a RP due to intermediate- to high-risk PCa a better sensitivity of PSMA-PET/MRI to detect extracapsular extension and infiltration of the seminal vesicle than with mpMRI [24]. Actually, despite the limited availability of comparative head-to-head studies with [^{18}F]F-PSMA-1007 and [^{68}Ga]Ga-PSMA-11 for prostate tumor staging, several studies have reported that both these tracers can equally identify predominant prostate lesions in patients with intermediate- or high-risk PCa, although the nonurinary excretion of [^{18}F]F-PSMA-1007 may help evaluate lesions near the urinary bladder better than [^{68}Ga]Ga-PSMA-11 [25–27].

Concerning the primary staging of the disease, results of the proPSMA study were recently published in Lancet (2020), which was a prospective multicenter two-arm randomized controlled trial including more than 300 patients with histologically confirmed high-risk PCa being considered for curative therapies [28]. One arm received CT and bone scans for the primary staging of their tumor, and the other arm received only PSMA PET scan. Results showed that PSMA PET staging is a suitable replacement for conventional imaging, providing superior accuracy to the combined findings of CT and bone scans, a higher proportion of management changes, and fewer equivocal results. In addition, the radiation exposure of the patients was significantly lower with PSMA PET scan than with the CT and bone scan. Concerning this topic and modifications in the therapy management of patients with primary PCa, Figure 1 presents the PSMA PET-CT images of a 56 year old man, newly diagnosed with PCa: GS (5 + 5) and a PSA value of 45.96 ng/mL at the time of diagnosis. The patient was scheduled for RP and extended pelvic lymphadenectomy. However, the PSMA-PET-CT examination prior to his surgical approach could detect two PSMA-avid bone lesions (frontoparietal in the left side of the skull and in the ninth thoracic vertebra left side), in addition to the primary tumor in the prostate and bilateral iliac lymph node metastases (Figure 1). Both these bone lesions were not visible on CT examination separately performed for the patient. The patient was then treated systemically with hormonal and chemotherapy instead of the primary surgical intent.

3.2. PSMA PET in Biochemical Recurrence PCa

Current guidelines of the European Association of Urology suggest a PSMA PET scan in any case of proven BCR, as well as in patients where the results of the examination will influence the subsequent treatment decisions. Indeed, the cutoff value for the definition of a PSA recurrence, which usually suggests the relapse of the disease, depends strongly on the type of primary tumor treatment. European Association of Urology defines a BCR of PCa as an increase in PSA to at least 0.2 ng/mL after RP with a cutoff value above 0.4 ng/mL most likely to predict further metastases. However, in patients treated primarily with local radiotherapy with or without hormonal therapy, a rise in PSA level of 2 ng/mL or more above the nadir level is determined as BCR based on RTOG-ASTRO Phoenix Consensus Conference in 2006 [29]. In this regard, identification of lesions responsible for the recurrence of the PCa in such a patient cohort, however, could sometimes be like searching for a needle in a haystack, especially in patients with very low PSA recurrence levels. Considering the role of imaging, conventional imaging with bone scan provides very low sensitivity for patients with a PSA level below 2 ng/mL, and positive bone scintigraphy can only be identified in approximately 9.4% of patients at that PSA level. Furthermore, findings indicated that the ability of CT scan to detect biochemical recurrence is only 11–14%. Kane et al., in a retrospective analysis of 132 men with biochemical relapse after RP, displayed a mean PSA level of 27.4 ng/mL, and a PSA velocity of 1.8 ng/mL/month was associated with a positive CT result [30]. Although [^{18}F]choline and [^{18}F]fluciclovine PET examinations have higher sensitivity (up to 90%) and specificity (up to 100%) than bone scans for finding bone metastases in patients with BCR [31,32], sensitivity of both these tracers is highly dependent on the PSA level and is less than 50% at PSA lower than 1.0 ng/mL [33], with poorer sensitivity of [^{18}F]choline for detecting lymph node metastases than [^{18}F]fluciclovine PET scan. However, if we compare PSMA PET with both these tracers, we can see that prospective and retrospective studies could demonstrate significantly higher sensitivity and detection rate of PSMA (58%) than choline (35%) or fluciclovine (23%), especially in patients with a PSA level lower than 0.5 ng/mL [34,35]. Regardless of the PSA level, PSMA PET can also detect small, choline PET-negative lymph node metastasis as could be demonstrated in previous studies, which showed PSMA-positive lymph nodes having diameters of only 6 mm [34]. Furthermore, PSMA PET offers not only higher sensitivity than choline PET but also higher contrast and higher specificity [34,36].

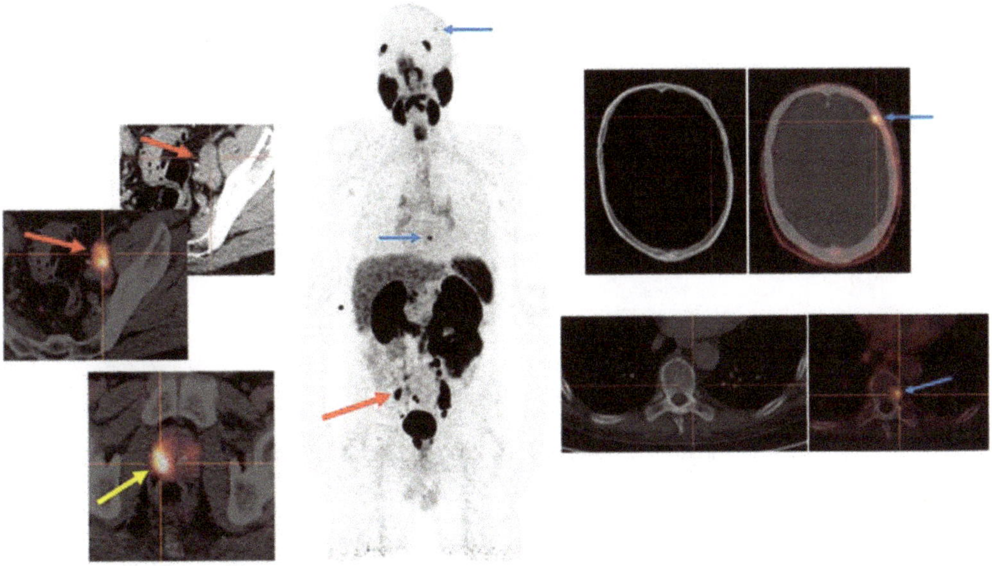

Figure 1. PSMA PET examination for primary staging of a high-risk prostate cancer. PSMA PET-CT Images for primary staging of a 56 year old man, newly diagnosed with PCa: Gleason score (5 + 5) and a PSA value of 45.96 ng/mL showing two PSMA-avid bone lesions frontoparietal in the left side of the skull and the ninth thoracic vertebra (blue arrows) in addition to the primary tumor in the right lobe of the prostate (yellow arrow) and bilateral iliac lymph node metastases (red arrows). Both these bone lesions were not visible on CT examination separately performed for the patient.

Furthermore, the presence and the timepoint of certain treatments, such as ADT, chemotherapy, and immunotherapy, could have an important influence on the sensitivity and specificity of the PSMA PET scan [37]. Although both in vivo and in vitro studies [38–40] have shown that ADT upregulates PSMA expression, with enhanced PSMA expression demonstrated in 55% of patients with high-grade and advanced PCa treated with ADT, it appears that PSMA expression rises with a short duration of ADT and declines with a longer duration of therapy [41,42]. Thus, it is not yet clear exactly how ADT affects the diagnostic value of this scan, necessitating further potential clinical prospective studies to address this issue, since the duration of ADT represents a critical factor.

In fact, the detection rate of PSMA PET increases with higher PSA levels and higher GS as can be seen in a study that included more than 200 patients with PCa. The sensitivity of PSMA PET imaging reached 46% even with a PSA level lower than 0.2 ng/mL [43]. This high sensitivity of PSMA to detect metastasis plays an important role in the treatment decision of prostate cancer patients. In a large meta-analysis that included 15 studies with more than 1100 patients mainly with BCR but also with primary PCa, the high impact of PSMA on changing the therapy and management of patients was demonstrated. With the help of the PSMA PET imaging, patients received more surgery and more radiation therapy than systemic treatments with hormonal or chemotherapies [44].

In this setting, it is further important to mention the study by Jentjens et al. [45] that included 34 patients with BCR, which prospectively underwent a PSMA PET-CT followed by PSMA PET-MRI scan. Results revealed significantly higher sensitivity of PET-CT and PET-MRI compared to CT and MRI. Although PET-MRI could detect more lesions than PET-CT, results indicated no significant differences between these two modalities in the detection of the presence of local recurrence or distant metastases. One other study could indicate the superiority of PET-MRI over PET-CT in the identification of local PCa recurrence in 119 patients with BCR due to the additional diagnostic information derived from MRI [46].

3.3. Evaluation of PCa Patients for Possibility of PSMA Radioligand Therapies

Another important clinical indication of PSMA PET examination in patients with PCa is to reveal whether their tumor-related metastases express PSMA for the purpose of potential PSMA radioligand therapies. The most common radionuclide for therapy is [^{177}Lu]Lutetium-labeled PSMA-617 ligand, which is generally abbreviated as PSMA-RLT. Although the rate of response to this therapy varies from 60–75% [47,48], treatment options for mCRPC patients have expanded since the introduction of this therapy, and numerous studies have verified the antitumor efficacy, as well as the good tolerability and favorable response rates, of PSMA-RLT [49–52]. However, it is currently used mainly as a last-line treatment when other available standard therapies fail, and the therapeutic protocols are very heterogeneous in terms of different treatment cycles, as well as different therapy doses and different inter-cycle intervals, depending on the general medical status of the patients. Multiple baseline characteristics such as the distribution of metastases, history of pretreatment with hormonal and chemotherapies, number and interval between PSMA-RLT, and the baseline serum levels of PSA, platelets, hemoglobin, alkaline phosphatase, and lactate dehydrogenase have been found to play an essential role in helping to anticipate the response to PSMA-RLT [47,53–56]. In addition, identification of the metabolic activity of the metastatic tumors using [^{18}F]fluorodeoxyglucose (FDG) aids in detecting the aggressive nature of these lesions [57] and allows prediction of their response to PSMA-RLT prior to therapy; lesions that are [^{18}F]FDG-positive but PSMA-negative on PET examination represent more aggressive and metabolically active lesions with worse outcome and response to PSMA-RLT than PSMA-positive but FDG-negative ones [58,59].

In this context, two significant studies in the field of PSMA-RLT are worth mentioning. One is the TheraP study, which was a multicenter, unblinded randomized phase II trial, involving 11 centers in Australia and including about 200 mCRPC patients; 98 of these patients were treated with the PSMA-RLT and 85 were treated with cabazitaxel chemotherapy at the same stage of the disease [60]. Results of the study revealed that patients who received PSMA-RLT hade a significantly higher percentage of PSA response and fewer serious adverse events than patients receiving cabazitaxel. Another important study is the VISION study sponsored by Novartis, which was an international prospective randomized open-label multicenter phase III study [14]. The study included two groups of patients with mCRPC; one group received PSMA RLT plus best standard of care (551) compared to a group of patients that received only best standard of care in the control arm. In this trial, all patients with one and/or more PSMA-negative lesions on PSMA PET scan were excluded from the study. Results showed that men in the investigational arm had a 38% reduction in risk of death and a 60% reduction in risk of radiographic disease progression or death compared to men in the control group, with a significant improvement in time to first symptomatic skeletal event, overall response rate, and disease control rate. Notably, the VISION study had 20% fewer patients with PSA reductions greater than 50% than the

TheraP trial. Factors that might contribute to these results were the higher exclusion rate of patients in the TheraP study compared with the VISION study, as the study excluded all patients with FDG-positive but PSMA-negative lesions on their PET scan, and the stricter selection criteria in the TheraP study than in the VISION study, as only patients with lesions with a maximum standardized PSMA uptake value of ≥20 were enrolled.

Figure 2 is an example of the role of PSMA PET examination in a patient with mCRPC, who had a PSA level of 65.74 ng/m before PSMA-RLT. After receiving six cycles of this therapy, the PSA level declined to 1.15 mg/mL, and the metastases largely disappeared on the PSMA PET scan.

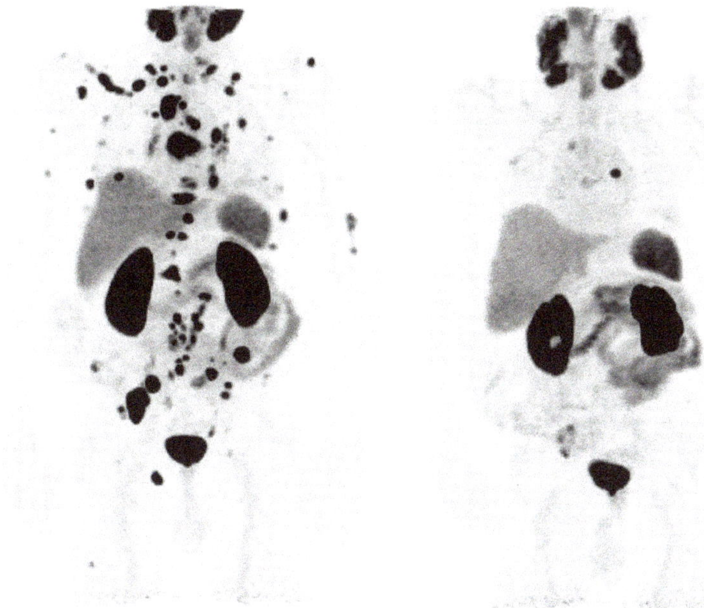

Figure 2. PSMA PET examination for evaluation of PSMA radioligand therapy. PSMA PET examination of a patient with metastatic castration-resistant prostate cancer and a PSA level of 65.74 ng/mL. Before PSMA-RLT (**left**), showing multiple bone and lymph node metastasis with highly increased PSMA-expression. After six cycles of PSMA-RLT and a PSA decline to only 1.15 ng/mL (**right**), showing a clear reduction in the number of PSMA-expressing bone and lymph node metastases.

4. Pitfalls and Limitation of PSMA PET Examination

As mentioned earlier, PSMA is not specific to prostate tissue, and physiologically PSMA-expressing areas or organs (lacrimal glands, parotid and salivary glands, brain, and proximal tubules of kidneys, small intestine, liver, and spleen) should be well distinguished from lesions that pathologically express PSMA. Furthermore, PSMA radioligand and especially [^{68}Ga]Ga-PSMA will be excreted via kidney and urinary bladder. Thus, an accumulation of this tracer throughout the urinary tract or in a specific segment of the ureter might be misinterpreted as a suspicious lymph node. Furthermore, [^{68}Ga]Ga-PSMA ligand uptake in sympathetic ganglia such as cervical, celiac, and sacral ganglia is very common should be carefully taken into consideration. These focal uptakes, however, are usually symmetrical and bilateral, and they could radiologically be differentiated from lymph nodes [61]. Moreover, some comparative studies demonstrated identification of nonspecific bone lesions with [^{18}F]PSMA-1007 but not with other PSMA PET tracers in patients with PCa [62,63].

Figure 3 displays the case of a 74 year old PCa patient treated with RP in 2011 with initial PSA of 14 ng/mL and GS 7 (3 + 4). After prostatectomy, the patient received no further therapies. Seven years later, the patient developed BCR with a PSA level of 0.7 ng/mL; a few months later, the PSA reached 1.22 ng/mL with a PSA doubling time of about 12 months. A PSMA PET examination was performed for the patient. The scan showed PSMA-positive bone lesions in the left iliac bone and in the vertebral arch of the fourth lumbar spine. In addition, PSMA-positive intrahepatic lesions were detected in liver segments IVa and IVb. However, no local recurrence and no suspicious lymph nodes were found. This case was discussed by our institutional tumor board, and a histological clarification of these intrahepatic PSMA-avid lesions was strongly recommended as PSA levels and the extent of liver metastases did not match well. Histopathology results demonstrated hepatocellular carcinoma rather than prostate cancer metastases in these lesions.

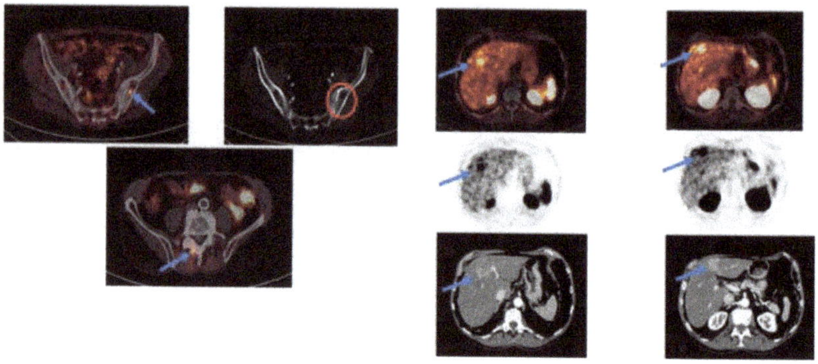

(**A**) (**B**)

Figure 3. PSMA PET examination for biochemical recurrence of prostate cancer. [^{68}Ga]Ga-PSMA PET-CT of a 74 year old patient with biochemical recurrence of prostate cancer revealed (**A**) PSMA-positive bone lesions in the left iliac bone with corresponding bone changes on the CT scan (red circle) and the vertebral arch of the fourth lumbar spine. (**B**) Additional PSMA-avid and contrast-enhanced intrahepatic lesions in liver segments IVa and IVb. Histological examination of these hepatic lesions demonstrated hepatocellular carcinoma.

Indeed, expression of PSMA is not specific to prostate tissue and PCa, as it might be increasingly expressed in the process of neovascularization of numerous solid tumors such as renal cell cancer [64], hepatocellular carcinoma, and squamous cell carcinoma of head and neck [65] (see Figure 4). In a prospective pilot study by Kessler et al. including 37 suspected malignant hepatic lesions of seven patients, results revealed that more than 95% hepatocellular carcinoma lesions expressed PSMA [66]. Consequently, in cases of low PSA levels and large PSMA-expressing organ lesions, possible etiologies other than PCa should be considered for the lesion. In this sense, in a review publication by Osmany et al. [67], the authors could show a list of benign and malignant conditions that might be associated with the increased expression of PSMA, including hepatic hemangioma, fractures, neuroendocrine tumor, lymphoma, multiple myeloma, osteosarcoma, breast cancer, and adenocarcinoma of the lung.

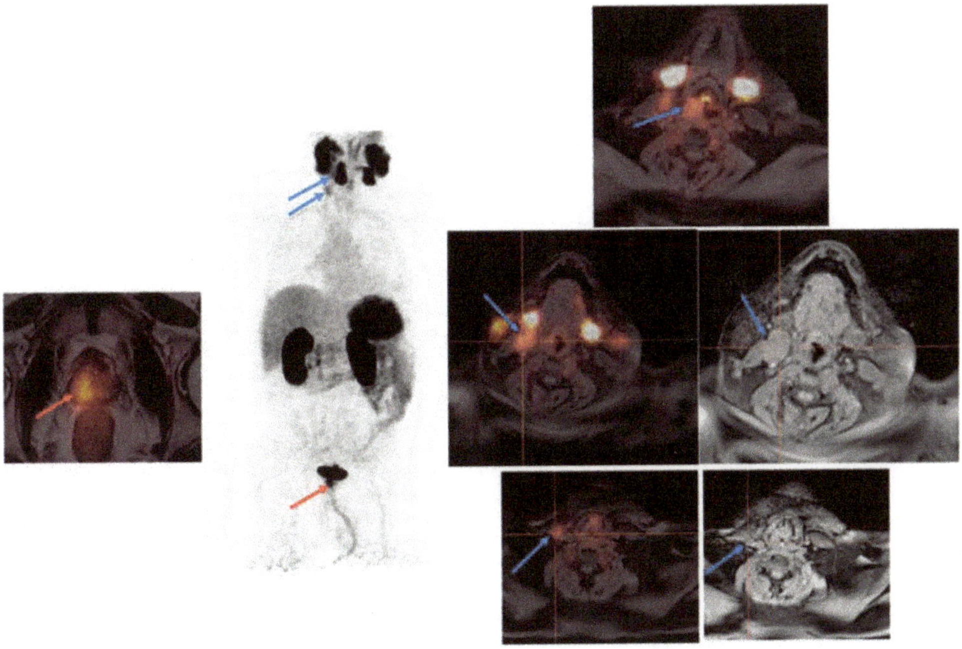

Figure 4. PSMA PET examination for primary staging of a high-risk prostate cancer. PSMA PET examination for primary staging of a 60 year old patient with a newly diagnosed high-risk prostate cancer (PSA value: 30.9 ng/mL) revealed the primary tumor in the right lobe of the prostate with a clear PSMA-expression (red arrow) and an incidental detection of a suspicious lesion with marked PSMA expression in the right piriform sinus (area of the right vocal cord) with multiple suspicious lymph nodes in the right neck region (blue arrows). The final histological examination of this neck lesion confirmed squamous cell carcinoma of the larynx.

5. Conclusions and Outlooks

This review, which is descriptive rather than systematic, emphasized the usefulness of PSMA PET examination in the management of patients with various stages of PCa. It should provide clinicians with a concise overview of the clinical potential of this diagnostic imaging modality in patients with PCa. In addition, it was intended to help ascertain the pitfalls and vulnerabilities associated with this scan to avoid misinterpretation of the results and to facilitate the decision making process in relation to the patient's further treatment.

We concluded that PSMA PET examination in PCa patients has an essential role in the high-risk group. It is the new gold standard in patients with BCR and plays an important role in treatment decision and selection of possible therapy options such as surgery, radiation, hormone, or chemotherapy treatment. Additionally, PSMA PET is a gold standard for the evaluation of PSMA targeted therapies in patients having progress of the disease after obtaining other available standard therapies.

Future prospective studies, particularly on the impact of PSMA PET on therapy stratification, may further strengthen the role of PSMA in the treatment of PCa patients. Nevertheless, the main issue with this investigation lies in its limited availability and reimbursement in hospitals and clinical centers, which reduce patients' access to this valuable diagnostic imaging modality.

Author Contributions: S.R. wrote the manuscript, conceived the paper, collected and researched the data, contributed to discussion. A.R.H. contributed to discussion, reviewed and edited the manuscript. Both authors have read and approved the final version of the manuscript.

Funding: This review paper received no external funding.

Conflicts of Interest: The authors declare no conflict of interest.

References

1. Hull, G.W.; Rabbani, F.; Abbas, F.; Wheeler, T.M.; Kattan, M.W.; Scardino, P.T. Cancer control with radical prostatectomy alone in 1,000 consecutive patients. *J. Urol.* **2002**, *167*, 528–534. [CrossRef]
2. Kupelian, P.; Katcher, J.; Levin, H.; Zippe, C.; Klein, E. Correlation of clinical and pathologic factors with rising prostate-specific antigen profiles after radical prostatectomy alone for clinically localized prostate cancer. *Urology* **1996**, *48*, 249–260. [CrossRef]
3. Chang, S.S. Overview of prostate-specific membrane antigen. *Rev. Urol.* **2004**, *6* (Suppl. S10), S13–S18.
4. Silver, D.A.; Pellicer, I.; Fair, W.R.; Heston, W.D.; Cordon-Cardo, C. Prostate-specific membrane antigen expression in normal and malignant human tissues. *Clin. Cancer Res.* **1997**, *3*, 81–85.
5. Petronis, J.D.; Regan, F.; Lin, K. Indium-111 capromab pendetide (ProstaScint) imaging to detect recurrent and metastatic prostate cancer. *Clin. Nucl. Med.* **1998**, *23*, 672–677. [CrossRef]
6. Afshar-Oromieh, A.; Malcher, A.; Eder, M.; Eisenhut, M.; Linhart, H.G.; Hadaschik, B.A.; Holland-Letz, T.; Giesel, F.L.; Kratochwil, C.; Haufe, S.; et al. PET imaging with a [^{68}Ga] gallium-labelled PSMA ligand for the diagnosis of prostate cancer: Biodistribution in humans and first evaluation of tumour lesions. *Eur. J. Nucl. Med. Mol. Imaging* **2013**, *40*, 486–495. [CrossRef]
7. Grubmuller, B.; Baltzer, P.; Hartenbach, S.; D'Andrea, D.; Helbich, T.H.; Haug, A.R.; Goldner, G.M.; Wadsak, W.; Pfaff, S.; Mitterhauser, M.; et al. PSMA ligand PET/MRI for primary prostate cancer: Staging performance and clinical impact. *Clin. Cancer Res.* **2018**, *24*, 6300–6307. [CrossRef]
8. Grubmuller, B.; Baltzer, P.; D'Andrea, D.; Korn, S.; Haug, A.R.; Hacker, M.; Grubmuller, K.H.; Goldner, G.M.; Wadsak, W.; Pfaff, S.; et al. ^{68}Ga-PSMA 11 ligand PET imaging in patients with biochemical recurrence after radical prostatectomy-diagnostic performance and impact on therapeutic decision-making. *Eur. J. Nucl. Med. Mol. Imaging* **2018**, *45*, 235–242. [CrossRef]
9. Grubmuller, B.; Senn, D.; Kramer, G.; Baltzer, P.; D'Andrea, D.; Grubmuller, K.H.; Mitterhauser, M.; Eidherr, H.; Haug, A.R.; Wadsak, W.; et al. Response assessment using ^{68}Ga-PSMA ligand PET in patients undergoing ^{177}Lu-PSMA radioligand therapy for metastatic castration-resistant prostate cancer. *Eur. J. Nucl. Med. Mol. Imaging* **2019**, *46*, 1063–1072. [CrossRef]
10. Grubmuller, B.; Rasul, S.; Baltzer, P.; Fajkovic, H.; D'Andrea, D.; Berndl, F.; Maj-Hes, A.; Grubmuller, K.H.; Mitterhauser, M.; Wadsak, W.; et al. Response assessment using [^{68}Ga] Ga-PSMA ligand PET in patients undergoing systemic therapy for metastatic castration-resistant prostate cancer. *Prostate* **2020**, *80*, 74–82. [CrossRef]
11. Rowe, S.P.; Campbell, S.P.; Mana-Ay, M.; Szabo, Z.; Allaf, M.E.; Pienta, K.J.; Pomper, M.G.; Ross, A.E.; Gorin, M.A. Prospective evaluation of PSMA-targeted ^{18}F-DCFPyL PET/CT in men with biochemical failure after radical prostatectomy for prostate cancer. *J. Nucl. Med.* **2020**, *61*, 58–61. [CrossRef]
12. Rowe, S.P.; Li, X.; Trock, B.J.; Werner, R.A.; Frey, S.; DiGianvittorio, M.; Bleiler, J.K.; Reyes, D.K.; Abdallah, R.; Pienta, K.J.; et al. Prospective comparison of PET imaging with PSMA-targeted ^{18}F-DCFPyL versus Na^{18}F for bone lesion detection in patients with metastatic prostate cancer. *J. Nucl. Med.* **2020**, *61*, 183–188. [CrossRef]
13. Morris, M.J.; Rowe, S.P.; Gorin, M.A.; Saperstein, L.; Pouliot, F.; Josephson, D.; Wong, J.Y.C.; Pantel, A.R.; Cho, S.Y.; Gage, K.L.; et al. Diagnostic performance of ^{18}F-DCFPyL-PET/CT in men with biochemically recurrent prostate cancer: Results from the CONDOR phase III, multicenter study. *Clin. Cancer Res.* **2021**, *27*, 3674–3682. [CrossRef] [PubMed]
14. Sartor, O.; de Bono, J.; Chi, K.N.; Fizazi, K.; Herrmann, K.; Rahbar, K.; Tagawa, S.T.; Nordquist, L.T.; Vaishampayan, N.; El-Haddad, G.; et al. Lutetium-177–PSMA-617 for metastatic castration-resistant prostate cancer. *N. Engl. J. Med.* **2021**, *385*, 1091–1103. [CrossRef] [PubMed]
15. Kinoshita, Y.; Kuratsukuri, K.; Landas, S.; Imaida, K.; Rovito, P.M., Jr.; Wang, C.Y.; Haas, G.P. Expression of prostate-specific membrane antigen in normal and malignant human tissues. *World J. Surg.* **2006**, *30*, 628–636. [CrossRef]
16. Eiber, M.; Herrmann, K.; Calais, J.; Hadaschik, B.; Giesel, F.L.; Hartenbach, M.; Hope, T.; Reiter, R.; Maurer, T.; Weber, W.A.; et al. Prostate cancer molecular imaging standardized evaluation (PROMISE): Proposed miTNM classification for the interpretation of PSMA-ligand PET/CT. *J. Nucl. Med.* **2018**, *59*, 469–478. [CrossRef] [PubMed]
17. Maurer, T.; Gschwend, J.E.; Rauscher, I.; Souvatzoglou, M.; Haller, B.; Weirich, G.; Wester, H.J.; Heck, M.; Kubler, H.; Beer, A.J.; et al. Diagnostic efficacy of ^{68}gallium-PSMA positron emission tomography compared to conventional imaging for lymph node staging of 130 consecutive patients with intermediate to high risk prostate cancer. *J. Urol.* **2016**, *195*, 1436–1443. [CrossRef] [PubMed]
18. Fendler, W.P.; Schmidt, D.F.; Wenter, V.; Thierfelder, K.M.; Zach, C.; Stief, C.; Bartenstein, P.; Kirchner, T.; Gildehaus, F.J.; Gratzke, C.; et al. ^{68}Ga-PSMA PET/CT detects the location and extent of primary prostate cancer. *J. Nucl. Med.* **2016**, *57*, 1720–1725. [CrossRef]
19. Herlemann, A.; Wenter, V.; Kretschmer, A.; Thierfelder, K.M.; Bartenstein, P.; Faber, C.; Gildehaus, F.J.; Stief, C.G.; Gratzke, C.; Fendler, W.P. ^{68}Ga-PSMA positron emission tomography/computed tomography provides accurate staging of lymph node regions prior to lymph node dissection in patients with prostate cancer. *Eur. Urol.* **2016**, *70*, 553–557. [CrossRef] [PubMed]

20. Al-Bayati, M.; Grueneisen, J.; Lutje, S.; Sawicki, L.M.; Suntharalingam, S.; Tschirdewahn, S.; Forsting, M.; Rubben, H.; Herrmann, K.; Umutlu, L.; et al. Integrated ^{68}gallium labelled prostate-specific membrane antigen-11 positron emission tomography/magnetic resonance imaging enhances discriminatory power of multi-parametric prostate magnetic resonance imaging. *Urol. Int.* **2018**, *100*, 164–171. [CrossRef] [PubMed]
21. Park, S.Y.; Zacharias, C.; Harrison, C.; Fan, R.E.; Kunder, C.; Hatami, N.; Giesel, F.; Ghanouni, P.; Daniel, B.; Loening, A.M.; et al. Gallium 68 PSMA-11 PET/MR imaging in patients with intermediate-or high-risk prostate cancer. *Radiology* **2018**, *288*, 495–505. [CrossRef] [PubMed]
22. Eiber, M.; Weirich, G.; Holzapfel, K.; Souvatzoglou, M.; Haller, B.; Rauscher, I.; Beer, A.J.; Wester, H.J.; Gschwend, J.; Schwaiger, M.; et al. Simultaneous ^{68}Ga-PSMA HBED-CC PET/MRI improves the localization of primary prostate cancer. *Eur. Urol.* **2016**, *70*, 829–836. [CrossRef]
23. Kawada, T.; Yanagisawa, T.; Rajwa, P.; Sari Motlagh, R.; Mostafaei, H.; Quhal, F.; Laukhtina, E.; Aydh, A.; Konig, F.; Pallauf, M.; et al. Diagnostic performance of prostate-specific membrane antigen positron emission tomography-targeted biopsy for detection of clinically significant prostate cancer: A systematic review and meta-analysis. *Eur. Urol. Oncol.* **2022**, *5*, 390–400. [CrossRef] [PubMed]
24. Muehlematter, U.J.; Burger, I.A.; Becker, A.S.; Schawkat, K.; Hotker, A.M.; Reiner, C.S.; Muller, J.; Rupp, N.J.; Ruschoff, J.H.; Eberli, D.; et al. Diagnostic accuracy of multiparametric MRI versus Ga-PSMA-11 PET/MRI for extracapsular extension and seminal vesicle invasion in patients with prostate cancer. *Radiology* **2019**, *293*, 350–358. [CrossRef] [PubMed]
25. Hoberuck, S.; Löck, S.; Borkowetz, A.; Sommer, U.; Winzer, R.; Zophel, K.; Fedders, D.; Michler, E.; Kotzerke, J.; Kopka, K.; et al. Intraindividual comparison of [^{68}Ga]-Ga-PSMA-11 and [^{18}F]-F-PSMA-1007 in prostate cancer patients: A retrospective single-center analysis. *EJNMMI Res.* **2021**, *11*, 109. [CrossRef]
26. Kuten, J.; Fahoum, I.; Savin, Z.; Shamni, O.; Gitstein, G.; Hershkovitz, D.; Mabjeesh, N.J.; Yossepowitch, O.; Mishani, E.; Even-Sapir, E. Head-to-head comparison of ^{68}Ga-PSMA-11 with ^{18}F-PSMA-1007 PET/CT in staging prostate cancer using histopathology and immunohistochemical analysis as a reference standard. *J. Nucl. Med.* **2020**, *61*, 527–532. [CrossRef]
27. Pattison, D.A.; Debowski, M.; Gulhane, B.; Arnfield, E.G.; Pelecanos, A.M.; Garcia, P.L.; Latter, M.J.; Lin, C.Y.; Roberts, M.J.; Ramsay, S.C.; et al. Prospective intra-individual blinded comparison of [^{18}F] PSMA-1007 and [^{68}Ga]Ga-PSMA-11 PET/CT imaging in patients with confirmed prostate cancer. *Eur. J. Nucl. Med. Mol. Imaging* **2022**, *49*, 763–776. [CrossRef]
28. Hofman, M.S.; Lawrentschuk, N.; Francis, R.J.; Tang, C.; Vela, I.; Thomas, P.; Rutherford, N.; Martin, J.M.; Frydenberg, M.; Shakher, R.; et al. Prostate-specific membrane antigen PET-CT in patients with high-risk prostate cancer before curative-intent surgery or radiotherapy (proPSMA): A prospective, randomised, multicentre study. *Lancet* **2020**, *395*, 1208–1216. [CrossRef]
29. Roach, M., III; Hanks, G.; Thames, H., Jr.; Schellhammer, P.; Shipley, W.U.; Sokol, G.H.; Sandler, H. Defining biochemical failure following radiotherapy with or without hormonal therapy in men with clinically localized prostate cancer: Recommendations of the RTOG-ASTRO Phoenix Consensus Conference. *Int. J. Radiat. Oncol. Biol. Phys.* **2006**, *65*, 965–974. [CrossRef] [PubMed]
30. Kane, C.J.; Amling, C.L.; Johnstone, P.A.; Pak, N.; Lance, R.S.; Thrasher, J.B.; Foley, J.P.; Riffenburgh, R.H.; Moul, J.W. Limited value of bone scintigraphy and computed tomography in assessing biochemical failure after radical prostatectomy. *Urology* **2003**, *61*, 607–611. [CrossRef]
31. Fuccio, C.; Castellucci, P.; Schiavina, R.; Santi, I.; Allegri, V.; Pettinato, V.; Boschi, S.; Martorana, G.; Al-Nahhas, A.; Rubello, D.; et al. Role of ^{11}C-choline PET/CT in the restaging of prostate cancer patients showing a single lesion on bone scintigraphy. *Ann. Nucl. Med.* **2010**, *24*, 485–492. [CrossRef]
32. Chau, A.; Gardiner, P.; Colletti, P.M.; Jadvar, H. Diagnostic performance of 18F-fluciclovine in detection of prostate cancer bone metastases. *Clin. Nucl. Med.* **2018**, *43*, e226–e231. [CrossRef] [PubMed]
33. Nanni, C.; Zanoni, L.; Pultrone, C.; Schiavina, R.; Brunocilla, E.; Lodi, F.; Malizia, C.; Ferrari, M.; Rigatti, P.; Fonti, C.; et al. ^{18}F-FACBC (anti1-amino-3-^{18}F-fluorocyclobutane-1-carboxylic acid) versus ^{11}C-choline PET/CT in prostate cancer relapse: Results of a prospective trial. *Eur. J. Nucl. Med. Mol. Imaging* **2016**, *43*, 1601–1610. [CrossRef] [PubMed]
34. Morigi, J.J.; Stricker, P.D.; van Leeuwen, P.J.; Tang, R.; Ho, B.; Nguyen, Q.; Hruby, G.; Fogarty, G.; Jagavkar, R.; Kneebone, A.; et al. Prospective comparison of ^{18}F-fluoromethylcholine versus ^{68}Ga-PSMA PET/CT in prostate cancer patients who have rising PSA after curative treatment and are being considered for targeted therapy. *J. Nucl. Med* **2015**, *56*, 1185–1190. [CrossRef]
35. Wang, R.; Shen, G.; Huang, M.; Tian, R. The diagnostic role of ^{18}F-choline, ^{18}F-fluciclovine and ^{18}F-PSMA PET/CT in the detection of prostate cancer with biochemical recurrence: A meta-analysis. *Front. Oncol.* **2021**, *11*, 684629. [CrossRef]
36. Afshar-Oromieh, A.; Zechmann, C.M.; Malcher, A.; Eder, M.; Eisenhut, M.; Linhart, H.G.; Holland-Letz, T.; Hadaschik, B.A.; Giesel, F.L.; Debus, J.; et al. Comparison of PET imaging with a ^{68}Ga-labelled PSMA ligand and ^{18}F-choline-based PET/CT for the diagnosis of recurrent prostate cancer. *Eur. J. Nucl. Med. Mol. Imaging* **2014**, *41*, 11–20. [CrossRef]
37. Vaz, S.; Hadaschik, B.; Gabriel, M.; Herrmann, K.; Eiber, M.; Costa, D. Influence of androgen deprivation therapy on PSMA expression and PSMA-ligand PET imaging of prostate cancer patients. *Eur. J. Nucl. Med. Mol. Imaging* **2020**, *47*, 9–15. [CrossRef]
38. Wright, G.L., Jr.; Grob, B.M.; Haley, C.; Grossman, K.; Newhall, K.; Petrylak, D.; Troyer, J.; Konchuba, A.; Schellhammer, P.F.; Moriarty, R. Upregulation of prostate-specific membrane antigen after androgen-deprivation therapy. *Urology* **1996**, *48*, 326–334. [CrossRef]
39. Murga, J.D.; Moorji, S.M.; Han, A.Q.; Magargal, W.W.; DiPippo, V.A.; Olson, W.C. Synergistic co-targeting of prostate-specific membrane antigen and androgen receptor in prostate cancer. *Prostate* **2015**, *75*, 242–254. [CrossRef]

40. Evans, M.J.; Smith-Jones, P.M.; Wongvipat, J.; Navarro, V.; Kim, S.; Bander, N.H.; Larson, S.M.; Sawyers, C.L. Noninvasive measurement of androgen receptor signaling with a positron-emitting radiopharmaceutical that targets prostate-specific membrane antigen. *Proc. Natl. Acad. Sci. USA* **2011**, *108*, 9578–9582. [CrossRef]
41. Hope, T.A.; Truillet, C.; Ehman, E.C.; Afshar-Oromieh, A.; Aggarwal, R.; Ryan, C.J.; Carroll, P.R.; Small, E.J.; Evans, M.J. ^{68}Ga-PSMA-11 PET imaging of response to androgen receptor inhibition: First human experience. *J. Nucl. Med.* **2017**, *58*, 81–84. [CrossRef] [PubMed]
42. Liu, T.; Wu, L.Y.; Fulton, M.D.; Johnson, J.M.; Berkman, C.E. Prolonged androgen deprivation leads to downregulation of androgen receptor and prostate-specific membrane antigen in prostate cancer cells. *Int. J. Oncol.* **2012**, *41*, 2087–2092. [CrossRef] [PubMed]
43. Afshar-Oromieh, A.; Holland-Letz, T.; Giesel, F.L.; Kratochwil, C.; Mier, W.; Haufe, S.; Debus, N.; Eder, M.; Eisenhut, M.; Schafer, M.; et al. Diagnostic performance of ^{68}Ga-PSMA-11 (HBED-CC) PET/CT in patients with recurrent prostate cancer: Evaluation in 1007 patients. *Eur. J. Nucl. Med. Mol. Imaging* **2017**, *44*, 1258–1268. [CrossRef]
44. Han, S.; Woo, S.; Kim, Y.J.; Suh, C.H. Impact of ^{68}Ga-PSMA PET on the management of patients with prostate cancer: A systematic review and meta-analysis. *Eur. Urol.* **2018**, *74*, 179–190. [CrossRef]
45. Jentjens, S.; Mai, C.; Ahmadi Bidakhvidi, N.; De Coster, L.; Mertens, N.; Koole, M.; Everaerts, W.; Joniau, S.; Oyen, R.; Van Laere, K.; et al. Prospective comparison of simultaneous [^{68}Ga]Ga-PSMA-11 PET/MR versus PET/CT in patients with biochemically recurrent prostate cancer. *Eur. Radiol.* **2022**, *32*, 901–911. [CrossRef]
46. Freitag, M.T.; Radtke, J.P.; Afshar-Oromieh, A.; Roethke, M.C.; Hadaschik, B.A.; Gleave, M.; Bonekamp, D.; Kopka, K.; Eder, M.; Heusser, T.; et al. Local recurrence of prostate cancer after radical prostatectomy is at risk to be missed in ^{68}Ga-PSMA-11-PET of PET/CT and PET/MRI: Comparison with mpMRI integrated in simultaneous PET/MRI. *Eur. J. Nucl. Med. Mol. Imaging* **2017**, *44*, 776–787. [CrossRef]
47. Rasul, S.; Hartenbach, M.; Wollenweber, T.; Kretschmer-Chott, E.; Grubmuller, B.; Kramer, G.; Shariat, S.; Wadsak, W.; Mitterhauser, M.; Pichler, V.; et al. Prediction of response and survival after standardized treatment with 7400 MBq ^{177}Lu-PSMA-617 every 4 weeks in patients with metastatic castration-resistant prostate cancer. *Eur. J. Nucl. Med. Mol. Imaging* **2021**, *48*, 1650–1657. [CrossRef]
48. Ahmadzadehfar, H.; Eppard, E.; Kurpig, S.; Fimmers, R.; Yordanova, A.; Schlenkhoff, C.D.; Gartner, F.; Rogenhofer, S.; Essler, M. Therapeutic response and side effects of repeated radioligand therapy with ^{177}Lu-PSMA-DKFZ-617 of castrate-resistant metastatic prostate cancer. *Oncotarget* **2016**, *7*, 12477–12488. [CrossRef]
49. McBean, R.; O'Kane, B.; Parsons, R.; Wong, D. Lu177-PSMA therapy for men with advanced prostate cancer: Initial 18 months experience at a single Australian tertiary institution. *J. Med. Imaging Radiat. Oncol.* **2019**, *63*, 538–545. [CrossRef]
50. Ahmadzadehfar, H.; Rahbar, K.; Kurpig, S.; Bogemann, M.; Claesener, M.; Eppard, E.; Gartner, F.; Rogenhofer, S.; Schafers, M.; Essler, M. Early side effects and first results of radioligand therapy with ^{177}Lu-DKFZ-617 PSMA of castrate-resistant metastatic prostate cancer: A two-centre study. *EJNMMI Res.* **2015**, *5*, 36. [CrossRef] [PubMed]
51. Rahbar, K.; Bode, A.; Weckesser, M.; Avramovic, N.; Claesener, M.; Stegger, L.; Bogemann, M. Radioligand therapy with 177Lu-PSMA-617 as a novel therapeutic option in patients with metastatic castration resistant prostate cancer. *Clin. Nucl. Med.* **2016**, *41*, 522–528. [CrossRef] [PubMed]
52. Rasul, S.; Hacker, M.; Kretschmer-Chott, E.; Leisser, A.; Grubmuller, B.; Kramer, G.; Shariat, S.; Wadsak, W.; Mitterhauser, M.; Hartenbach, M.; et al. Clinical outcome of standardized ^{177}Lu-PSMA-617 therapy in metastatic prostate cancer patients receiving 7400 MBq every 4 weeks. *Eur. J. Nucl. Med. Mol. Imaging* **2020**, *47*, 713–720. [CrossRef] [PubMed]
53. Rathke, H.; Holland-Letz, T.; Mier, W.; Flechsig, P.; Mavriopoulou, E.; Rohrich, M.; Kopka, K.; Hohenfellner, M.; Giesel, F.L.; Haberkorn, U.; et al. Response prediction of ^{177}Lu-PSMA-617 radioligand therapy using prostate-specific antigen, chromogranin A, and lactate dehydrogenase. *J. Nucl. Med.* **2020**, *61*, 689–695. [CrossRef]
54. Ferdinandus, J.; Eppard, E.; Gaertner, F.C.; Kurpig, S.; Fimmers, R.; Yordanova, A.; Hauser, S.; Feldmann, G.; Essler, M.; Ahmadzadehfar, H. Predictors of response to radioligand therapy of metastatic castrate-resistant prostate cancer with ^{177}Lu-PSMA-617. *J. Nucl. Med.* **2017**, *58*, 312–319. [CrossRef] [PubMed]
55. Rahbar, K.; Bogeman, M.; Yordanova, A.; Eveslage, M.; Schafers, M.; Essler, M.; Ahmadzadehfar, H. Delayed response after repeated ^{177}Lu-PSMA-617 radioligand therapy in patients with metastatic castration resistant prostate cancer. *Eur. J. Nucl. Med. Mol. Imaging* **2018**, *45*, 243–246. [CrossRef] [PubMed]
56. Kessel, K.; Seifert, R.; Schafers, M.; Weckesser, M.; Schlack, K.; Boegemann, M.; Rahbar, K. Second line chemotherapy and visceral metastases are associated with poor survival in patients with mCRPC receiving ^{177}Lu-PSMA-617. *Theranostics* **2019**, *9*, 4841–4848. [CrossRef] [PubMed]
57. Chen, R.; Wang, Y.; Zhu, Y.; Shi, Y.; Xu, L.; Huang, G.; Liu, J. The added value of ^{18}F-FDG PET/CT compared to ^{68}Ga-PSMA PET/CT in patients with castration-resistant prostate cancer. *J. Nucl. Med.* **2021**, *63*, 69–75. [CrossRef] [PubMed]
58. Hotta, M.; Gafita, A.; Czernin, J.; Calais, J. Outcome of patients with PSMA-PET/CT screen failure by VISION criteria and treated with ^{177}Lu-PSMA therapy: A multicenter retrospective analysis. *J. Nucl. Med.* **2022**. [CrossRef]
59. Thang, S.P.; Violet, J.; Sandhu, S.; Iravani, A.; Akhurst, T.; Kong, G.; Ravi Kumar, A.; Murphy, D.G.; Williams, S.G.; Hicks, R.J.; et al. Poor outcomes for patients with metastatic castration-resistant prostate cancer with low prostate-specific membrane antigen (PSMA) expression deemed ineligible for ^{177}Lu-labelled PSMA radioligand therapy. *Eur. Urol. Oncol.* **2019**, *2*, 670–676. [CrossRef]

60. Hofman, M.S.; Emmett, L.; Sandhu, S.; Iravani, A.; Joshua, A.M.; Goh, J.C.; Pattison, D.A.; Tan, T.H.; Kirkwood, I.D.; Ng, S.; et al. [^{177}Lu] Lu-PSMA-617 versus cabazitaxel in patients with metastatic castration-resistant prostate cancer (TheraP): A randomised, open-label, phase 2 trial. *Lancet* **2021**, *397*, 797–804. [CrossRef]
61. Rischpler, C.; Beck, T.I.; Okamoto, S.; Schlitter, A.M.; Knorr, K.; Schwaiger, M.; Gschwend, J.; Maurer, T.; Meyer, P.T.; Eiber, M. ^{68}Ga-PSMA-HBED-CC uptake in cervical, celiac, and sacral ganglia as an important pitfall in prostate cancer PET imaging. *J. Nucl. Med.* **2018**, *59*, 1406–1411. [CrossRef] [PubMed]
62. Rauscher, I.; Kronke, M.; Konig, M.; Gafita, A.; Maurer, T.; Horn, T.; Schiller, K.; Weber, W.; Eiber, M. Matched-pair comparison of ^{68}Ga-PSMA-11 PET/CT and ^{18}F-PSMA-1007 PET/CT: Frequency of pitfalls and detection efficacy in biochemical recurrence after radical prostatectomy. *J. Nucl. Med.* **2020**, *61*, 51–57. [CrossRef] [PubMed]
63. Dietlein, F.; Kobe, C.; Hohberg, M.; Zlatopolskiy, B.D.; Krapf, P.; Endepols, H.; Tager, P.; Hammes, J.; Heidenreich, A.; Persigehl, T.; et al. Intraindividual comparison of ^{18}F-PSMA-1007 with renally excreted PSMA ligands for PSMA PET imaging in patients with relapsed prostate cancer. *J. Nucl. Med.* **2020**, *61*, 729–734. [CrossRef] [PubMed]
64. Pozzessere, C.; Bassanelli, M.; Ceribelli, A.; Rasul, S.; Li, S.; Prior, J.O.; Cicone, F. Renal cell carcinoma: The oncologist asks, can PSMA PET/CT answer? *Curr. Urol. Rep.* **2019**, *20*, 68. [CrossRef] [PubMed]
65. Lawhn-Heath, C.; Flavell, R.R.; Glastonbury, C.; Hope, T.A.; Behr, S.C. Incidental detection of head and neck squamous cell carcinoma on 68Ga-PSMA-11 PET/CT. *Clin. Nucl. Med.* **2017**, *42*, e218–e220. [CrossRef] [PubMed]
66. Kesler, M.; Levine, C.; Hershkovitz, D.; Mishani, E.; Menachem, Y.; Lerman, H.; Zohar, Y.; Shibolet, O.; Even-Sapir, E. ^{68}Ga-PSMA is a novel PET-CT tracer for imaging of hepatocellular carcinoma: A prospective pilot study. *J. Nucl. Med.* **2019**, *60*, 185–191. [CrossRef] [PubMed]
67. Osmany, S.; Zaheer, S.; Bartel, T.B.; Johnston, M.; Peh, W.M.; Barmaky, S.; Jadvar, H. Gallium-68–labeled prostate-specific membrane antigen-11 PET/CT of prostate and nonprostate cancers. *Am. J. Roentgenol.* **2019**, *213*, 286–299. [CrossRef]

cancers

Article

Safety and Efficacy of Carbon-Ion Radiotherapy for Elderly Patients with High-Risk Prostate Cancer

Yuichi Hiroshima [1,2], Hitoshi Ishikawa [1,2,*], Yuma Iwai [3], Masaru Wakatsuki [1], Takanobu Utsumi [4], Hiroyoshi Suzuki [4], Koichiro Akakura [5], Masaoki Harada [1], Hideyuki Sakurai [2], Tomohiko Ichikawa [6] and Hiroshi Tsuji [1]

1 QST Hospital, National Institutes for Quantum Science and Technology, Chiba 263-8555, Japan
2 Department of Radiation Oncology, Proton Medical Research Center, University of Tsukuba Hospital, Tsukuba 305-8576, Japan
3 Department of Radiology, Chiba University Graduate School of Medicine, Chiba 260-8670, Japan
4 Department of Urology, Toho University Sakura Medical Center, Chiba 285-8741, Japan
5 Department of Urology, Japan Community Health-Care Organization Tokyo Shinjuku Medical Center, Tokyo 162-8543, Japan
6 Department of Urology, Chiba University Graduate School of Medicine, Chiba 260-8670, Japan
* Correspondence: ishikawa.hitoshi@qst.go.jp

Simple Summary: As the population ages, the number of elderly prostate cancer patients is increasing, and the choice of treatment options for elderly patients with poor general condition is becoming more difficult. The aim of our retrospective study is to assess the clinical outcomes after carbon-ion radiotherapy for elderly patients with high-risk prostate cancer. We compared 173 patients ≥75 years as the elderly group and 754 patients <75 years as the young group. Disease-specific and biochemical relapse-free survivals did not differ significantly between the young and elderly groups, and there were also no significant differences in adverse events between the two groups. Although this study is retrospective, carbon-ion radiotherapy may be a safe and effective treatment for elderly high-risk prostate cancer patients.

Abstract: Carbon-ion radiotherapy (CIRT) is a high-dose intensive treatment, whose safety and efficacy have been proven for prostate cancer. This study aims to evaluate the outcomes of CIRT in elderly patients with prostate cancer. Patients aged 75 years or above at the initiation of CIRT were designated as the elderly group, and younger than 75 years as the young group. The overall survival (OS), disease-specific survival (DSS), biochemical control rate (BCR), biochemical relapse-free survival (BRFS), and adverse events were compared between the elderly and young patients with high-risk prostate cancer treated with CIRT. The elderly group comprised 173 of 927 patients treated for high-risk prostate cancer between April 2000 and May 2018. The overall median age was 69 (range: 45–92) years. The median follow-up period was 91.9 (range: 12.6–232.3) months. The 10-year OS, DSS, BCR, and BRFS rates in the young and elderly groups were 86.9%/71.5%, 96.6%/96.8%, 76.8%/88.1%, and 68.6%/64.3%, respectively. The OS ($p < 0.001$) was longer in the younger group and the BCR was better in the elderly group ($p = 0.008$). The DSS and BRFS did not differ significantly between the two groups. The rates of adverse events between the two groups did not differ significantly and no patient had an adverse event of Grade 4 or higher during the study period. CIRT may be as effective and safe in elderly patients as the treatment for high-risk prostate cancer.

Keywords: carbon-ion radiotherapy; elderly patients; high-risk prostate cancer; radiotherapy; particle beam therapy

Citation: Hiroshima, Y.; Ishikawa, H.; Iwai, Y.; Wakatsuki, M.; Utsumi, T.; Suzuki, H.; Akakura, K.; Harada, M.; Sakurai, H.; Ichikawa, T.; et al. Safety and Efficacy of Carbon-Ion Radiotherapy for Elderly Patients with High-Risk Prostate Cancer. *Cancers* **2022**, *14*, 4015. https://doi.org/10.3390/cancers14164015

Academic Editors: José I. López and Claudia Manini

Received: 14 July 2022
Accepted: 17 August 2022
Published: 19 August 2022

1. Introduction

In Japan, the incidence of prostate cancer has increased in recent years, partially due to the popularity of prostate-specific antigen (PSA) screening; the reported incidence was 92,021 in 2018, making it the most commonly diagnosed cancer, surpassing stomach, colorectal, and lung cancers [1]. The incidence rate increased with age, with 23.3 cases per 100,000 persons in the 50–54 age years group and 75.1 cases in the 55–59 age years group, with a particularly sharp increase in those aged 60 and over, and a high rate of about 650 cases in those aged 75 and over [1]. Japan is one of the fastest aging countries in the world, and it is clear that the incidence of prostate cancer will rise with the increase in the number of elderly individuals. On the other hand, the decline in the performance status (PS) and activities of daily living (ADLs) in elderly individuals prior to treatment often makes healthcare providers, patients, and their families hesitant to consider curative treatment even after the diagnosis of malignancy [2–4].

There are two types of treatment for prostate cancer: curative and noncurative. For the elderly, androgen deprivation therapy (ADT) may be chosen as noncurative treatment based on the predicted prognosis. However, with the increase in life expectancy, prostate cancer may become resistant to castration and, progress and metastasize before death from other diseases. In particular, prostate cancer is known to metastasize to bones, and bone fractures in the elderly people have been reported to reduce prognosis as well as quality of life [5,6]. Against this background, minimally invasive curative treatment that can be undergone by the elderly is desired.

The two principal types of curative treatment for localized prostate cancer are surgery and radiotherapy. The use of robot-assisted laparoscopic prostatectomy (RALP) has become more widespread, while the precision of radiotherapy, especially external-beam radiotherapy (EBRT), has increased with the development of image-guided radiotherapy [7–9]. Surgery is usually not performed for men over 75 years of age due to the high risk of perioperative mortality, whereas radiotherpay is minimally invasive and can be performed for those over 75 years of age if they are in a good general condition [10]. Therefore, the patient population undergoing radiotherapy tends to be relatively older compared to that undergoing surgery, and there is potential for further growth in the number of patients, owing to the aging of the population. However, radiotherapy in the elderly is characterized by a greater likelihood of deterioration in the quality of life affecting the performance of daily activities, especially during long treatment periods, which is exacerbated by adverse events, such as frequent urination during treatment and hematuria and hematochezia after treatment [11].

Particle beam therapy, an EBRT technique, uses atoms accelerated at a high speed by an accelerator. The efficacy of particle beam therapy has also been reported for the treatiment of pancreatic, liver, and other cancers [12–19]. Particle beam therapy can be further categorized into proton beam therapy and carbon-ion radiotherapy (CIRT). CIRT is known to have a smaller peripheral effect and greater biological effectiveness compared to proton beam therapy. Therefore, a previous study reported that CIRT can facilitate a further reduction in the irradiation to the surrounding organs, including the rectum, thereby reducing adverse events and enabling the effective treatment of prostate cancer over a shorter therapeutic period [20,21].

CIRT has been used for the treatment of prostate cancer since 1995 [20]. Although the previous studies have reported on the treatment results and toxicity, none have focused on elderly patients [22–34]. Therefore, this study aims to assess and report the outcomes and toxicity of CIRT for prostate cancer in patients aged 75 years and older, compared to those in their younger counterparts.

2. Materials and Methods

2.1. Patients

We retrospectively searched for patients who underwent CIRT for prostate cancer at our institution between April 2000 and May 2018. The eligibility criteria for CIRT at our institution have been described in previous studies [28,29,32,33]. The inclusion criteria for this study were as follows: (1) prostate cancer diagnosis confirmed by biopsy, whose Gleason score (GS) was evaluated by a pathologist; (2) patients who met the diagnostic criteria for high-risk disease based on the National Comprehensive Cancer Network risk classification: initial PSA $\geq$ 20 ng/mL, or T3a-T3bN0M0, or GS $\geq$ 8 [35]; (3) endocrine therapy started 3–6 months before the initiation of CIRT; (4) patients aged 20 years or older at the start of therapy; and (5) PS in the range of 0–2. The exclusion criteria for this study included the following: (1) lymph node or distant metastasis prior to the initiation of treatment, (2) previous treatment of prostate cancer with a modality other than ADT, (3) active and refractory infections of the prostate, and (4) presence of other active cancers. The participants provided informed consent for participation or had the opportunity to opt-out of the study.

2.2. ADT

ADT was based on the luteinizing hormone-releasing hormone (LHRH) analog or castration combined with anti-androgen therapy. The LHRH analog alone was also acceptable in case of liver dysfunction. ADT was to be started at least 3 months prior to the start of CIRT and continued for the duration of treatment and for at least 1 year after completion. However, in cases where it was difficult to continue the treatment due to side effects or other reasons, discontinuation of the treatment was allowed.

2.3. CIRT

The details of the CIRT regimen for prostate cancer have been reported previously [20,23]. In the past, thermoplastics were used to fix the head, trunk, and pelvis, but due to advances in image guidance equipment, treatment is gradually being performed using only a pelvic fixture. CT is obtained in 2 mm slices and combined with MRI to improve prostate positioning accuracy. In addition, about 100 mL of urine was stored in the bladder at the time of CT imaging, and ultrasound was used to confirm that the same amount of urine was stored at the time of treatment. Laxatives were also occasionally used to keep the rectum as empty as possible. Briefly, the entire prostate and proximal third or half of the seminal vesicle (SV) for T1–T3a and as much of the SV as possible for T3b were designated as the clinical target volume. The radiation dose of CIRT was expressed as the photon equivalent dose (Gy) (i.e., the relative biological effect (RBE)-weighted absorbed dose) and defined as the physical dose multiplied by the carbon-ion RBE [36]. Total dose and number of fractions (fr) were treated at 66 Gy/20 fr from January 1998 to August 2005, 63 Gy/20 fr from September 2005 to August 2007, 57.6 Gy/16 fr from September 2007 to March 2013, 54 Gy/16 fr from September 2011 to July 2012, and 51.6 Gy/12 fr from April 2013 to the present, respectively, as the clinical trials were conducted at different times when treatment was performed [20]. In April 2013, CIRT using the scanning irradiation method was started instead of the passive irradiation method [23]. Treatment planning for CIRT was performed with HIPLAN (NIRS, Chiba, Japan) until March 2013, and after that, XiO-N (ELEKTA, Stockholm, Sweden). Patient background and treatment details and their respective numbers and percentages are shown in Table 1.

2.4. Evaluation

After the completion of CIRT, patients were required to undergo a follow-up examination at 3-month intervals for 2 years initally, and at 6-month intervals thereafter. Blood samples were collected during each visit. After the completion of CIRT, patients were required to undergo a follow-up examination at 3–6-month intervals for the first 5 years, and at 6–12-month intervals thereafter if no sign of the recurrences were observed. Even after a 10-year follow-up examination, patients basically came to the hospital as long as they could in order to evaluate the risk of subsequent primary second cancers in patients with prostate cancer after CIRT [37], but patients whose regular follow-up was interrupted were contacted directly by mail or telephone to check on their condition. Blood samples were collected during each visit. Post-treatment biochemical recurrence (BR) was defined as a PSA value exceeding 2.0 ng/mL over the minimum value during treatment according to the Phoenix definition [35]. In the event of BR, bone scintigraphy, computed tomography

(CT), and magnetic resonance imaging (MRI) were performed to confirm the presence of recurrence. Local recurrence was defined as regrowth, as observed on CT and MRI. Adverse events were assessed in accordance with the Common Terminology Criteria for Adverse Events (CTCAE) version 4.0, and the previous studies [38–40]. Acute-phase adverse events were defined as those occurring within 3 months of treatment; and late-phase adverse events were defined as those occurring 3 months after treatment.

Table 1. Patient, tumor, and treatment characteristics.

	Overall Study Population	Young Group	Elderly Group	p-Value
age	Median: 69 (range 45–92)	Median: 67 (range: 45–74)	Median: 77 (range: 75–92)	
Initial PSA (ng/mL)	Median: 16.7 (range 2.1–329.7)	Median: 16.92 (range: 2.1–320.0)	Median: 16.0 (range: 3.3–329.7)	0.590
Clinical T stage 1c/2a/2b/2c/3	139/159/22/149/458 15.0%/17.2%/2.4%/16.1%/49.4%	112/122/20/115/385 14.9%/16.2%/2.7%/15.3%/51.1%	27/37/2/34/73 15.6%/21.4%/1.2%/19.7%/42.2%	0.110
GS sum 5/6/7/8/9/10	2/45/301/212/364/3 0.2%/4.9%/32.5%/22.9%/39.3%/0.3%	1/37/258/163/292/3 0.1%/4.9%/34.2%/21.6%/38.7%/0.4%	1/8/43/49/72/0 5.8%/4.6%/24.9%/28.3%/41.6%/0%	0.114
Primary GS ~3/4/5	255/621/51 27.5%/67.0%/5.5%	211/502/41 28.0%/66.6%/5.4%	44/119/10 25.4%/68.8%/5.8%	0.792
Total ADT duration (months)	Median: 20.9 (range: 0.6–63.6)	Median: 22.0 (range: 0.6–63.5)	Median: 18.2 (range: 0.6–63.6)	0.265
Total dose (Gy)/fraction 51.6/20, 54/16, 57.6/16, 63/20, 66/20	311, 4, 331, 128, 153 33.5%, 0.4%, 35.7%, 13.8%, 16.5%	251, 3, 267, 104, 129 33.3%, 0.4%, 35.4%, 13.8%, 17.1%	60, 1, 64, 24, 24 34.7%, 0.6%, 37.0%, 13.9%, 13.9%	0.881
Treatment planning system HIPLAN/Xio-N	686/231 74.8%/25.2%	570/184 75.6%/24.4%	116/47 71.2%/28.8%	0.360

Abbreviation: PSA: prostate-specific antigen, GS: Gleason score, ADT: androgen-deprivation therapy, Gy: Gray.

2.5. Statistical Analysis

The Mann–Whitney U and χ^2 tests were used for the statistical analysis of the continuous and categorical variables, respectively. The Kaplan–Meier method was used to analyze overall survival (OS), disease-specific survival (DSS), biochemical control rate (BCR), and biochemical relapse-free survival (BRFS), and the log-rank test was used for a comparison of these parameters. SPSS software version 28.0 (IBM Inc., Armonk, NY, USA) was used for all analyses. p-values < 0.05 were considered statistically significant.

2.6. Ethics Approval

This research was conducted in accordance with the Declaration of Helsinki (1964) and the subsequent Code of Ethics, and the study design was approved by the Institutional Review Board (permit number: N21-005).

3. Results

3.1. Patient Characteristics

A total of 927 patients with prostate cancer who started CIRT between April 2000 and May 2018 met the eligibility criteria. Of these, 173 were aged 75 years or older and placed in the elderly group, and others who were younger than 75 years were defined as the young group. The median age of the entire patient population was 69 years (range: 45–92 years), and the median PSA level before treatment initiation was 16.7 ng/mL (range: 2.1–329.7). The other patient characteristics are presented in Table 1. There were no significant differences in the patient factors between the young and elderly groups.

The median follow-up duration for the entire patient population and elderly group was 91.9 (range: 12.6–232.3 months) and 84.4 (range: 13.5–214.8 months) months, respectively. During the observation period, there were 147 and 38 instances of all-cause motality in the entire patient population and elderly group, respectively, including 31 and 4 deaths due to prostate cancer, respectively. Recurrences during the observation period were as follows: BCR was observed in 161 patients (17.4%) in the entire patient population and 15 patients

(8.7%) in the elderly group. The distributions of local, regional, and distant metastases are shown in Figure 1. One patient developed regional recurrence and another developed regional and distant recurrences concurrently, neither of whom had BCR.

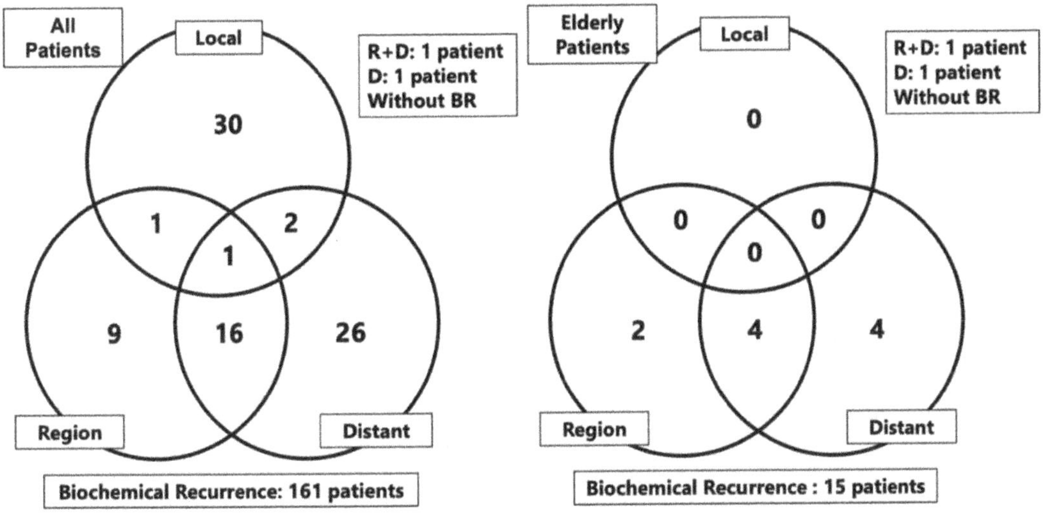

Figure 1. Form of recurrences after carbon-ion radiotherapy in the overall study population and elderly group. Abbreviation: R: region, D; distant, BR: biochemical recurrence.

3.2. Adverse Events

The gastrointestinal (GI) and genitourinary (GU) adverse events that occured during the observation period are shown in Table 2. CTCAE grades 4 or 5 adverse events were not observed in any patient during the study period. The occurence of adverse events, especially the higher grades, in the elderly group was similar or lower than that in the young group.

Table 2. Adverse events after carbon-ion radiotherapy in patients with high-risk prostate cancer.

	Grade 1			Grade 2			Grade 3		
	Overall Study Population	Young Group	Elderly Group	Overall Study Population	Young Group	Elderly Group	Overall Study Population	Young Group	Elderly Group
Radiation dermatitis	17 1.8%	15 2.0%	2 1.2%	1 0.1%	0 0%	1 0.6%	0	0	0
Rectal bleeding	110 11.8%	93 12.3%	17 9.8%	15 1.6%	14 1.9%	1 0.6%	0	0	0
Genitourinary events (Except hematuria)	503 54.1%	399 52.9%	104 60.1%	80 8.6%	65 8.6%	15 8.7%	12 1.3%	11 1.5%	1 0.6%
Hematuria (Number of patients with bladder cancer)	4 0.4%	4 0.5%	0 0%	45 4.8% (6 0.6%)	38 5.0% (5 0.7%)	7 4.0% (1 0.5%)	12 1.3%	11 1.5%	1 0.6%

3.3. Outcomes

The OS, DSS, BCR, and BRFS in the young group at 5 and 10 years were 96.3% and 86.9%, 99.0% and 96.6%, 89.7% and 76.8%, and 87.1% and 68.6%, respectively (Figure 2). The corresponding rates in the elderly group were 92.6% and 71.5%, 98.2% and 96.8%, 93.4% and 88.1%, and 87.4% and 64.3%, respectively.

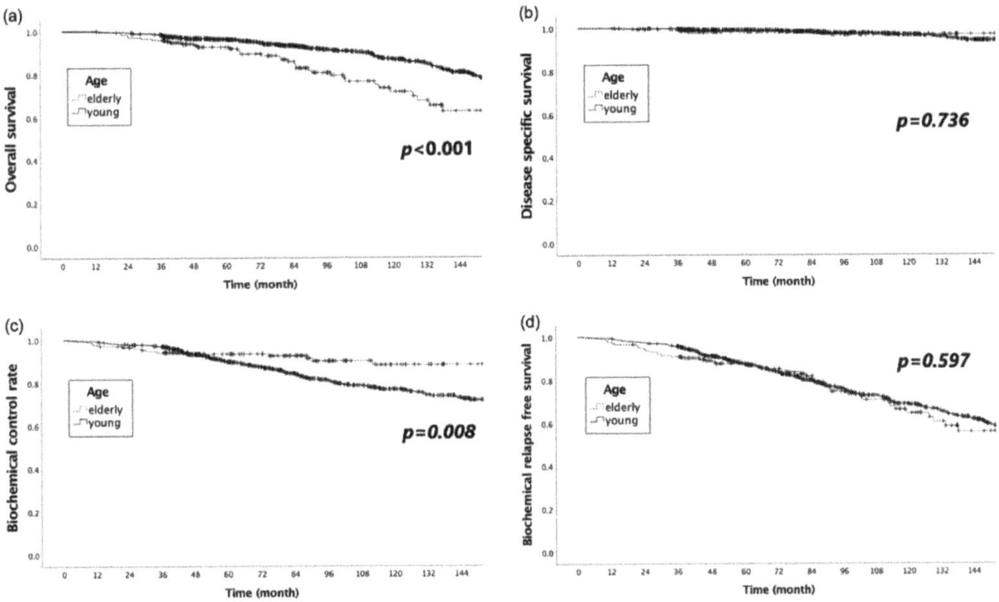

Figure 2. Comparison of the Kaplan–Meier curves for (**a**) overall survival, (**b**) disease-specific survival, (**c**) biochemical recurrence control rate, and (**d**) biochemical relapse-free survival for the young and elderly groups.

The OS was significantly longer in the younger group, while the BCR was significantly better in the elderly group ($p < 0.001$ and $p = 0.008$, respectively). The DSS and BRFS did not significantly differ between the two groups ($p = 0.736$ and $p = 0.597$, respectively).

The survival times according to age in the present study were compared with the expected life expectancy by age for Japanese men in 2020 [41]. No obvious differences were observed in the two groups under 74 years of age. However, the survival time was better for those aged 75–79 and 83 years in the group treated with CIRT (Figure 3).

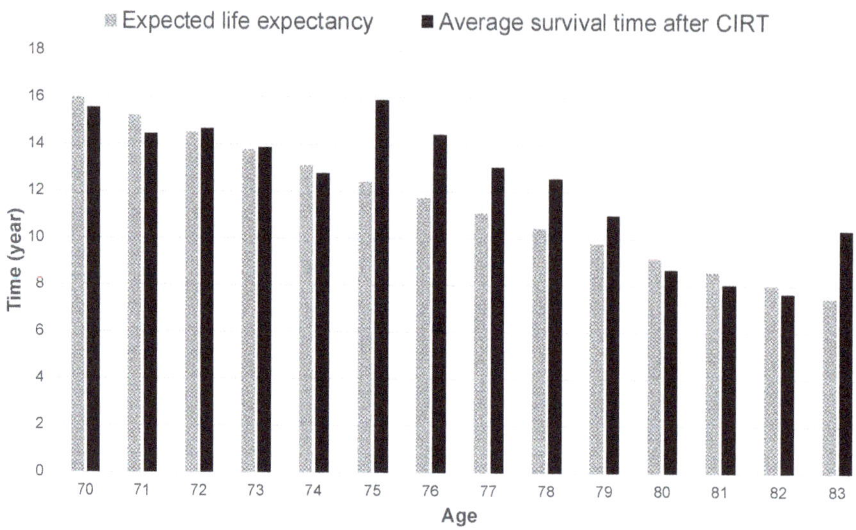

Figure 3. Expected life expectancy by age in Japan in 2020 compared to the average overall survival time by age at the start of carbon-ion radiotherapy.

4. Discussion

This study examined the efficacy and toxicity of CIRT for high-risk prostate cancer in patients aged above and below 75 years. The OS was significantly better in the young group, while the DSS and BRFS did not differ significantly between the two groups. This suggests that while the curative effect of CIRT for prostate cancer was equivalent for both groups, the OS was shortened by other diseases that occur more frequently in the elderly population, such as death due to other cancers and cardiovascular events. On the other hand, the BCR was significantly better in the elderly, although there were no differences in the prostate cancer risk category and treatment methods between the two groups. The difference in BCR may be attributed to the age-related decline in androgen secretion. Androgens are among the factors that promote the progression of prostate cancer, and their secretion is also known to decrease with age [42–44]. Thus, there may be background factors that may decrease the likelihood of prostate cancer recurrences in the elderly, or it may take a long time for the PSA to rise sufficiently to the diagnostic range.

More hypofractionated radiotherapy for prostate cancer is becoming more common with intensity-modulated radiotherapy and stereotactic body radiotherapy. A meta-analysis, including a phase III prospective study, found 14.6% of GI adverse events and 19.4% of GU adverse events to be grade 2 or higher, which is a higher percentage than either group in this study [45]. Another meta-analysis of prospective studies found 1.1% of GI adverse events and 2.0% of GU adverse events to be grade 3 or higher, also a higher rate than in the present study. Thus, CIRT may be safer than existing EBRT [46]. No grade 4 or higher adverse events were observed in this study, and the occurence of adverse events in each category was either similar or, in some cases, slightly higher in young group. Although the OS of the elderly group was shorter than that of the young group, the duration of follow-up was comparable. These results suggest that the frequency and severity of adverse events may be similar between the elderly and young groups. The occurence of cardiovascular disease, such as atrial fibrillation, is often higher in elderly patients, who are frequently prescibed anticoagulants, exposing to a higher risk for bleeding-related adverse events [47–49]. These medications were not withdrawn or discontinued at the time of CIRT, and patients were allowed to continue taking them afterwards. The fact that there was no apparent difference in the frequency of adverse events, especially the GU and GI adverse events, suggests that CIRT is a safe treatment modality for elderly as well as young patients.

A comparison of the mean OS rates and expected life expectancy of the patient groups included in the study with the general population revealed that OS for patients treated with CIRT was similar or superior, especially in the 75–79-year-old group. This suggests that even elderly patients with high-risk prostate cancer can survive up to their respective age-adjusted mean life expectancies after radical treatment with CIRT. The encouraging results in the 76–80-year-old age group in the present study may be partially attributed to the fact that, historically, age was one of the criteria for surgery in Japan. Urologists were first consulted when the PSA screening results required further investigation; moreover, urologists often prescribed surgery for patients up to 75 years of age, when the expected life expectancy was 10 years or less, before the widespread use of RALP [7]. Therefore, patients aged older than 76 years who underwent CIRT were referred to our hospital because they were not considered as suitable surgical candidates due to their advanced age. However, it is possible that the results of the present study are the product of enrolling patients whose PS and ADLs are better than those of their (age-matched) contemporaries. At present, RALP is widely used, and is performed even for patients aged 76–80 years if their PS and ADLs are good. The accumulation of patients in the future will clarify whether this is the reason for good life expectancy in the 76–80-year-old age group observed in the study.

The elderly population continues to grow in Japan, with a corresponding rise in the incidence of prostate cancer. In the real-world clinical setting, younger patients may be ready for radical treatment, while elderly patients are often hesitant because of concerns about worsening of the general condition and adverse events after treatment, which may discourage healthcare providers, the patient, and family members from proceeding with treatment. In this context, guidelines and reports have been devised for elderly cancer patients [2,4,11]. The common desires binding all patients is the wish to complete the treatment in as short a time as possible with high precision, and in as few days as possible, especially for elderly patients. In this regard, CIRT is considered to be a guideline-compliant treatment, which offers highly precise physical dose distribution and allows for hypofractionation. Further advancements in hypofractionation in the future may enable even more elderly patient-friendly treatment.

The limitations of this study included its retrospective design, bias arising from the patients' background, relatively small sample size of the elderly group (n = 173), single-center setting, variation in the time of CIRT initiation, and differences in the total dose and fractionation. On the other hand, no other study has examined the prognosis after CIRT by exclusively focusing on elderly patients with prostate cancer. Furthermore, the GS, which varied depending on each individual pathologist, was evaluated by only one pathologist at our institution for more than 25 years, rendering the reliability of risk classification to be high. In addition, this study included missing data from 83 (8.9%) of the 927 patients (43 in the young group and 40 in the elderly group) because of being lost to follow-up until 10 years after CIRT. However, the median follow-up times for the 43 young and 40 elderly patients were 81.3 and 85.3 months, respectively. Since these follow-up times were similar to those of the entire cohort in the present study, the impact of the missing data on comparing treatment outcomes of the young and elderly groups may be limited. Further studies with a longer follow-up period, larger sample size, and prospective design are needed to validate the results of this study.

5. Conclusions

The evaluation of the efficacy and toxicity of CIRT in elderly patients with prostate cancer revealed that CIRT may yield a good therapeutic effect on these patients, comparable to that of their younger counterparts.

Author Contributions: Conceptualization, Y.H. and H.I.; methodology, Y.H.; validation, Y.I. and Y.H.; formal analysis, Y.H.; investigation, M.W., T.U., H.S. and K.A.; resources, Y.H., H.I. and M.H.; data curation, Y.H. and T.U.; writing—original draft preparation, Y.H.; writing—review and editing, H.I., M.W., T.U., H.S. (Hiroyoshi Suzuki), K.A. and H.T.; visualization, Y.H.; supervision, H.S. (Hideyuki

Sakurai), T.I. and H.T.; project administration, H.T. All authors have read and agreed to the published version of the manuscript.

Funding: This work was supported by Grants-in-Aid for Scientific Research (C) (21K07715) from the Ministry of Education, Culture, Sports, Science and Technology of Japan.

Institutional Review Board Statement: The study was conducted in accordance with the Declaration of Helsinki and approved by the Institutional Review Board of National Institutes for Quantum Science and Technology (permit number: N21-005).

Informed Consent Statement: Informed consent was obtained from all participants involved in the study.

Data Availability Statement: The data are not publicly available due to the facility's privacy policy.

Conflicts of Interest: The authors declare no conflict of interest.

References

1. Cancer Information Service, National Cancer Center, Japan (Vital Statistics of Japan). Cancer Registry and Statistics. Available online: https://ganjoho.jp/reg_stat/statistics/dl/index.html (accessed on 14 May 2022).
2. Popescu, T.; Karlsson, U.; Vinh-Hung, V.; Trigo, L.; Thariat, J.; Vuong, T.; Baumert, B.G.; Motta, M.; Zamagni, A.; Bonet, M.; et al. Challenges Facing Radiation Oncologists in The Management of Older Cancer Patients: Consensus of The International Geriatric Radiotherapy Group. *Cancers* **2019**, *11*, 371.
3. Keenan, L.G.; O'Brien, M.; Ryan, T.; Dunne, M.; McArdle, O. Assessment of older patients with cancer: Edmonton Frail Scale (EFS) as a predictor of adverse outcomes in older patients undergoing radiotherapy. *J. Geriatr. Oncol.* **2017**, *8*, 206–210. [PubMed]
4. VanderWalde, N.; Jagsi, R.; Dotan, E.; Baumgartner, J.; Browner, I.S.; Burhenn, P.; Cohen, H.J.; Edil, B.H.; Edwards, B.; Extermann, M.; et al. NCCN Guidelines Insights: Older Adult Oncology, Version 2.2016. *J. Natl. Compr. Cancer Netw.* **2016**, *14*, 1357–1370.
5. Lavallée, L.T.; McLarty, R.; Tran, C.; Breauj, R.H.; Richard, P.; Shayegan, B.; Danielson, B.; Jammal, M.-P.; Saad, F. Canadian Urologic Association best practice report: Bone health in prostate cancer. *Can. Urol. Assoc. J.* **2021**, *15*, 375–382. [PubMed]
6. von Friesendorff, M.; McGuigan, F.E.; Wizert, A.; Rogmark, C.; Holmberg, A.H.; Woolf, A.D.; Akesson, K. Hip fracture, mortality risk, and cause of death over two decades. *Osteoporos. Int.* **2016**, *27*, 2945–2953.
7. Mottet, N.; Bellmunt, J.; Bolla, M.; Briers, E.; Cumberbatch, M.G.; De Santis, M.; Fossati, N.; Gross, T.; Henry, A.M.; Joniau, S.; et al. EAU-ESTRO-SIOG Guidelines on Prostate Cancer. Part 1: Screening, Diagnosis, and Local Treatment with Curative Intent. *Eur. Urol.* **2017**, *71*, 618–629.
8. Sandhu, S.; Moore, C.M.; Chiong, E.; Beltran, H.; Bristow, R.G.; Williams, S.G. Prostate cancer. *Lancet* **2021**, *398*, 1075–1090.
9. Lantz, A.; Bock, D.; Akre, O.; Angenete, E.; Bjartell, A.; Carlsson, S.; Modig, K.K.; Nyberg, M.; Kollberg, K.S.; Steineck, G.; et al. Functional and Oncological Outcomes After Open Versus Robot-assisted Laparoscopic Radical Prostatectomy for Localised Prostate Cancer: 8-Year Follow-up. *Eur. Urol.* **2021**, *80*, 650–660.
10. Fung, C.; Dale, W.; Mohile, S.G. Prostate Cancer in the Elderly Patient. *J. Clin. Oncol.* **2014**, *32*, 2523–2530.
11. Droz, J.-P.; Albrand, G.; Gillessen, S.; Hughes, S.; Mottet, N.; Oudard, S.; Payne, H.; Puts, M.; Zulian, G.; Balducci, L.; et al. Management of Prostate Cancer in Elderly Patients: Recommendations of a Task Force of the International Society of Geriatric Oncology. *Eur. Urol.* **2017**, *72*, 521–531.
12. Hiroshima, Y.; Ishikawa, H.; Murakami, M.; Nakamura, M.; Shimizu, S.; Enomoto, T.; Oda, T.; Mizumoto, M.; Nakai, K.; Okumura, T.; et al. Proton Beam Therapy for Local Recurrence of Rectal Cancer. *Anticancer Res.* **2021**, *41*, 3589–3595. [CrossRef]
13. Igaki, H.; Mizumoto, M.; Okumura, T.; Hasegawa, K.; Kokudo, N.; Sakurai, H. A systematic review of publications on charged particle therapy for hepatocellular carcinoma. *Int. J. Clin. Oncol.* **2017**, *23*, 423–433. [CrossRef] [PubMed]
14. Hiroshima, Y.; Fukumitsu, N.; Saito, T.; Numajiri, H.; Murofushi, K.N.; Ohnishi, K.; Nonaka, T.; Ishikawa, H.; Okumura, T.; Sakurai, H. Concurrent chemoradiotherapy using proton beams for unresectable locally advanced pancreatic cancer. *Radiother. Oncol.* **2019**, *136*, 37–43. [CrossRef] [PubMed]
15. Komatsu, S.; Fukumoto, T.; Demizu, Y.; Miyawaki, D.; Terashima, K.; Sasaki, R.; Hori, Y.; Hishikawa, Y.; Ku, Y.; Murakami, M. Clinical results and risk factors of proton and carbon ion therapy for hepatocellular carcinoma. *Cancer* **2011**, *117*, 4890–4904. [CrossRef] [PubMed]
16. Kawashiro, S.; Yamada, S.; Okamoto, M.; Ohno, T.; Nakano, T.; Shinoto, M.; Shioyama, Y.; Nemoto, K.; Isozaki, Y.; Tsuji, H.; et al. Multi-institutional Study of Carbon-ion Radiotherapy for Locally Advanced Pancreatic Cancer: Japan Carbon-ion Radiation Oncology Study Group (J-CROS) Study 1403 Pancreas. *Int. J. Radiat. Oncol.* **2018**, *101*, 1212–1221. [CrossRef]
17. Shinoto, M.; Yamada, S.; Okamoto, M.; Shioyama, Y.; Ohno, T.; Nakano, T.; Nemoto, K.; Isozaki, Y.; Kawashiro, S.; Tsuji, H.; et al. Carbon-ion radiotherapy for locally recurrent rectal cancer: Japan Carbon-ion Radiation Oncology Study Group (J-CROS) Study 1404 Rectum. *Radiother. Oncol.* **2019**, *132*, 236–240. [CrossRef]
18. McDonald, M.W.; Liu, Y.; Moore, M.G.; Johnstone, P.A.S. Acute toxicity in comprehensive head and neck radiation for nasopharynx and paranasal sinus cancers: Cohort comparison of 3D conformal proton therapy and intensity modulated radiation therapy. *Radiat. Oncol.* **2016**, *11*, 1–10. [CrossRef]

19. Baba, K.; Mizumoto, M.; Oshiro, Y.; Shimizu, S.; Nakamura, M.; Hiroshima, Y.; Iizumi, T.; Saito, T.; Numajiri, H.; Nakai, K.; et al. An Analysis of Vertebral Body Growth after Proton Beam Therapy for Pediatric Cancer. *Cancers* **2021**, *13*, 349. [CrossRef]
20. Ishikawa, H.; Hiroshima, Y.; Kanematsu, N.; Inaniwa, T.; Shirai, T.; Imai, R.; Suzuki, H.; Akakura, K.; Wakatsuki, M.; Ichikawa, T.; et al. Carbon-ion radiotherapy for urological cancers. *Int. J. Urol.* **2022**, *52*, 942–945. [CrossRef]
21. Kanai, T.; Endo, M.; Minohara, S.; Miyahara, N.; Koyama-Ito, H.; Tomura, H.; Matsufuji, N.; Futami, Y.; Fukumura, A.; Hiraoka, T.; et al. Biophysical characteristics of HIMAC clinical irradiation system for heavy-ion radiation therapy. *Int. J. Radiat. Oncol.* **1999**, *44*, 201–210. [CrossRef]
22. Utsumi, T.; Suzuki, H.; Ishikawa, H.; Hiroshima, Y.; Wakatsuki, M.; Harada, M.; Ichikawa, T.; Akakura, K.; Tsuji, H. External validation of the Candiolo nomogram for high-risk prostate cancer patients treated with carbon ion radiotherapy plus androgen deprivation therapy: A retrospective cohort study. *Jpn. J. Clin. Oncol.* **2022**, *52*, 942–945. [CrossRef] [PubMed]
23. Sato, H.; Kasuya, G.; Ishikawa, H.; Nomoto, A.; Ono, T.; Nakajima, M.; Isozaki, Y.; Yamamoto, N.; Iwai, Y.; Nemoto, K.; et al. Long-term clinical outcomes after 12-fractionated carbon-ion radiotherapy for localized prostate cancer. *Cancer Sci.* **2021**, *112*, 3598–3606. [CrossRef] [PubMed]
24. Ishikawa, H.; Tsuji, H.; Murayama, S.; Sugimoto, M.; Shinohara, N.; Maruyama, S.; Murakami, M.; Shirato, H.; Sakurai, H. Particle therapy for prostate cancer: The past, present and future. *Int. J. Urol.* **2019**, *26*, 971–979. [CrossRef] [PubMed]
25. Kasuya, G.; Ishikawa, H.; Tsuji, H.; Haruyama, Y.; Kobashi, G.; Ebner, D.K.; Akakura, K.; Suzuki, H.; Ichikawa, T.; Shimazaki, J.; et al. Cancer-specific mortality of high-risk prostate cancer after carbon-ion radiotherapy plus long-term androgen deprivation therapy. *Cancer Sci.* **2017**, *108*, 2422–2429. [CrossRef] [PubMed]
26. Nomiya, T.; Tsuji, H.; Kawamura, H.; Ohno, T.; Toyama, S.; Shioyama, Y.; Nakayama, Y.; Nemoto, K.; Tsujii, H.; Kamada, T. A multi-institutional analysis of prospective studies of carbon ion radiotherapy for prostate cancer: A report from the Japan Carbon ion Radiation Oncology Study Group (J-CROS). *Radiother. Oncol.* **2016**, *121*, 288–293. [CrossRef] [PubMed]
27. Kasuya, G.; Ishikawa, H.; Tsuji, H.; Nomiya, T.; Makishima, H.; Kamada, T.; Akakura, K.; Suzuki, H.; Shimazaki, J.; Haruyama, Y.; et al. Significant impact of biochemical recurrence on overall mortality in patients with high-risk prostate cancer after carbon-ion radiotherapy combined with androgen deprivation therapy. *Cancer* **2016**, *122*, 3225–3231. [CrossRef] [PubMed]
28. Nomiya, T.; Tsuji, H.; Maruyama, K.; Toyama, S.; Suzuki, H.; Akakura, K.; Shimazaki, J.; Nemoto, K.; Kamada, T.; Tsujii, H. Phase I/II trial of definitive carbon ion radiotherapy for prostate cancer: Evaluation of shortening of treatment period to 3 weeks. *Br. J. Cancer* **2014**, *110*, 2389–2395. [CrossRef]
29. Okada, T.; Tsuji, H.; Kamada, T.; Akakura, K.; Suzuki, H.; Shimazaki, J.; Tsujii, H.; Working Group for Genitourinary, T. Carbon Ion Radiotherapy in Advanced Hypofractionated Regimens for Prostate Cancer: From 20 to 16 Fractions. *Int. J. Radiat. Oncol. Biol. Phys.* **2012**, *84*, 968–972. [CrossRef]
30. Ishikawa, H.; Tsuji, H.; Kamada, T.; Akakura, K.; Suzuki, H.; Shimazaki, J.; Tsujii, H. The Working Group for Genitourinary Tumors Carbon-ion radiation therapy for prostate cancer. *Int. J. Urol.* **2012**, *19*, 296–305. [CrossRef]
31. Wakatsuki, M.; Tsuji, H.; Ishikawa, H.; Yanagi, T.; Kamada, T.; Nakano, T.; Suzuki, H.; Akakura, K.; Shimazaki, J.; Tsujii, H. Quality of Life in Men Treated With Carbon Ion Therapy for Prostate Cancer. *Int. J. Radiat. Oncol. Biol. Phys.* **2008**, *72*, 1010–1015. [CrossRef]
32. Ishikawa, H.; Tsuji, H.; Kamada, T.; Yanagi, T.; Mizoe, J.-E.; Kanai, T.; Morita, S.; Wakatsuki, M.; Shimazaki, J.; Tsujii, H. Carbon ion radiation therapy for prostate cancer: Results of a prospective phase II study. *Radiother. Oncol.* **2006**, *81*, 57–64. [CrossRef] [PubMed]
33. Tsuji, H.; Yanagi, T.; Ishikawa, H.; Kamada, T.; Mizoe, J.-E.; Kanai, T.; Morita, S.; Tsujii, H. Hypofractionated radiotherapy with carbon ion beams for prostate cancer. *Int. J. Radiat. Oncol. Biol. Phys.* **2005**, *63*, 1153–1160. [CrossRef] [PubMed]
34. Akakura, K.; Tsujii, H.; Morita, S.; Tsuji, H.; Yagishita, T.; Isaka, S.; Ito, H.; Akaza, H.; Hata, M.; Fujime, M.; et al. Phase I/II clinical trials of carbon ion therapy for prostate cancer. *Prostate* **2003**, *58*, 252–258. [CrossRef]
35. National Comprehensive Cancer Network. NCCN Clinical Practice Guidelines in Oncology Prostate Cancer (Version 3.2022). 2022. Available online: https://www.nccn.org/professionals/physician_gls/pdf/prostate.pdf (accessed on 8 May 2022).
36. Inaniwa, T.; Kanematsu, N.; Matsufuji, N.; Kanai, T.; Shirai, T.; Noda, K.; Tsuji, H.; Kamada, T.; Tsujii, H. Reformulation of a clinical-dose system for carbon-ion radiotherapy treatment planning at the National Institute of Radiological Sciences, Japan. *Phys. Med. Biol.* **2015**, *60*, 3271–3286. [CrossRef] [PubMed]
37. Mohamad, O.; Tabuchi, T.; Nitta, Y.; Nomoto, A.; Sato, A.; Kasuya, G.; Makishima, H.; Choy, H.; Yamada, S.; Morishima, T.; et al. Risk of subsequent primary cancers after carbon ion radiotherapy, photon radiotherapy, or surgery for localised prostate cancer: A propensity score-weighted, retrospective, cohort study. *Lancet Oncol.* **2019**, *20*, 674–685. [CrossRef]
38. US Department of Health Human Services. *Common Terminology Criteria for Adverse Events (CTCAE) Version 4.0*; National Cancer Institute: Bethesda, MD, USA, 2009; Volume 4.
39. Trotti, A.; Colevas, A.; Setser, A.; Rusch, V.; Jaques, D.; Budach, V.; Langer, C.; Murphy, B.; Cumberlin, R.; Coleman, C. CTCAE v3.0: Development of a comprehensive grading system for the adverse effects of cancer treatment. In *Seminars in Radiation Oncology*; Elsevier: Amsterdam, The Netherlands, 2003; pp. 176–181.
40. Trotti, A.; Byhardt, R.; Stetz, J.; Gwede, C.; Corn, B.; Fu, K.; Gunderson, L.; McCormick, B.; Morris∫, M.; Rich, T.; et al. Common toxicity criteria: Version 2.0. an improved reference for grading the acute effects of cancer treatment: Impact on radiotherapy. *Int. J. Radiat. Oncol. Biol. Phys.* **2000**, *47*, 13–47. [CrossRef]

41. Ministry of Health, Labour and Welfare of Japan. Simplified Life Chart in 2020. Available online: https://www.mhlw.go.jp/toukei/saikin/hw/life/life20/index.html (accessed on 16 May 2022).
42. Glantz, G.M. Cirrhosis and Carcinoma of the Prostate Gland. *J. Urol.* **1964**, *91*, 291–293. [CrossRef]
43. Gann, P.H.; Hennekens, C.H.; Ma, J.; Longcope, C.; Stampfer, M.J. Prospective Study of Sex Hormone Levels and Risk of Prostate Cancer. *JNCI J. Natl. Cancer Inst.* **1996**, *88*, 1118–1126. [CrossRef]
44. Iwamoto, T.; Yanase, T.; Koh, E.; Horie, H.; Baba, K.; Namiki, M.; Nawata, H. Reference ranges of serum total and free testosterone in japanese male adults. *Jpn. J. Urol.* **2004**, *95*, 751–760. [CrossRef]
45. Lehrer, E.J.; Kishan, A.U.; Yu, J.B.; Trifiletti, D.M.; Showalter, T.N.; Ellis, R.; Zaorsky, N.G. Ultrahypofractionated versus hypofractionated and conventionally fractionated radiation therapy for localized prostate cancer: A systematic review and meta-analysis of phase III randomized trials. *Radiother. Oncol.* **2020**, *148*, 235–242. [CrossRef]
46. Jackson, W.C.; Silva, J.; Hartman, H.E.; Dess, R.T.; Kishan, A.U.; Beeler, W.H.; Gharzai, L.A.; Jaworski, E.M.; Mehra, R.; Hearn, J.W.D.; et al. Stereotactic body radiation therapy for localized prostate cancer: A systematic review and meta-analysis of over 6000 patients treated on prospective studies. *Int. J. Radiat. Oncol. Biol. Phys.* **2019**, *104*, 778–789. [CrossRef] [PubMed]
47. Rutherford, O.-C.W.; Jonasson, C.; Ghanima, W.; Söderdahl, F.; Halvorsen, S. Effectiveness and safety of oral anticoagulants in elderly patients with atrial fibrillation. *Heart* **2021**, *108*, 345–352. [CrossRef] [PubMed]
48. Akao, M.; Shimizu, W.; Atarashi, H.; Ikeda, T.; Inoue, H.; Okumura, K.; Koretsune, Y.; Tsutsui, H.; Toyoda, K.; Hirayama, A.; et al. Oral Anticoagulant Use in Elderly Japanese Patients with Non-Valvular Atrial Fibrillation—Subanalysis of the ANAFIE Registry. *Circ. Rep.* **2020**, *2*, 552–559. [CrossRef] [PubMed]
49. Cavallari, I. Efficacy and safety of oral anticoagulation in elderly patients with atrial fibrillation. *Anatol. J. Cardiol.* **2018**, *19*, 67–71. [CrossRef]

cancers

MDPI

Article

A Matched-Pair Analysis after Robotic and Retropubic Radical Prostatectomy: A New Definition of Continence and the Impact of Different Surgical Techniques

Nicola d'Altilia [1], Vito Mancini [1], Ugo Giovanni Falagario [1], Leonardo Martino [1], Michele Di Nauta [1], Beppe Calò [2], Francesco Del Giudice [3], Satvir Basran [4], Benjamin I. Chung [4], Angelo Porreca [5], Lorenzo Bianchi [6], Riccardo Schiavina [6], Eugenio Brunocilla [6], Gian Maria Busetto [1,*], Carlo Bettocchi [1], Pasquale Annese [1], Luigi Cormio [1,2] and Giuseppe Carrieri [1]

1 Department of Urology and Renal Transplantation, Policlinico Riuniti di Foggia, University of Foggia, 71122 Foggia, Italy
2 Department of Urology, Bonomo Teaching Hospital, 76123 Andria, Italy
3 Department of Urology, Sapienza Rome University, 00185 Rome, Italy
4 Department of Urology, Stanford University School of Medicine, Stanford, CA 94305, USA
5 Oncological Urology, Veneto Institute of Oncology (IOV), Istituto di Ricovero e Cura a Carattere Scientifico (IRCCS), 37138 Padua, Italy
6 Department of Urology, University of Bologna, 40126 Bologna, Italy
* Correspondence: gianmaria.busetto@unifg.it; Tel.: +39-0881-733856

Simple Summary: The primary goal of radical prostatectomy is oncologic efficacy, but mitigation of the side effects related to prostatectomy is also important to preserve quality of life in these patients. In recent decades, robotic prostatectomy has become increasingly popular due to advantages related to better visualization of the anatomical structures, surgical precision, reduced blood loss, and shorter hospital stay. However, the available literature shows no difference in oncologic outcomes between robotic and open. Some authors suggest differences in its potential to improve functional outcomes, i.e., erectile function and urinary continence. The first is certainly more common but more accepted than urinary incontinence, considering that some patients already suffer from it before surgery and others are not particularly interested in recovering from it, unlike urinary incontinence, which causes psychophysical changes that are often difficult to resolve.

Abstract: (1) Background: Radical prostatectomy is considered the gold-standard treatment for patients with localized prostate cancer. The literature suggests there is no difference in oncological and functional outcomes between robotic-assisted radical prostatectomy (RARP) and open (RRP). (2) Methods: The aim of this study was to compare continence recovery rates after RARP and RRP measured with 24 h pad weights and the International Consultation on Incontinence Questionnaire—Short Form (ICIQ-SF). After matching the population (1:1), 482 met the inclusion criteria, 241 patients per group. Continent patients with a 24 h pad test showing <20 g of urinary leakage were considered, despite severe incontinence, and categorized as having >200 g of urinary leakage. (3) Results: There was no difference between preoperative data. As for urinary continence (UC) and incontinence (UI) rates, RARP performed significantly better than RRP based on objective and subjective results at all evaluations. Univariable and multivariable Cox Regression Analysis pointed out that the only significant predictors of continence rates were the bilateral nerve sparing technique (1.25 (CI 1.02,1.54), $p = 0.03$) and the robotic surgical approach (1.42 (CI 1.18,1.69) $p \leq 0.001$). (4) Conclusions: The literature reports different incidences of UC depending on assessment and definition of continence "without pads" or "social continence" based on number of used pads per day. In this, our first evaluation, the advantage of objective measurement through the weight of the 24 h and subjective measurement with the ICIQ-SF questionnaire best demonstrates the difference between the two surgical techniques by enhancing the use of robotic surgery over traditional surgery.

Keywords: prostate cancer; male incontinence; robotic prostatectomy

Citation: d'Altilia, N.; Mancini, V.; Falagario, U.G.; Martino, L.; Di Nauta, M.; Calò, B.; Del Giudice, F.; Basran, S.; Chung, B.I.; Porreca, A.; et al. A Matched-Pair Analysis after Robotic and Retropubic Radical Prostatectomy: A New Definition of Continence and the Impact of Different Surgical Techniques. *Cancers* **2022**, *14*, 4350. https://doi.org/10.3390/cancers14184350

Academic Editor: Ana Faustino

Received: 8 August 2022
Accepted: 4 September 2022
Published: 7 September 2022

1. Introduction

RP is considered the gold-standard treatment for patients with localized prostate cancer (PCa) and a life expectancy of more than 10 years [1]. The main goal of RP is oncological efficacy, but care is taken to preserve the anatomical structures surrounding the prostate to reduce the receipt of side effects, without compromising oncologic outcomes. Over the last twenty years, RARP has gained popularity and is now used on more than half of patients undergoing surgery for PCa [2]. The robotic platform allows an immersive surgical experience with magnified 3D high-definition vision and depth perception. The surgeon can see tissue planes clearly, identify structures, and dissect with a higher level of precision. This is expected to result in better oncological and functional outcomes [3].

However, the available literature suggests that there is no difference in oncological outcomes between RARP and RRP. The robotic approach has undeniably reduced blood loss and length of stay, but questions remain as to its ability to improve functional outcomes, such as erectile function and UC. Erectile dysfunction (ED) is a common side effect and those who experience ED after surgery and are motivated to regain erectile function can rely on several effective non-surgical and surgical treatments. Conversely, UI is less common but can be more psychologically challenging due to the fact that very few patients have underlying pre-RP UI. Due to this impact of de novo post-RP incontinence, all are strongly motivated to regain continence after surgery, but treatment options can often require additional surgical procedures, which can be a daunting prospect.

Prospective randomized trials comparing open and robotic surgery demonstrated comparable results in terms of early (85% vs. 84%, $p = 0.8$) [4] and late (91% vs. 90%, $p = 0.6$) [5] continence recovery following both techniques, whereas systematic reviews and meta-analyses showed better continence recovery at 12 months in favor of RARP (OR: 1.53, $p = 0.03$) (R: 2.84, $p = 0.002$) [6,7].

Therefore, due these above conflicting data, the aim of this study was to compare the continence recovery rates after RARP and RRP using pad weights over a 24 h period and to determine predictive factors for continence recovery.

2. Materials and Methods

2.1. Eligibility Criteria

The Da Vinci Xi surgical system has been available at our institution since January 2016; since then, RARP has become the preferred approach for RP, with RRP being performed only in the few patients who refuse or are not considered fit for robotic surgery due to their comorbidities.

For the present study, we queried our Internal-Review-Board-approved prospective database on PCa to identify patients that underwent RARP (n = 600), had a minimum follow-up of 12 months, and a matched population of patients that underwent RRP (n = 945) between January 2010 and December 2015. Patients with incomplete clinical data or those who had adjuvant treatment after surgery were excluded. Additionally, the first 100 patients who underwent RARP were excluded due to the possible impact of the learning curve on functional outcomes [8].

According to our policy [9–11], all patients had uroflowmetry with evaluation of peak flow rate (Q-max) and a post-void residual urinary volume (PVR). Transrectal ultrasound was used to measure both Pvol and to calculate PSA density.

For RRP, we used the technique described by Walsh [12], whereas for the RARP we used an intraperitoneal access and an anterograde dissection through the space of Retzius [13,14]. Pelvic lymph node dissection was carried out according to Briganti nomogram indications using an extended template as recommended by EAU guidelines [1]. The criteria for neurovascular bundle preservation were identical for both procedures and were carefully balanced against the risk of positive surgical margins [15–17]. The vesicourethral anastomosis, always preceded by posterior reconstruction [18], required 6 stitches in RRP, while RARP required 2 barbed running sutures. A drain was always placed at the end of each procedure and all patients underwent a retrograde cystogram prior to bladder catheter removal.

UC was assessed and discussed by two dedicated "Continence Nurses" before surgery and then 7 days after catheter removal. Patients were seen weekly for the first post-operative month in a formal pelvic floor rehabilitation program and then at 3, 6, and 12 months. During all visits, the ability to perform Kegel exercises was assessed; in addition, data on 24 h pad test and ICIQ-7 questionnaire were recorded.

2.2. Statistical Analysis

The primary study objective was to compare the rates of patients continent at 1 week, 1 month, 6 months, and 1 year after the two procedures.

To control for measured potential confounders in the data set, a 1:1 propensity score (PS)-matched population of patients undergoing RP was selected using age at surgery, PSA, Biopsy Gleason Grade Group (GGG), and digital rectal exam (DRE). After PS matching, the quality of covariate balance was checked by assessment of standardized differences. Continuous variables were reported as median and interquartile range and compared by the Mann–Whitney U-test, whereas categorical variables were reported as rates and tested by the Fisher's exact test or the chi-square test, as appropriate. A Kaplan–Meier curve of the probability of urinary continence recovery following RP over time was created. Finally, uni- and multivariable Cox regression analysis, involving the entire study cohort, was used to evaluate predictors of continence. Statistical analyses were performed using Stata-SE 15 (StataCorp LP, College Station, TX, USA). All tests were 2-sided with a significance level set at $p < 0.05$.

3. Results

After propensity score matching of the patient population, 482 met the inclusion criteria, 241 patients per group; Table 1 shows the preoperative characteristics of the study population. Of note, there were no differences in age, DRE, PSA, preoperative IIEF-5 score, IPSS, maximum peak flow rate (Q-max), PVR, Pvol, and biopsy GGG.

Table 1. Preoperative characteristics of the study population after propensity matching.

Characteristics	RRP [1] n = 241 (50%)	RARP [1] n = 241 (50%)	p-Value [2]
Age at surgery (y)	66.0 (62.0, 71.0)	66.0 (61.0, 70.0)	0.2
PSA, (ng/mL)	6.2 (4.6, 9.1)	6.2 (4.8, 9.2)	0.6
DRE, n (%)			
Negative	125 (52%)	119 (49%)	0.6
Positive	116 (48%)	122 (51%)	
Biopsy GGG, n (%)			
1	110 (46%)	110 (46%)	0.9
2	57 (24%)	64 (27%)	
3	42 (17%)	41 (17%)	
4	22 (9%)	19 (8%)	
5	10 (4%)	7 (3%)	
Prostate volume (cm^3)	40.0 (31.0, 56.0)	40.0 (32.0, 55.0)	1
Q-max (mL/s)	15.6 (12.0, 18.6)	14.0 (11.7, 21.0)	0.7
PVR (mL)	20.0 (1.0, 40.0)	30.0 (1.0, 50.0)	0.060
Pre-operative IPSS	7.0 (4.0, 12.0)	7.0 (3.0, 12.0)	0.3
Pre-operative IIEF	19.0 (11.0, 22.0)	20.0 (15.0, 23.0)	0.067
Nerve Sparing			
Non nerve sparing	172 (71.4%)	141 (58.5%)	0.002
Unilateral	26 (10.8%)	23 (9.5%)	
Bilateral	43 (17.8%)	77 (32.0%)	

[1] Median (IQR); n (%). [2] Wilcoxon rank-sum test, Pearson's Chi-square test.

As for UC rates (Table 2), RARP performed significantly better than RRP at all time points. The continence rate for RARP and RRP was 58% vs. 47% at 1 week ($p = 0.018$), 82% vs. 61% at 1 month ($p < 0.0001$), 92% vs. 75% at 6 months ($p < 0.0001$), and 94% vs. 80% at 1 year ($p < 0.0001$), respectively.

Table 2. Continence recovery rates (pad 0–20 gr) in patients undergoing RRP and RARP at 1 week, 1 month, 6 month, and 1-year follow-up.

Characteristic	RRP [1] n = 241 (50%)	RARP [1] n = 241 (50%)	p-Value [2]
1 week			
Pad < 20 gr, n (%)	114 (47%)	140 (58%)	0.018
ICIQ score *	9.0 (5.0, 13.0)	7.0 (4.0, 10.0)	0.011
1 month			
Pad < 20 gr, n (%)	148 (61%)	198 (82%)	<0.0001
ICIQ score *	8 (5, 21)	7 (0, 21.0)	0.001
6 months			
Pad < 20 gr, n (%)	180 (75%)	222 (92%)	<0.0001
ICIQ score *	5 (0, 10)	4 (0, 6)	0.029
1 year			
Pad < 20 gr, n (%)	193 (80%)	227 (94%)	<0.0001
ICIQ score *	5 (0, 8)	3 (0, 6)	0.09

[1] Median (IQR); n (%). [2] Wilcoxon rank-sum test, Pearson's Chi-square test. * ICIQ: International Consultation Incontinence Questionnaire—short form (ICIQ-short form).

Further, RARP performed significantly better than RRP in UI (Table 3), since severe leakage (>200 gr/24 h) was 40.9% vs. 15.8% at 6 months ($p = 0.04$) and 39.6% vs. 7.1% at 1 year ($p = 0.02$), in favor of RARP. Furthermore, Table 3 shows the trends in the score on the ICIQ questionnaire, which reflects the trend in the overall severity of urinary incontinence.

Table 3. Incontinence recovery rates (pad > 20 gr) in patients undergoing RRP and RARP at 1 week, 1 month, 6 month, and 1-year follow-up.

Characteristic	RRP [1] n = 241 (50%)	RARP [1] n = 241 (50%)	p-Value [2]
1 Week	**n = 127**	**n = 101**	
21–200 gr, n (%)	63 (49.6%)	82 (81.2%)	0.00001
ICIQ score *	20.5 (18, 21)	20.5 (18, 21)	0.27
>200 gr, n (%)	64 (50.4%)	19 (18.8%)	0.00001
ICIQ score *	21 (21, 21)	21 (21, 21)	0.72
1 Month	**n = 93**	**n = 43**	
21–100 gr, n (%)	44 (47.3%)	35 (81.4%)	0.00001
ICIQ score *	15 (9, 21)	10 (7, 15)	0.27
>200 gr, n (%)	51 (54.8%)	8 (18.6%)	0.00001
ICIQ score *	21 (21, 21)	21 (21, 21)	0.4
6 Month	**n = 61**	**n = 19**	
21–100 gr, n (%)	36 (59.1%)	16 (84.2%)	0.04
ICIQ score *	10 (7, 17)	10 (6, 16)	0.52
>200 gr, n (%)	25 (40.9%)	3 (15.8%)	0.04
ICIQ score *	21 (18, 21)	15 (10, 16)	0.58
1 Year	**n = 48**	**n = 14**	
21–100 gr, n (%)	29 (60.4%)	13 (82.9%)	0.02
ICIQ score *	10 (8, 15)	10 (7, 10)	0.51
>200 gr, n (%)	19 (39.6%)	1 (7.1%)	0.02
ICIQ score *	21 (21, 21)	21 (21, 21)	0.52

[1] Median (IQR); n (%). [2] Wilcoxon rank-sum test, Pearson's Chi-square test. * ICIQ: International Consultation Incontinence Questionnaire - short form (ICIQ-short form).

Kaplan–Maier curves (Figure 1) confirmed a significant difference in continence rates between techniques (Figure 1a), especially in those who received a bilateral nerve sparing (Figure 1b).

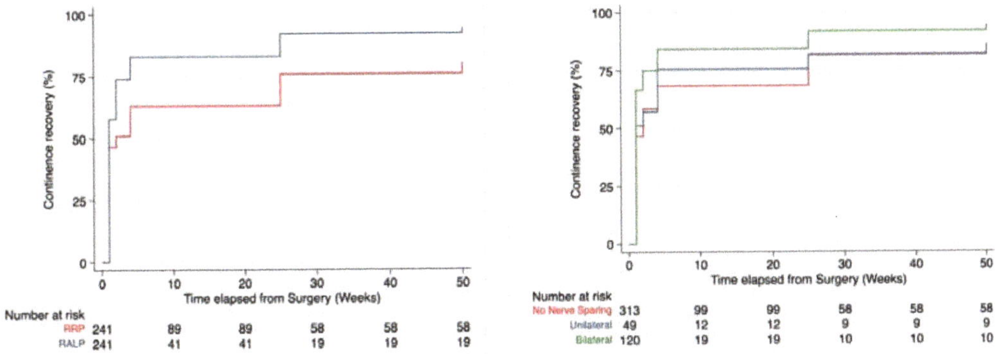

(a) (b)

Figure 1. Continence recovery rates between RRP and RARP at follow-up after propensity matching, Pr > chi2 $\leq$ 0.0001 (**a**). Continence recovery rates divided by nerve sparing grade on the entire population study, Pr > chi2 = 0.0012 (**b**).

In Table 4, the analysis on the entire study population (n = 482 patients) identified the bilateral nerve-sparing technique and robotic surgical approach as having significant predictor value of continence rates (1.25 (CI 1.02,1.54), $p = 0.030$; 1.42 (CI 1.18,1.69), $p \leq 0.001$).

Table 4. Univariable and multivariable Cox Regression Analysis to predict continence recovery in the entire population.

	Univariable Cox Regression Analysis			Multivariable Cox Regression Analysis		
Covariate	**Haz. Ratio**	**95% CI**	***p* Value**	**Haz. Ratio**	**95% CI**	***p* Value**
Age	0.99	0.98, 1.00	0.174			
GGG biopsy						
1	Ref.			Ref.		
2	1.05	0.84, 1.31	0.661	1.02	0.82, 1.28	0.857
3	0.87	0.67, 1.13	0.301	0.90	0.69, 1.17	0.430
4	0.88	0.68, 1.14	0.329	0.98	0.75, 1.28	0.876
5	1.03	0.72, 1.47	0.877	1.15	0.80, 1.65	0.462
Prostate volume	1.00	0.99, 1.00	0.231			
Q-max	1.00	1.00, 1.01	0.464			
PVR	1.00	1.00, 1.01	0.397			
IPSS	0.99	0.97, 1.00	0.155			
IIEF	1.01	1.00, 1.02	0.163			
Nerve sparing						
No	Ref.			Ref.		
Unilateral	1.06	0.79, 1.43	0.711	1.01	0.75, 1.37	0.932
Bilateral	1.35	1.10, 1.64	0.003	1.25	1.02, 1.54	0.030
Catheters days						
≥8	Ref.					
<8	0.85	0.72, 1.00	0.056	0.88	0.75, 1.05	0.159

Table 4. *Cont.*

	Univariable Cox Regression Analysis			Multivariable Cox Regression Analysis		
Covariate	**Haz. Ratio**	**95% CI**	***p* Value**	**Haz. Ratio**	**95% CI**	***p* Value**
Surgery type						
RRP	Ref.			Ref.		
RARP	1.46	1.23, 1.73	<0.001	1.42	1.18, 1.69	<0.001

4. Discussion

This study demonstrates better early and late recovery continence and lower rates of UI favoring RARP. Further, we show an overall benefit of the bilateral nerve-sparing approach as a factor in facilitating urinary recovery. Further, regarding time to continence, our data showed not only better but also faster continence recovery with RARP than with RRP. Indeed, RARP performed better than RRP at all evaluation times, considering continence rates were 47% vs. 58% at 1 week (p = 0.018), 61% vs. 82% at 1 month (p = 0.001), 75% vs. 92 % at 6 months (p = 0.018), and 80% vs. 94% at 1 year (p = 0.001).

UC rates after open and robotic prostatectomy vary widely in the available literature and several factors may explain these differences. The robotic approach has clear advantages, including better visualization of delicate structures involved in continence recovery [14–20] and higher surgical precision in the reconstructive phase. However, on the flip side, in reported series, up to 31% might have UI (pad > 0) at 1 year [7]. Patient and cancer characteristics, operative techniques, surgical experience, surgeon dexterity, high-volume centers, post-operative urinary rehabilitation, and adjuvant treatments are all factors to be consider when comparing post-prostatectomy results [21,22]. In this setting, the only existing randomized prospective study showed similar and comparable results in oncological and functional aspects, both in the short and long term [4–7,21,22], although the functional domain was judged on data acquired from questionnaires administered at follow-up and with no mention of postoperative rehabilitation.

The level of evidence of most of the published data is retrospective and, by nature, limited by selection bias. Furthermore, in the first ten years after the introduction of robotic technology, many studies were characterized by a significant publication bias, with post-RARP continence rates (pad 0) reported up to 50–80%, 90–96%, and 89% to 98%, interesting values after catheter removal compared with open at 6 and 12 months, respectively [23]. Further, Patel et al. [24] reported his technique of RARP provided a 1-year continence rate (pad 0) of 96.4% and Menon et al., a 6-month continence rate (pads $\leq$ 1) of 92% [13].

Subsequent data have been conflicting in nature regarding any advantages in continence with the robotic approach. An early meta-analysis compared the two techniques in terms of recovery of UC at 12 months and showed significant advantages to a robotic approach (OR: 1.53; 95% CI, 1.04–2.25; p = 0.03). Indeed, after RARP, continence recovery ranged from 89% to 100% compared to RRP, which ranges from 80% to 97% [7,25–27]. However, this new technology still exposed patients to IU, though with less impact (11% vs. 7%). Similarly, Seo et al. compared 10 matching studies between open and robotic RP with 2214 patients and reported a lower incontinence rate after RARP (RR 0.66, 95% CI 0.45–0.99, I2 = 45%; p = 0.040) [28].

Subsequently, several authors began to present data that demonstrated minimal differences in continence between the open and robotic approaches. A 2017 meta-analysis included 78 studies, in which Kun Tang showed no statistical differences in the 3-month and 12-month UC within the two surgical techniques (3 months: OR:1.54; 95% CI: 0.92 to 2.58; p = 0.10; 12 months: OR:1.03; 95% CI: 0.84 to 1.27; p = 0.75, respectively) [29]. Similarly, Lan Cao et al. published a systematic review and meta-analysis in 2019 with 8522 patients and found no difference in the 12-month continence rates among the two procedures. UC was defined as either no pad use or no leakage (71.6% vs. 70.8%; OR 1.14, 95% CI 0.99 to 1.31, p = 0.08) [30]. In a 1:2 matching population series between 294 and 588 patients undergoing RARP and RRP, one-year continence recovery rate (security pad) was overlapping

(93.7% vs. 91.8%; $p = 0.34$) also [31]. Several meta-analyses pointed out that the data were insufficient to define a technique as superior, given variables related to surgeon experience, patient history, and oncologic finding.

Quantifying the degree and severity of urinary incontinence has been an area of ongoing debate. Even though pad weight and pad usage are well correlated with quality of life [32], the variability in the definitions used in the literature makes it difficult to objectify an outcome and, therefore, make a real comparison among the results of traditional and technological surgery. The above-mentioned studies in recent years have only assessed incontinence by interviewing the patient based on the number of pads used per day (0, 0–1, one safety pad). Therefore, the differences between the two techniques are not highlighted, emphasizing the use of technology but not for functional advantages.

The rate of functional recovery also varies widely in proportion to pad use. As a result, one study pointed out that the continence rate after RP at 3 and 12 months was 44% and 68% when defined as "pad 0" and increased to 71% and 90% when defined as "pad 0–1" [33]. In addition, a systematic review, reported average values of post-RARP incontinence at 1 year of 16% and 9% according to the definition of "dry" or "a safe pad" [7]. Although Holze et al. argues zero pads best reflects the perception of "dry" [33], Moore et al., suggests a minimum value greater than 4 g/day of leakage as incontinence after prostatectomy, but greater than 20 g/day is characterized as symptomatic incontinence [34]. We, therefore, chose the threshold of a 24 h pad weight test greater than 20 g as a reasonable value for assessing symptomatic male incontinence after an RP. However, a 0 g pad in 24 h is not easy to obtain, as sweat can also increase the weight of the pad and can classify a continent patient as incontinent. For this reason, other definitions are established, such as socially continent. With this, 20 g as a threshold seems reasonable and easy to assess to define a patient as socially continent.

Based on our experience, we have historically chosen to avoid testing UC through interviews about pad count, which is strongly related to subjective feelings of "wetness". Instead, we choose "third-party" assessment by a urinary specialist nurse who objectively records the 24 h pad weight test at follow-up. We believe it is a reliable tool for assessing the presence and degree of urinary leakage and combining it with those obtained from the recognized ICIQ-7 subjective questionnaire for UI [35].

Others have used 24 h pad weight as an objective measurement of the degree of incontinence. In 2010, Dubbelman was the first to present functional outcomes after RRP using the 24 h pad weight test at 6 months in a series of 66 patients. He reported recovery of continence, defined as 4 g per day, in 44% of patients who were receiving a "Continence Nurses" or folder-guided approach [36]. In 2017, Sathianathen was the first to present functional outcomes after RARP using 24 h pad weight test. Using zero grams as a definition, Sathianathen reported continence recovery after RARP in 20% and 38% of patients, respectively, after 4 and 12 weeks, with average continence recovery at 10 weeks. They also pointed out the fact that if the number of pads had been used as the definition, it would probably have produced better results than UC [37].

However, currently, the 24 h pad test is used by less than 1 out of 10 clinicians [38]. It is considered a controversial way of objectively reporting a measure of male post-prostatectomy UI [39], because the weight of the 24 h pad is directly influenced by daily activity and fluid intake and the same weight can vary by 51 g (95% CI 30.3–72.1, $p < 0.001$) in those who do not change their activity level. On the other hand, what is perceived to be a single wet pad can range from 23 to 121 g and it is these values that make an objective comparison of the degree of leakage possible. More importantly, for each 1-unit increase in pads used, the average pad weight increased by 114.1 g (95% CI 74.8–153.4, $p < 0.001$) [40]. With this in mind, the definition of "Pad 0–1" may not reflect continence status as it may include symptomatic urinary leakage [34]. The added value of our study is to better stratify patients and define already symptomatic incontinent patients with >20 g or more of leakage who, in other definitions, would be continent.

Additional studies suggest that characteristics, such as age, body mass index (BMI), Pvol, and lower urinary tract disorders (LUTS), influence functional outcomes [40–42]. As we have previously shown, age did not influence continence recovery in RRP [43], even though Jeong found advantages in young patients, both in early and long-term follow-up [44] and Becker claims that beyond the age of 50, the recovery rate drops from 97.4% to 91.6% [45].

Another interesting insight of the analysis suggests that bilateral sparing of the neurovascular bundles is advantageous in facilitating recovery of UC on an entire study population, especially in RARP, as shown in multivariate analysis (Table 4) and Kaplan–Meier curves (Figure 1b). In this context, the benefits of nerve sparing on continence recovery, both in the short term and long term for RRP and RARP, have already been reported [46–48]. These findings are so strong that Steineck continues to recommend nerve sparing, regardless of age and preoperative erectile function, as the benefits are significant [49].

One of the strengths of our study was the management and evaluation of the prostatectomized matching pair population that was entrusted to the "urinary nurse" who trained and collected the data ensuring an unbiased evaluation. The definition was based on the objective measurement of a 24 h pad test in grams rather than on the pad use together with the subjective data obtained from the ICIQ-SF questionnaire. Having stated that, the results obtained justify the use of robotic surgery, not only for less blood loss and hospitalization but also for its impact on the recovery of UC and quality of life.

This work certainly has limitations; first, this is a retrospective analysis of prospectively collected data. Further, body mass index (BMI) and an assessment of preoperative continence were not assessed. However, in the second case, we have a subjective comparison of LUTS through the IPSS questionnaire with overlapping results between the two groups.

5. Conclusions

In urologic oncologic surgery, to date, robotic surgery has brought advantages in terms of decreases in blood loss, shorter hospitalization, and, possibly, improved quality of life. When it comes to treating localized prostate cancer, a prostatectomy not only has oncologic but also functional implications. In terms of continence recovery, our study demonstrates that, with a rigorous objective evaluation of postoperative urinary leakage, robotic prostatectomy has advantages over traditional surgery, in both short- and long-term measures of continence.

Author Contributions: Conceptualization, N.d., V.M., L.C. and U.G.F.; Data curation, N.d., G.M.B., L.M., M.D.N. and B.C.; Formal analysis, U.G.F., L.M., L.C. and V.M.; Investigation, N.d., V.M., G.M.B., L.B., L.C., G.C., C.B., P.A. and B.C.; Methodology, N.d., G.M.B., L.M., M.D.N., L.C., G.C. and C.B.; Project ad-ministration, N.d., G.M.B., L.M., M.D.N., L.C., G.C. and C.B.; Resources, F.D.G., S.B. and L.M.; Software, U.G.F., F.D.G. and S.B.; Supervision, G.C., L.C., E.B. and R.S.; Validation, C.B., F.D.G., S.B., L.B., L.C., G.C., C.B., P.A. and G.M.B.; Visualization, F.D.G., S.B. and C.B.; Writing—original draft, N.d.; Writing—review and editing, N.d., V.M., G.M.B., L.B., L.C., G.C., C.B., P.A., A.P. and B.I.C. All authors have read and agreed to the published version of the manuscript.

Funding: This research received no external funding.

Institutional Review Board Statement: All procedures performed in studies involving human participants were in accordance with the ethical standards of the institutional and/or national research committee and with the 1964 Helsinki Declaration and its later amendments or comparable ethical standards. This study was approved by the Department of Urology and Renal Transplantation, Policlinico Riuniti di Foggia, University of Foggia Department of Urology, University of Foggia, approval code: 143/CE/2020 and approved on DDG n.696 01.12.2020.

Informed Consent Statement: Informed consent was obtained from all subjects involved in the study.

Data Availability Statement: The data presented in this study are available on request from the corresponding author. The data are not publicly available.

Conflicts of Interest: The authors declare no conflict of interest.

References

1. Mottet, N.; van den Bergh, R.C.N.; Briers, E.; Van den Broeck, T.; Cumberbatch, M.G.; De Santis, M.; Fanti, S.; Fossati, N.; Gandaglia, G.; Gillessen, S.; et al. EAU-EANM-ESTRO-ESUR-SIOG Guidelines on Prostate Cancer-2020 Update. Part 1: Screening, Diagnosis, and Local Treatment with Curative Intent. *Eur. Urol.* **2021**, *79*, 243–262. [CrossRef] [PubMed]
2. Mottrie, A.; Larcher, A.; Patel, V. The Past, the Present, and the Future of Robotic Urology: Robot-assisted Surgery and Human-assisted Robots. *Eur. Urol. Focus* **2018**, *4*, 629–631. [CrossRef] [PubMed]
3. Leal Ghezzi, T.; Campos Corleta, O. 30 Years of Robotic Surgery. *World J. Surg.* **2016**, *40*, 2550–2557. [CrossRef] [PubMed]
4. Yaxley, J.W.; Coughlin, G.D.; Chambers, S.K.; Occhipinti, S.; Samaratunga, H.; Zajdlewicz, L.; Dunglison, N.; Carter, R.; Williams, S.; Payton, D.J.; et al. Robot-assisted laparoscopic prostatectomy versus open radical retropubic prostatectomy: Early outcomes from a randomised controlled phase 3 study. *Lancet* **2016**, *388*, 1057–1066. [CrossRef]
5. Coughlin, G.D.; Yaxley, J.W.; Chambers, S.K.; Occhipinti, S.; Samaratunga, H.; Zajdlewicz, L.; Teloken, P.; Dunglison, N.; Williams, S.; Lavin, M.F.; et al. Robot-assisted laparoscopic prostatectomy versus open radical retropubic prostatectomy: 24-month outcomes from a randomised controlled study. *Lancet Oncol.* **2018**, *19*, 1051–1060. [CrossRef]
6. Ficarra, V.; Novara, G.; Ahlering, T.E.; Costello, A.; Eastham, J.A.; Graefen, M.; Guazzoni, G.; Menon, M.; Mottrie, A.; Patel, V.R.; et al. Systematic review and meta-analysis of studies reporting potency rates after robot-assisted radical prostatectomy. *Eur. Urol.* **2012**, *62*, 418–430. [CrossRef]
7. Ficarra, V.; Novara, G.; Rosen, R.C.; Artibani, W.; Carroll, P.R.; Costello, A.; Menon, M.; Montorsi, F.; Patel, V.R.; Stolzenburg, J.U.; et al. Systematic review and meta-analysis of studies reporting urinary continence recovery after robot-assisted radical prostatectomy. *Eur. Urol.* **2012**, *62*, 405–417. [CrossRef]
8. Ou, Y.C.; Yang, C.K.; Chang, K.S.; Wang, J.; Hung, S.W.; Tung, M.C.; Tewari, A.K.; Patel, V.R. The surgical learning curve for robotic-assisted laparoscopic radical prostatectomy: Experience of a single surgeon with 500 cases in Taiwan, China. *Asian J. Androl.* **2014**, *16*, 728–734. [CrossRef]
9. Cormio, L.; Lucarelli, G.; Selvaggio, O.; Di Fino, G.; Mancini, V.; Massenio, P.; Troiano, F.; Sanguedolce, F.; Bufo, P.; Carrieri, G. Absence of Bladder Outlet Obstruction Is an Independent Risk Factor for Prostate Cancer in Men Undergoing Prostate Biopsy. *Medicine* **2016**, *95*, e2551. [CrossRef]
10. Cormio, L.; Lucarelli, G.; Netti, G.S.; Stallone, G.; Selvaggio, O.; Troiano, F.; Di Fino, G.; Sanguedolce, F.; Bufo, P.; Grandaliano, G.; et al. Post-void residual urinary volume is an independent predictor of biopsy results in men at risk for prostate cancer. *Anticancer Res.* **2015**, *35*, 2175–2182.
11. Cormio, L.; Cindolo, L.; Troiano, F.; Marchioni, M.; Di Fino, G.; Mancini, V.; Falagario, U.; Selvaggio, O.; Sanguedolce, F.; Fortunato, F.; et al. Development and Internal Validation of Novel Nomograms Based on Benign Prostatic Obstruction-Related Parameters to Predict the Risk of Prostate Cancer at First Prostate Biopsy. *Front. Oncol.* **2018**, *16*, 438. [CrossRef] [PubMed]
12. Walsh, P.C. Anatomic radical prostatectomy: Evolution of the surgical technique. *J. Urol.* **1998**, *160 Pt 2*, 2418–2424. [CrossRef]
13. Menon, M.; Tewari, A.; Peabody, J.; VIP Team. Vattikuti Institute prostatectomy: Technique. *J. Urol.* **2003**, *169*, 2289–2292. [CrossRef]
14. Martini, A.; Falagario, U.G.; Villers, A.; Dell'Oglio, P.; Mazzone, E.; Autorino, R.; Moschovas, M.C.; Buscarini, M.; Bravi, C.A.; Briganti, A.; et al. Contemporary Techniques of Prostate Dissection for Robot-assisted Prostatectomy. *Eur. Urol.* **2020**, *78*, 583–591. [CrossRef] [PubMed]
15. Falagario, U.G.; Jambor, I.; Ratnani, P.; Martini, A.; Treacy, P.J.; Wajswol, E.; Lantz, A.; Papastefanou, G.; Weil, R.; Phillip, D.; et al. Performance of prostate multiparametric MRI for prediction of prostate cancer extra-prostatic extension according to NCCN risk categories: Implication for surgical planning. *Minerva Urol. Nefrol.* **2020**, *72*, 746–754. [CrossRef] [PubMed]
16. Martini, A.; Cumarasamy, S.; Gupta, A.; Falagario, U.G.; Shah, Q.N.; Beksac, A.T.; Haines, K.G.; Tewari, A.K. Clinical implications of prostatic capsular abutment or bulging on multiparametric magnetic resonance imaging. *Minerva Urol. Nefrol.* **2019**, *71*, 502–507. [CrossRef]
17. Martini, A.; Marqueen, K.E.; Falagario, U.G.; Waingankar, N.; Wajswol, E.; Khan, F.; Fossati, N.; Briganti, A.; Montorsi, F.; Tewari, A.K.; et al. Estimated Costs Associated with Radiation Therapy for Positive Surgical Margins During Radical Prostatectomy. *JAMA Netw. Open* **2020**, *3*, e201913. [CrossRef]
18. Rocco, B.; Gregori, A.; Stener, S.; Santoro, L.; Bozzola, A.; Galli, S.; Knez, R.; Scieri, F.; Scaburri, A.; Gaboardi, F. Posterior reconstruction of the rhabdosphincter allows a rapid recovery of continence after transperitoneal videolaparoscopic radical prostatectomy. *Eur. Urol.* **2007**, *51*, 996–1003. [CrossRef]
19. Sheetz, K.H.; Claflin, J.; Dimick, J.B. Trends in the Adoption of Robotic Surgery for Common Surgical Procedures. *JAMA Netw. Open* **2020**, *3*, e1918911. [CrossRef]
20. Haglind, E.; Carlsson, S.; Stranne, J.; Wallerstedt, A.; Wilderäng, U.; Thorsteinsdottir, T.; Lagerkvist, M.; Damber, J.E.; Bjartell, A.; Hugosson, J.; et al. Urinary Incontinence and Erectile Dysfunction after Robotic Versus Open Radical Prostatectomy: A Prospective, Controlled, Nonrandomised Trial. *Eur. Urol.* **2015**, *68*, 216–225. [CrossRef]
21. Kim, S.C.; Song, C.; Kim, W.; Kang, T.; Park, J.; Jeong, I.G.; Lee, S.; Cho, Y.M.; Ahn, H. Factors determining functional outcomes after radical prostatectomy: Robot-assisted versus retropubic. *Eur. Urol.* **2011**, *60*, 413–419. [CrossRef] [PubMed]

22. Trieu, D.; Ju, I.E.; Chang, S.B.; Mungovan, S.F.; Patel, M.I. Surgeon case volume and continence recovery following radical prostatectomy: A systematic review. *ANZ J. Surg.* **2021**, *91*, 521–529. [CrossRef] [PubMed]
23. Singh, I.; Hemal, A.K. Robotic-assisted radical prostatectomy in 2010. *Expert Rev. Anticancer Ther.* **2010**, *10*, 671–682. [CrossRef] [PubMed]
24. Patel, V.R.; Sivaraman, A.; Coelho, R.F.; Chauhan, S.; Palmer, K.J.; Orvieto, M.A.; Camacho, I.; Coughlin, G.; Rocco, B. Pentafecta: A new concept for reporting outcomes of robot-assisted laparoscopic radical prostatectomy. *Eur. Urol.* **2011**, *59*, 702–707. [CrossRef]
25. Ficarra, V.; Novara, G.; Fracalanza, S.; D'Elia, C.; Secco, S.; Iafrate, M.; Cavalleri, S.; Artibani, W. A prospective, non-randomized trial comparing robot-assisted laparoscopic and retropubic radical prostatectomy in one European institution. *BJU Int.* **2009**, *104*, 534–539. [CrossRef]
26. Di Pierro, G.B.; Baumeister, P.; Stucki, P.; Beatrice, J.; Danuser, H.; Mattei, A. A prospective trial comparing consecutive series of open retropubic and robot-assisted laparoscopic radical prostatectomy in a centre with a limited caseload. *Eur. Urol.* **2011**, *59*, 1–6. [CrossRef]
27. Ou, Y.C.; Yang, C.R.; Wang, J.; Cheng, C.L.; Patel, V.R. Comparison of robotic-assisted versus retropubic radical prostatectomy performed by a single surgeon. *Anticancer Res.* **2009**, *29*, 1637–1642. [CrossRef]
28. Seo, H.-J.; Lee, N.R.; Son, S.K.; Kim, D.K.; Rha, K.H.; Lee, S.H. Comparison of Robot-Assisted Radical Prostatectomy and Open Radical Prostatectomy Outcomes: A Systematic Review and Meta-Analysis. *Yonsei Med. J.* **2016**, *57*, 1165–1177. [CrossRef]
29. Tang, K.; Jiang, K.; Chen, H.; Chen, Z.; Xu, H.; Ye, Z. Robotic vs. Retropubic radical prostatectomy in prostate cancer: A systematic review and an meta-analysis update. *Oncotarget* **2017**, *8*, 32237–32257. [CrossRef]
30. Cao, L.; Yang, Z.; Qi, L.; Chen, M. Robot-assisted and laparoscopic vs open radical prostatectomy in clinically localized prostate cancer: Perioperative, functional, and oncological outcomes: A Systematic review and meta-analysis. *Medicine* **2019**, *98*, e15770. [CrossRef]
31. Krambeck, A.E.; DiMarco, D.S.; Rangel, L.J.; Bergstralh, E.J.; Myers, R.P.; Blute, M.L.; Gettman, M.T. Radical prostatectomy for prostatic adenocarcinoma: A matched comparison of open retropubic and robot-assisted techniques. *BJU Int.* **2009**, *103*, 448–453. [CrossRef]
32. Nitti, V.W.; Mourtzinos, A.; Brucker, B.M.; SUFU Pad Test Study Group. Correlation of patient perception of pad use with objective degree of incontinence measured by pad test in men with post-prostatectomy incontinence: The SUFU Pad Test Study. *J. Urol.* **2014**, *192*, 836–842. [CrossRef] [PubMed]
33. Holze, S.; Mende, M.; Healy, K.V.; Koehler, N.; Gansera, L.; Truss, M.C.; Rebmann, U.; Degener, S.; Stolzenburg, J.U. Comparison of various continence definitions in a large group of patients undergoing radical prostatectomy: A multicentre, prospective study. *BMC Urol.* **2019**, *19*, 70. [CrossRef] [PubMed]
34. Moore, K.N.; Truong, V.; Estey, E.; Voaklander, D.C. Urinary incontinence after radical prostatectomy: Can men at risk be identified preoperatively? *J. Wound Ostomy Cont. Nurs.* **2007**, *34*, 270–279. [CrossRef] [PubMed]
35. Karantanis, E.; Fynes, M.; Moore, K.H.; Stanton, S.L. Comparison of the ICIQ-SF and 24-hour pad test with other measures for evaluating the severity of urodynamic stress incontinence. *Int. Urogynecol. J. Pelvic Floor Dysfunct.* **2004**, *15*, 111–116. [CrossRef]
36. Dubbelman, Y.; Groen, J.; Wildhagen, M.; Rikken, B.; Bosch, R. The recovery of urinary continence after radical retropubic prostatectomy: A randomized trial comparing the effect of physiotherapist-guided pelvic floor muscle exercises with guidance by an instruction folder only. *BJU Int.* **2010**, *106*, 515–522. [CrossRef]
37. Sathianathen, N.J.; Johnson, L.; Bolton, D.; Lawrentschuk, N.L. An objective measurement of urinary continence recovery with pelvic floor physiotherapy following robotic assisted radical prostatectomy. *Transl. Androl. Urol.* **2017**, *6* (Suppl. S2), S59–S63. [CrossRef]
38. Zimmern, P.; Kobashi, K.; Lemack, G. Outcome measure for stress urinary incontinence treatment (OMIT): Results of two society of urodynamics and female urology (SUFU) surveys. *Neurourol. Urodyn.* **2010**, *29*, 715–718. [CrossRef]
39. Rasmussen, A.; Mouritsen, L.; Dalgaard, A.; Frimodt-Møller, C. Twenty-four hour pad weighing test: Reproducibility and dependency of activity level and fluid intake. *Neurourol. Urodyn.* **1994**, *13*, 261–265. [CrossRef]
40. Malik, R.D.; Cohn, J.A.; Fedunok, P.A.; Chung, D.E.; Bales, G.T. Assessing variability of the 24-hour pad weight test in men with post-prostatectomy incontinence. *Int. Braz. J. Urol.* **2016**, *42*, 327–333. [CrossRef]
41. Novara, G.; Ficarra, V.; D'elia, C.; Secco, S.; Cioffi, A.; Cavalleri, S.; Artibani, W. Evaluating urinary continence and preoperative predictors of urinary continence after robot assisted laparoscopic radical prostatectomy. *J. Urol.* **2010**, *184*, 1028–1033. [CrossRef] [PubMed]
42. Shikanov, S.; Desai, V.; Razmaria, A.; Zagaja, G.P.; Shalhav, A.L. Robotic radical prostatectomy for elderly patients: Probability of achieving continence and potency 1 year after surgery. *J. Urol.* **2010**, *183*, 1803–1807. [CrossRef] [PubMed]
43. d'Altilia, N.; Di Nauta, M.; Falagario, U.G.; Calò, B.; Selvaggio, O.; Sanguedolce, F.; Mancini, V.; Stallone, G.; Barret, E.; Cormio, L.; et al. Elderly patients are not at higher risk of urinary incontinence after radical prostatectomy. *JGG* **2018**, *66*, 168–172.
44. Jeong, S.J.; Yeon, J.S.; Lee, J.K.; Cha, W.H.; Jeong, J.W.; Lee, B.K.; Lee, S.C.; Jeong, C.W.; Kim, J.H.; Hong, S.K.; et al. Development and validation of nomograms to predict the recovery of urinary continence after radical prostatectomy: Comparisons between immediate, early, and late continence. *World J. Urol.* **2014**, *32*, 437–444. [CrossRef]
45. Becker, A.; Tennstedt, P.; Hansen, J.; Trinh, Q.D.; Kluth, L.; Atassi, N.; Schlomm, T.; Salomon, G.; Haese, A.; Budaeus, L.; et al. Functional and oncological outcomes of patients aged <50 years treated with radical prostatectomy for localised prostate cancer in a European population. *BJU Int.* **2014**, *114*, 38–45. [CrossRef]

46. Michl, U.; Tennstedt, P.; Feldmeier, L.; Mandel, P.; Oh, S.J.; Ahyai, S.; Budäus, L.; Chun, F.K.H.; Haese, A.; Heinzer, H.; et al. Nerve-sparing Surgery Technique, Not the Preservation of the Neurovascular Bundles, Leads to Improved Long-term Continence Rates After Radical Prostatectomy. *Eur. Urol.* **2016**, *69*, 584–589. [CrossRef]
47. Salazar, A.; Regis, L.; Planas, J.; Celma, A.; Díaz, F.; Gallardo, I.; Trilla, E.; Morote, J. Early continence after radical prostatectomy: A systematic review. *Actas Urol. Esp.* **2019**, *43*, 526–535. [CrossRef]
48. Reeves, F.; Preece, P.; Kapoor, J.; Everaerts, W.; Murphy, D.G.; Corcoran, N.M.; Costello, A.J. Preservation of the neurovascular bundles is associated with improved time to continence after radical prostatectomy but not long-term continence rates: Results of a systematic review and meta-analysis. *Eur. Urol.* **2015**, *68*, 692–704. [CrossRef]
49. Steineck, G.; Bjartell, A.; Hugosson, J.; Axén, E.; Carlsson, S.; Stranne, J.; Wallerstedt, A.; Persson, J.; Wilderäng, U.; Thorsteinsdottir, T.; et al. Degree of preservation of the neurovascular bundles during radical prostatectomy and urinary continence 1 year after surgery. *Eur. Urol.* **2015**, *67*, 559–568. [CrossRef]

cancers

MDPI

Article

Clinical Benefits of Indocyanine Green Fluorescence in Robot-Assisted Partial Nephrectomy

Yu-Kuan Yang [1], Ming-Li Hsieh [1,2], Sy-Yuan Chen [1], Chung-Yi Liu [2,3], Po-Hung Lin [1,2], Hung-Cheng Kan [1,2], See-Tong Pang [1,2,*,†] and Kai-Jie Yu [1,2,4,*,†]

1 Division of Urology, Department of Surgery, Linkou Chang Gung Memorial Hospital, Taoyuan 333, Taiwan; luke820737@cgmh.org.tw (Y.-K.Y.); h0810@cgmh.org.tw (M.-L.H.); mr1711@cgmh.org.tw (S.-Y.C.); m7587@cgmh.org.tw (P.-H.L.); m0320@cgmh.org.tw (H.-C.K.)

2 College of Medicine, Graduate Institute of Clinical Medical Sciences, Chang Gung University, Taoyuan 333, Taiwan; 8902087@cgmh.org.tw

3 Department of Urology, New Taipei Municipal Tucheng Chang Gung Memorial Hospital, New Taipei City 236, Taiwan

4 Department of Chemical Engineering and Biotechnology, Graduate Institute of Biochemical and Biomedical Engineering, National Taipei University of Technology, Taipei 106, Taiwan

* Correspondence: jacobpang@cgmh.org.tw (S.-T.P.); m7398@cgmh.org.tw (K.-J.Y.); Tel.: +886-3-3281200 (ext. 2103) (K.-J.Y.)

† These authors contributed equally to this work.

Simple Summary: Indocyanine green (ICG) administration in robot-assisted partial nephrectomy (RAPN) can minimize warm ischemia time and preserve more parenchyma resulting in exceptional preservation of renal function and reduced incidence of postoperative complications. However, previous studies have seldom compared how ICG-RAPN use differs when used to treat benign versus malignant renal tumors because the baseline patient and tumor characteristics as well as treatment goals are completely different. The aim of our retrospective study was to compare the intraoperative and postoperative outcomes and the differences in the results of ICG administration between patients with benign and malignant tumors. We have demonstrated that ICG-RAPN yielded superior preservation of short-term renal function. Of the patients with malignant renal tumors, it had less operative blood loss without a more positive margin rate than standard RAPN.

Abstract: Background: To compare the intraoperative and postoperative outcomes of indocyanine green (ICG) administration in robot-assisted partial nephrectomy (RAPN) and report the differences in the results between patients with benign and malignant renal tumors. Methods: From 2017 to 2020, 132 patients underwent RAPN at our institution, including 21 patients with ICG administration. Clinical data obtained from our institution's RAPN database were retrospectively reviewed. Intraoperative, postoperative, pathological, and functional outcomes of RAPN were assessed. Results: The pathological results indicated that among the 127 patients, 38 and 89 had received diagnoses of benign and malignant tumors, respectively. A longer operative time (311 vs. 271 min; $p = 0.006$) but superior preservation of estimated glomerular filtration rate (eGFR) at 3-month follow-up (90% vs. 85%; $p = 0.031$) were observed in the ICG-RAPN group. Less estimated blood loss, shorter warm ischemia time, and superior preservation of eGFR at postoperative day 1 and 6-month follow-up were also noted, despite no significant differences. Among the patients with malignant tumors, less estimated blood loss (30 vs. 100 mL; $p < 0.001$) was reported in the ICG-RAPN subgroup. Conclusions: Patients with ICG-RAPN exhibited superior short-term renal function outcomes compared with the standard RAPN group. Of the patients with malignant tumors, ICG-RAPN was associated with less blood loss than standard RAPN without a more positive margin rate. Further studies with larger cohorts and prospective designs are necessary to verify the intraoperative and functional advantages of the green dye.

Keywords: robot-assisted partial nephrectomy; renal cell carcinoma; indocyanine green; nephron sparing

Citation: Yang, Y.-K.; Hsieh, M.-L.; Chen, S.-Y.; Liu, C.-Y.; Lin, P.-H.; Kan, H.-C.; Pang, S.-T.; Yu, K.-J. Clinical Benefits of Indocyanine Green Fluorescence in Robot-Assisted Partial Nephrectomy. *Cancers* **2022**, *14*, 3032. https://doi.org/10.3390/cancers14123032

Academic Editor: Alchiede Simonato

Received: 24 May 2022
Accepted: 12 June 2022
Published: 20 June 2022

1. Introduction

Partial nephrectomy (PN) is the standard surgical treatment for small renal masses, as it has exhibited superior renal functional preservation and long-term patient survival relative to radical nephrectomy [1–4]. Compared with other methods for PN, robot-assisted PN (RAPN), which has become increasingly prominent in recent years, yields fewer intraoperative and postoperative complications, superior renal functional outcomes, and lower rates of positive margins and conversion to radical nephrectomy than open and laparoscopic PN [5–7]. Near-infrared fluorescence (NIRF) using indocyanine green (ICG) has been adopted as a safe and practical tool to identify anatomical structures in both oncological and non-oncological surgeries, helping verify devascularization and indicate resection margins, offering a benefit in terms of preserving short-term renal function [8–21]. Studies have mainly discussed the outcomes of ICG in aiding selective clamping without ischemia time or with reduced warm ischemia time (WIT) [14,17–19,22,23]; however, they have seldom compared how ICG-RAPN use differs when used to treat benign versus malignant renal tumors because the baseline patient and tumor characteristics as well as treatment goals are completely different [3,4,24,25]. Therefore, in this study, we not only compared the outcomes of ICG-RAPN and standard RAPN but also reported the differences in the results between patients with benign and malignant tumors.

2. Materials and Methods

2.1. Patients

From January 2017 through December 2020, 132 patients received RAPN for kidney tumor at a tertiary center and 5 patients were excluded due to off-clamp method; the group included 21 patients who underwent intraoperative administration of ICG. All the patients had completed either a preoperative computed tomography (CT) scan or magnetic resonance imaging to allow health care personnel to examine the anatomical structures of their kidneys and the characteristics of their kidney tumors. We retrospectively reviewed the demographic data, tumor complexity (using RENAL nephrometry scoring), preoperative hemoglobin (Hb), renal function (using serum estimated glomerular filtration rate (eGFR) determined using the CDK-EPI equation), operative variables (WIT, operative time, and estimated blood loss), postoperative outcomes (pathological results, length of admission in days, intraoperative and postoperative complications according to Clavien–Dindo classification (CD)), 1-year recurrence, trifecta achievement (defined as WIT $\leq$ 25 min, negative surgical margins, and the absence of $\geq$3a CD complications), Hb at postoperative day 1, and renal function at postoperative day 1 as well as 3- and 6-month follow-ups.

2.2. ICG Injection

RAPN was performed using either a fluorescence-capable da Vinci Si system or the Xi surgical system (Intuitive Surgical, Sunnyvale, CA, USA), and the surgeon determined whether to use ICG depending on the appearance of an unfavorable kidney anatomical structure or high RENAL score. After the renal pedicle was controlled with Bulldogs, 3 to 5 mL of ICG (25 mg, dissolved in 10 mL of distilled water, with a final concentration of 2.5 mg/mL) was applied intravenously to confirm renal ischemia with NIRF imaging. The fluorescence could be seen in the main renal vessels after 75 s. Thereafter, unclamped the renal artery for margin identification then clamped the renal artery again before resection.

2.3. Statistical Analysis

Statistical analysis was performed with SPSS for Mac, version 25.0 (Chicago, IL, USA). Continuous variables are presented using median values with interquartile ranges (IQR), and categorical variables are presented as counts and proportions (percentages). The Mann–Whitney *U* test and Fisher's exact test or chi-square analysis were employed as appropriate to compare the intraoperative and postoperative data, with statistical significance defined at $p < 0.05$.

3. Results

3.1. Patients Characteristics

Among the 132 patients who had undergone RAPN, 5 were excluded due to the use of off-clamp surgery. A total of 127 patients were included in the final analysis; 38 (30%) and 89 (70%) patients had received diagnoses of benign and malignant renal tumors, respectively. Significant differences between the benign and malignant tumor groups in terms of the preoperative and postoperative patient characteristics at baseline were observed. More male patients ($p < 0.001$), more hypertension cases ($p = 0.003$), higher preoperative and postoperative Hb levels, lower preoperative and postoperative eGFR, and smaller tumor size ($p = 0.001$) were recorded in the malignant renal tumors group. Significantly longer operative time (311 vs. 271 min; $p = 0.006$) and superior eGFR preservation rate at 3-month follow-up (90% vs. 85%; $p = 0.031$) were observed in the ICG-RAPN groups. Less estimated blood loss (50 vs. 100 mL; $p = 0.09$), shorter WIT (21 vs. 24 min; $p = 0.25$), and higher postoperative eGFR preservation rate at day 1 and 6-month follow-up were noted in the ICG-RAPN groups, despite no significant differences. Furthermore, no other significant differences were noted in terms of the postoperative outcomes, and no ICG-related adverse effects were recorded. Demographic and preoperative characteristics of our study participants are listed in Table 1, and intraoperative and postoperative variables are summarized in Table 2.

Table 1. Baseline and clinical data of participants.

Variable	Total (n = 127)		
Patients, n	ICG (21)	No ICG (106)	p-Value
Age (years), median (IQR)	58 (42–67)	57 (49–66)	0.343
Male, n (%)	12 (57)	60 (57)	0.964
BMI (kg/m^2), median (IQR)	26.5 (23.2–29.2)	25.1 (23.1–28.1)	0.219
Hypertension, n (%)	8 (38)	51 (48)	0.4
Diabetes, n (%)	4 (19)	30 (28)	0.59
ASA score, median (IQR)	2 (2–2)	2 (2–2)	0.819
ASA score, n (%)			0.933
1	4 (19)	25 (24)	
2	15 (71)	69 (65)	
3	2 (10)	12 (11)	
Left side, n (%)	8 (38)	43 (41)	0.833
RENAL score, median (IQR)	8 (6–8)	8 (6–9)	0.763
Tumor complexity, n (%)			0.78
Low (4–6)	7 (33)	28 (26.9)	
Intermediate (7–9)	12 (57)	63 (60.6)	
High (10–12)	2 (10)	13 (12.5)	
Tumor diameter in CT/MRI (cm), median (IQR)	3.5 (2.7–7.3)	3.2 (2.5–4.8)	0.239
Preoperative Hb (g/dL), median (IQR)	14.4 (12.7–16.0)	13.7 (12.5–15.0)	0.188
Preoperative eGFR (mL/min/1.73 m^2), median (IQR)	91.9 (74.3–109.1)	91.5 (74.4–110.5)	0.731

Abbreviations: BMI = body mass index; CT = computed tomography; eGFR = estimated glomerular filtration rate; Hb = hemoglobin; ICG = indocyanine green; IQR = interquartile range; n = number; MRI = magnetic resonance imaging; RAPN = robot-assisted partial nephrectomy.

Table 2. Intraoperative and postoperative data of participants.

Variable	Total (n = 127)		
Patients, n	ICG (21)	No ICG (106)	p-Value
Operative time (min), median (IQR)	311 (263–360)	271 (217–310)	0.006
Estimated blood loss (mL), median (IQR)	50 (30–200)	100 (50–200)	0.09
Warm ischemia time (min), median (IQR)	21 (16–27)	24 (17–35)	0.25
Tumor size (cm), median (IQR)	3.3 (2.5–5.8)	2.9 (2.3–4.1)	0.174

Table 2. *Cont.*

Variable	Total (n = 127)		
Stay length (day), median (IQR)	7 (6–8)	7 (6–8)	0.545
Positive surgical margins, n (%)	2 (11)	8 (8)	0.66
Postoperative complications, n (%)	4 (19)	25 (24)	0.781
Minor (Clavien-Dindo I–II)	3 (14)	20 (19)	0.902
Major (Clavien-Dindo III–IV)	1 (5)	5 (5)	
Clavien-Dindo III ≥ 3, n (%)	1 (5)	5 (5)	1
One-year recurrence, n (%)	0 (0)	3 (4)	1
Trifecta achievement, n (%)	9 (56)	44 (46)	0.462
Post-op Hb at day one (g/dL), median (IQR)	12.9 (11.1–14.4)	12.2 (10.8–13.6)	0.223
Preservation rate of post-op Hb at day one (%), median (IQR)	89 (84–92)	89 (86–95)	0.398
Post-op eGFR at day one (mL/min/1.73 m^3), median (IQR)	81.0 (64.0–93.8)	68.6 (51.5–93.9)	0.296
Preservation rate of post-op eGFR at day one (%), median (IQR)	84 (70–96)	79 (68–90)	0.2
Post-op eGFR at 3-months (mL/min/1.73 m^3), median (IQR)	74.3 (62.8–91.7)	77.7 (65.1–94.0)	0.762
Preservation rate of post-op eGFR at 3-months (%), median (IQR)	90 (85–97)	85 (77–91)	0.031
Post-op eGFR at 6-months (mL/min/1.73 m^3), median (IQR)	71.5 (65.6–94.6)	76.1 (64.9–90.1)	0.735
Preservation rate of post-op eGFR at 6-months (%), median (IQR)	83 (80–91)	81 (74–94)	0.346

Abbreviations: eGFR = estimated glomerular filtration rate; Hb = hemoglobin; ICG = indocyanine green; IQR = interquartile range; n = number; RAPN = robot-assisted partial nephrectomy; RCC = renal cell carcinoma.

3.2. Patients with Benign Renal Tumors

Among the 38 patients with benign renal tumors, 10 received ICG-RAPN, and 28 received standard RAPN. No significant differences in terms of baseline patient characteristics, tumor complexity, or laboratory data were noted between these two groups. However, larger tumor sizes in preoperative images (7.3 cm vs. 4.4 cm; p = 0.172) were observed in the ICG-RAPN group despite the difference being nonsignificant. ICG-RAPN entailed a longer operation time (325 vs. 228 min; p < 0.001). Greater estimated blood loss was noted in ICG-RAPN, but the difference was nonsignificant. The WIT was 20 min for both groups. Two (20% [of subgroup]) and four (14%) patients reported postoperative complications in the ICG-RAPN and standard RAPN groups, respectively, but none of these complications had a CD classification of ≥3. The ICG-RAPN group exhibited lower day 1 postoperative Hb preservation (87% vs. 89%; p = 0.351) but superior postoperative eGFR preservation at day 1 as well as 3- and 6-month follow-ups; nevertheless, these differences were all nonsignificant. Angiomyolipoma (AML) was the most common benign kidney tumor in both the ICG and standard groups, accounting for 90% and 96% of such tumors, respectively. Intraoperative and postoperative variables of benign tumors group are summarized in Table 3.

Table 3. Intraoperative and postoperative data of benign and malignant tumors group.

Variable	Benign (n = 38)			Malignant (n = 89)		Variable
Patients, n	ICG (10)	No ICG (28)	p	ICG (11)	No ICG (78)	p Value
Operative time (min), median (IQR)	325 (275–396)	228 (187–291)	0.001	298 (247–350)	280 (222–325)	0.25
Estimated blood loss (mL), median (IQR)	200 (50–350)	90 (50–200)	0.272	30 (20–50)	100 (50–200)	<0.001
Warm ischemia time (min), median (IQR)	20 (16–25)	20 (15–29)	0.722	24 (15–28)	26 (18–36)	0.207
Tumor size (cm), median (IQR)	5.6 (2.3–9.2)	3.8 (2.7–5.5)	0.36	2.9 (2.5–3.7)	2.7 (2.2–3.7)	0.447
Stay length (day), median (IQR)	8 (6–8)	7 (6–8)	0.317	6 (5–8)	7 (6–8)	0.074
Positive surgical margins, n (%)	1 (13)	4 (18)	1	1 (9)	4 (5)	0.491
Post-op complications, n (%)	2 (20)	4 (14)	0.644	2 (18)	21 (27)	0.721
Minor (Clavien-Dindo I–II)	2 (20)	4 (14)	0.644	1 (9)	16 (21)	0.627
Major (Clavien-Dindo III–IV)	0 (0)	0 (0)		1 (9)	5 (6)	
Clavien-Dindo III ≥ 3, n (%)	0 (0)	0 (0)		1 (9)	5 (6)	0.558
One-year recurrence, n (%)	0 (0)	2 (8)	1	0 (0)	1 (2)	1
Trifecta achievement, n (%)	6 (67)	10 (48)	0.44	3 (43)	34 (46)	1

Table 3. *Cont.*

Variable	Benign (n = 38)			Malignant (n = 89)		Variable
Post-op Hb at day one (g/dL), median (IQR)	11.5 (10.9–13.3)	11.5 (11.0–12.2)	0.66	13.7 (11.4–15.2)	12.5 (10.8–14.0)	0.1
Preservation rate of post-op Hb at day one (%), median (IQR)	87 (83–90)	89 (86–93)	0.351	91 (84–96)	90 (86–95)	0.985
Post-op eGFR at day one (mL/min/1.73 m^3), median (IQR)	80.0 (66.0–119.1)	89.2 (71.7–114.1)	0.66	81.0 (39.9–85.1)	64.2 (45.0–84.4)	0.48
Preservation rate of post-op eGFR at day one (%), median (IQR)	93 (75–98)	85 (79–99)	0.961	77 (68–97)	76 (65–87)	0.313
Post-op eGFR at 3-months (mL/min/1.73 m^3), median (IQR)	73.5 (65.1–139.6)	98.6 (83.7–121.3)	0.775	75.0 (56.3–84.4)	73.9 (61.2–90.1)	0.736
Preservation rate of post-op eGFR at 3-months (%), median (IQR)	93 (86–111)	86 (80–89)	0.095	88 (81–94)	85 (76–92)	0.201
Post-op eGFR at 6-months (mL/min/1.73 m^3), median (IQR)	97.1 (69.2–98.8)	85.9 (67.3–100.0)	0.787	68.5 (54.3–83.3)	73.8 (64.2–88.1)	0.348
Preservation rate of post-op eGFR at 6-months (%), median (IQR)	84 (79–91)	80 (72–96)	0.516	83 (80–95)	81 (74–94)	0.448

Abbreviations: eGFR = estimated glomerular filtration rate; Hb = hemoglobin; ICG = indocyanine green; IQR = interquartile range; n = number; post-op = postoperative.

3.3. Patients with Malignant Renal Tumors

Among the 89 patients with malignant renal tumors, 11 received ICG-RAPN, and 78 received standard RAPN. The ICG-RAPN group was 11 years younger, on average, than the standard group, but the age difference was nonsignificant (p = 0.059). No significant differences were evident in body mass index (BMI), underlying diseases, preoperative tumor size (in images), RENAL score, or preoperative Hb and eGFR. The ICG-RAPN group exhibited longer operation times (298 vs. 280 min; p = 0.25) and shorter WIT (24 vs. 26; p = 0.207), but neither difference was significant. Furthermore, estimated blood loss was significantly less in the ICG-RAPN group (30 vs. 100 mL; p < 0.001) than in the standard one. Two patients (18%) reported postoperative complications in the ICG-RAPN group, one of which a CD classification of ≥3 (postoperative renal pseudoaneurysm was discovered, and the patient underwent transarterial embolization). Twenty-one patients (27%) reported postoperative complications in the standard RAPN group, five (6%) of whom had complications with a CD classification of ≥3 (one patient underwent reopen surgery after the initial RAPN due to hemogenic shock). The patients with high-grade complications and subsequent interventions are listed in Table 4. The preservation rate of postoperative Hb and eGFR at day 1, 3 months, and 6 months were higher in the ICG-RAPN group but with a nonsignificant difference. No statistically significant differences in terms of length of hospital stay, positive surgical margin rate, or trifecta achievement were observed between these two groups. For the histopathologic type, both the ICG and standard RAPN groups had renal cell carcinoma (RCC), and the clear cell subtype was the most prevalent, appearing in 46% and 73% of patients, respectively. The pathologic stage distribution was pT1a (91%), pT1b (9%) in ICG-RAPN and pT1a (77%), pT1b (17%), pT2a (1%) and pT3a (5%) in standard-RAPN. Intraoperative and postoperative variables of the malignant tumors group are summarized in Table 3.

Table 4. High-grade complications and subsequent interventions.

Complication (Clavien-Dindo ≥ 3)	Benign		Malignant		Intervention
Patients, n	ICG (0)	No ICG (0)	ICG (1)	No ICG (5)	ICG (0)
Renal pseudoaneurysm, n (%)	0	0	1 (100)	3 (60)	Transarterial embolization
Hemogenic shock, n (%)	0	0	0	1 (20)	Emergent re-open surgery
Urine leakage with urinoma and UPJ obstruction, n (%)	0	0	0	1 (20)	Stent placement

4. Discussion

The main goal of PN is to achieve favorable functional outcomes with oncological safety equivalent to radical nephrectomy when treating small renal tumors [1–4,24].

The previous study revealed that preoperative renal function, the volume of preserved parenchyma and WIT effect the postoperative short-term renal function outcomes after PN [19]. With the help of ICG-RAPN, it can minimize WIT and preserve more parenchyma resulting in exceptional preservation of renal function and reduced incidence of postoperative complications [12–15,17–23,26] Our study also compared the differences in the results of ICG administration between patients with benign and malignant tumors. We have demonstrated that ICG-RAPN yielded superior preservation of short-term renal function. Of the patients with malignant renal tumors, it had less operative blood loss without a more positive margin rate than standard RAPN.

The preoperative characteristics of benign and malignant tumors differ; for instance, RCC, which is the most common malignant tumor, is more common in men and is associated with obesity, hypertension, and chronic kidney disease; our previous study results affirmed these associations [27,28]. Furthermore, the indicators and treatment goals of benign and malignant renal tumors are entirely different. Malignant renal tumors such as RCC are mostly found by accident in cross-sectional images prior to symptom manifestation; once clinical T1 cancer is suspected, the latest guidelines recommend surgical intervention (especially PN) as the first option [3,4]. In addition to preserving renal function, negative surgical margins should always be a priority, and enucleation should even be considered in some situations [3,4].

However, for benign renal tumor, active surveillance represents a reasonable strategy, and surgical intervention is usually performed when malignancy is suspected, increasing the risk of hemorrhage in patients with symptoms or large tumors [29–31]. PN remains the first option of surgical intervention, but negative surgical margins are not emphasized because of the low recurrence rate after surgery [29,31]. Therefore, when evaluating the benefit of ICG-RAPN on surgical margins, benign renal tumors should be excluded from the analysis, with many of them having unrecognizable surgical margins, as in the case of 8 of the 38 (21%) such patients enrolled in our study.

ICG-RAPN can assist surgeons with the following three tasks. First, it can be employed for identifying the arterial blood supply of renal tumors and can even be used with selective clamping to minimize the ischemia of normal renal parenchyma [8,9,16]. This is because ICG binds with plasma proteins rapidly after its intravenous application, rendering the perfusion of renal arteries easily visible [16]. Treatment of larger tumors, especially malignant ones, can benefit from this method substantially because such tumors tend to have more accessory and complex blood supplies that are difficult to thoroughly identify with preoperative imaging [15]. In our study's malignant tumor group, the estimated blood loss in the ICG-RAPN subgroup (30 mL) was significantly less than that of the benign tumor group and was much less than reported in previous studies (range: 75–300 mL) [14–16,18,19,22,23]. For the high-grade postoperative complications like reopen surgery due to hemorrhagic shock or postoperative urinary leakage, they were only seen in our malignant group with standard-RAPN. Though there was no statistical difference in complications just like in the previous studies, it may be explained by the already low rate of complications in RAPN [12,32]. Furthermore, other cases with complicated anatomic renal structures, such as horseshoe, solitary, or pelvic kidneys, or duplex kidneys with non-functioning symptomatic upper or lower moiety in pediatric patients, can also warrant ICG administration [12,21].

Second, ICG-RAPN can render renal tumors hypofluorescent (i.e., the tumor tissue is darker than normal) in near-infrared light, which allows for easy identification of the margin of the tumor and enables precision resection in real-time [8,9,16,20,21]. This method complements perioperative ultrasound, which is often inadequate because it cannot provide a real-time image due to the probe interfering when resection is performed [11]. The mechanism of hypofluorescence is based on ICG binding to a transmembrane protein called bilitranslocase (BLT), which expresses high concentrations in normal renal proximal and distal tubules (but not in glomeruli) and emits fluorescent light [33,34]. Previous studies have suggested that cells of AML and RCC do not express BLT and do not store ICG intracellularly [8,16]. Thus, their hypofluorescence is visible under NIRF, simplifying the differentiation of cancer cells from normal renal parenchyma. The aforementioned advantages have allowed ICG-RAPN to significantly shorten the WIT in most studies (WIT range: 11.5–24 min), and our study's malignant group also displayed this trend finding despite the difference with the standard RAPN group being nonsignificant (24 vs. 26 min) [12]. This was critical because a WIT longer than 25 min is an independent predictor of newly diagnosed chronic kidney disease (CKD) in patients with a solitary kidney [26]. Finally, ICG-RAPN allows the blood supply of the remaining renal parenchyma to be evaluated after renorrhaphy because renorrhaphy itself can result in ischemic injury, especially in patients who already have CKD [11,15,35,36].

Nevertheless, ICG-RAPN has some limitations, such as being more suitable for superficially localized tumors due to its limited tissue penetration [16,23]. To increase its penetration depth and allow it to be performed in endophytic renal masses, the ICG was mixed with lipiodol, which was used to avoid a quick ICG washout from the renal tumor, in a previous study. Preoperative superselective transarterial delivery of the lipiodol–ICG mixture was initially employed, and then a postprocedural CT scan was completed for localization [11]. This method of using RAPN to deal with a completely endophytic tumor demonstrated acceptable renal functional outcomes and a lack of intraoperative and postoperative complications [11]. Another limitation of ICG-RAPN is that it offers the surgeon only one opportunity to evaluate the renal blood supply because once ICG is applied, it remains in the circulation system [15]. Furthermore, although none of our patients presented adverse effects from ICG usage, a previous study reported some rare allergic reactions associated with ICG usage in a population without known allergy-related vulnerabilities [37].

Our results suggested that ICG-RAPN's advantage over standard RAPN in functional outcomes seemed to decrease with time because the postoperative eGFR preservation at 6 months was lower than at 3 months in both the benign and malignant groups. A similar phenomenon was also reported in a previous study [19]. Two possible mechanisms may explain the significant long-term decrease in renal function observed in our study. First, the greater recovery from kidney injury in patients who received standard RAPN (because they are more susceptible to acute tubular necrosis owing to increased renal ischemia) compared with the ICG-RAPN group at 6-month follow-up may explain their greater decrease from day 1 postoperative eGFR. Second, the compensation of the normal contralateral kidney (a feature in most patients of both groups and none of our patients had solitary kidney) for loss of renal function may have partially normalized the eGFR over the time leading up to the 6-month follow-up. Although superior short-term preservation of eGFR in the ICG-RAPN group may have implied less renal parenchyma ischemia, this benefit could have been masked over time by the compensation of the normal contralateral kidney, minimizing the difference between ICG and standard groups [38]. Therefore, for patients with a normal contralateral kidney, serum eGFR may not be a sensitive measure of ICG-RAPN's value. Some studies have suggested the use of renal scan, CT scan with 3D volume rendering, or magnetic resonance renography to estimate the single residual renal volume more accurately [14,15,19].

Our study entailed several limitations. First, the relatively small number of patients and the retrospective analysis may have made the study susceptible to underrepresentation bias. In both the benign and malignant tumor groups, ICG-RAPN exhibited superior short- and long-term renal functional outcomes relative to standard RAPN. Although comparing the benign and malignant tumor groups one by one suggested only a nonsignificant advantage of ICG-RAPN over standard RAPN, a significant advantage in short-term renal functional outcomes was noted when we combined the benign and malignant tumor groups together. Therefore, if the number of patients is increased, the intraoperative and postoperative outcomes may reach significance. Second, because the individual surgeons determined whether to perform ICG-RAPN, selection would have been strongly biased toward standard RAPN. Third, serum eGFR could not reveal the exact residual kidney function and the contralateral normal kidney may have hampered the estimation of residual kidney function. Incomplete laboratory data and image examinations during follow-up might also have occurred, and different routines of the various surgeons may have resulted in a failure to accurately reflect all the outcomes.

5. Conclusions

In summary, although patients who underwent ICG-RAPN had longer operative times, they demonstrated superior short-term renal functional outcomes relative to those who received standard RAPN. For patients with malignant renal tumors, ICG-RAPN resulted in less operative blood loss than standard RAPN without increased positive surgical margin rates. Therefore, ICG-RAPN is an ostensibly safe procedure with potentially superior short-term renal functional outcomes. Further prospective randomized controlled trials are required to confirm whether this technique effectively provides the discussed intraoperative and functional advantages.

Author Contributions: Conceptualization, Y.-K.Y., C.-Y.L. and M.-L.H.; methodology, K.-J.Y. and S.-Y.C.; software, Y.-K.Y. and H.-C.K.; validation, K.-J.Y. and S.-T.P.; formal analysis, S.-Y.C. and Y.-K.Y.; investigation, C.-Y.L. and P.-H.L.; resources, S.-T.P., K.-J.Y., P.-H.L., C.-Y.L. and H.-C.K.; writing—review and editing, M.-L.H.; visualization, S.-T.P. and K.-J.Y.; supervision, C.-Y.L. All authors have read and agreed to the published version of the manuscript.

Funding: This research received no external funding.

Institutional Review Board Statement: The study was conducted in accordance with the Declaration of Helsinki, and approved by the Institutional Review Board of Linkou Chang Gung Memorial Hospital (CMRPG3K0581-2).

Informed Consent Statement: Informed consent was obtained from all subjects involved in the study.

Data Availability Statement: All data can be found in the text.

Acknowledgments: The authors would like to thank Linkou Chang Gung Memorial Hospital for support via grant CMRPG3K0581-2.

Conflicts of Interest: The authors declare no conflict of interest.

References

1. Mir, M.C.; Derweesh, I.; Porpiglia, F.; Zargar, H.; Mottrie, A.; Autorino, R. Partial Nephrectomy Versus Radical Nephrectomy for Clinical T1b and T2 Renal Tumors: A Systematic Review and Meta-analysis of Comparative Studies. *Eur. Urol.* **2017**, *71*, 606–617. [CrossRef]
2. Gershman, B.; Leibovich, B.C.; Kim, S.P. Partial Versus Radical Nephrectomy for the Clinical T1a Renal Mass. *Eur. Urol. Focus* **2019**, *5*, 970–972. [CrossRef] [PubMed]
3. Ljungberg, B.; Albiges, L.; Abu-Ghanem, Y.; Bensalah, K.; Dabestani, S.; Fernández-Pello, S.; Giles, R.H.; Hofmann, F.; Hora, M.; Kuczyk, M.A.; et al. European Association of Urology Guidelines on Renal Cell Carcinoma: The 2019 Update. *Eur. Urol.* **2019**, *75*, 799–810. [CrossRef] [PubMed]
4. Campbell, S.C.; Clark, P.E.; Chang, S.S.; Karam, J.A.; Souter, L.; Uzzo, R.G. Renal Mass and Localized Renal Cancer: Evaluation, Management, and Follow-Up: AUA Guideline: Part I. *J. Urol.* **2021**, *206*, 199–208. [CrossRef] [PubMed]
5. Cacciamani, G.E.; Medina, L.G.; Gill, T.; Abreu, A.; Sotelo, R.; Artibani, W.; Gill, I.S. Impact of Surgical Factors on Robotic Partial Nephrectomy Outcomes: Comprehensive Systematic Review and Meta-Analysis. *J. Urol.* **2018**, *200*, 258–274. [CrossRef]
6. Carbonara, U.; Simone, G.; Minervini, A.; Sundaram, C.P.; Larcher, A.; Lee, J.; Checcucci, E.; Fiori, C.; Patel, D.; Meagher, M.; et al. Outcomes of robot-assisted partial nephrectomy for completely endophytic renal tumors: A multicenter analysis. *Eur. J. Surg. Oncol.* **2021**, *47*, 1179–1186. [CrossRef]
7. Shatagopam, K.; Bahler, C.D.; Sundaram, C.P. Renorrhaphy techniques and effect on renal function with robotic partial nephrectomy. *World J. Urol.* **2020**, *38*, 1109–1112. [CrossRef]
8. Ferroni, M.C.; Sentell, K.; Abaza, R. Current Role and Indications for the Use of Indocyanine Green in Robot-assisted Urologic Surgery. *Eur. Urol. Focus* **2018**, *4*, 648–651. [CrossRef]

9. Hekman, M.C.H.; Rijpkema, M.; Langenhuijsen, J.F.; Boerman, O.C.; Oosterwijk, E.; Mulders, P.F.A. Intraoperative Imaging Techniques to Support Complete Tumor Resection in Partial Nephrectomy. *Eur. Urol. Focus* **2018**, *4*, 960–968. [CrossRef]
10. Sentell, K.T.; Ferroni, M.C.; Abaza, R. Near-infrared fluorescence imaging for intraoperative margin assessment during robot-assisted partial nephrectomy. *BJU Int.* **2020**, *126*, 259–264. [CrossRef]
11. Simone, G.; Tuderti, G.; Anceschi, U.; Ferriero, M.; Costantini, M.; Minisola, F.; Vallati, G.; Pizzi, G.; Guaglianone, S.; Misuraca, L.; et al. "Ride the Green Light": Indocyanine Green—Marked Off-clamp Robotic Partial Nephrectomy for Totally Endophytic Renal Masses. *Eur. Urol.* **2019**, *75*, 1008–1014. [CrossRef] [PubMed]
12. Veccia, A.; Antonelli, A.; Hampton, L.J.; Greco, F.; Perdonà, S.; Lima, E.; Hemal, A.K.; Derweesh, I.; Porpiglia, F.; Autorino, R. Near-infrared Fluorescence Imaging with Indocyanine Green in Robot-assisted Partial Nephrectomy: Pooled Analysis of Comparative Studies. *Eur. Urol. Focus* **2020**, *6*, 505–512. [CrossRef] [PubMed]
13. Cacciamani, G.E.; Shakir, A.; Tafuri, A.; Gill, K.; Han, J.; Ahmadi, N.; Hueber, P.A.; Gallucci, M.; Simone, G.; Campi, R.; et al. Best practices in near-infrared fluorescence imaging with indocyanine green (NIRF/ICG)-guided robotic urologic surgery: A systematic review-based expert consensus. *World J. Urol.* **2020**, *38*, 883–896. [CrossRef]
14. Mattevi, D.; Luciani, L.G.; Mantovani, W.; Cai, T.; Chiodini, S.; Vattovani, V.; Puglisi, M.; Malossini, G. Fluorescence-guided selective arterial clamping during RAPN provides better early functional outcomes based on renal scan compared to standard clamping. *J. Robot. Surg.* **2019**, *13*, 391–396. [CrossRef] [PubMed]
15. Diana, P.; Buffi, N.M.; Lughezzani, G.; Dell'Oglio, P.; Mazzone, E.; Porter, J.; Mottrie, A. The Role of Intraoperative Indocyanine Green in Robot-assisted Partial Nephrectomy: Results from a Large, Multi-institutional Series. *Eur. Urol.* **2020**, *78*, 743–749. [CrossRef]
16. Lukas Gadus, J.K.; Chmelik, F.; Matejkova, M.; Heracek, J. Robotic Partial Nephrectomy with Indocyanine Green Fluorescence Navigation. *Contrast Media Mol. Imaging* **2020**, *2020*, 8.
17. Krane, L.S.; Manny, T.B.; Hemal, A.K. Is near infrared fluorescence imaging using indocyanine green dye useful in robotic partial nephrectomy: A prospective comparative study of 94 patients. *Urology* **2012**, *80*, 110–116. [CrossRef]
18. Borofsky, M.S.; Gill, I.S.; Hemal, A.K.; Marien, T.P.; Jayaratna, I.; Krane, L.S.; Stifelman, M.D. Near-infrared fluorescence imaging to facilitate super-selective arterial clamping during zero-ischaemia robotic partial nephrectomy. *BJU Int.* **2013**, *111*, 604–610. [CrossRef]
19. McClintock, T.R.; Bjurlin, M.A.; Wysock, J.S.; Borofsky, M.S.; Marien, T.P.; Okoro, C.; Stifelman, M.D. Can selective arterial clamping with fluorescence imaging preserve kidney function during robotic partial nephrectomy? *Urology* **2014**, *84*, 327–332. [CrossRef]
20. Esposito, C.; Settimi, A.; Del Conte, F.; Cerulo, M.; Coppola, V.; Farina, A.; Crocetto, F.; Ricciardi, E.; Esposito, G.; Escolino, M. Image-Guided Pediatric Surgery Using Indocyanine Green (ICG) Fluorescence in Laparoscopic and Robotic Surgery. *Front. Pediatrics* **2020**, *8*, 314. [CrossRef]
21. Esposito, C.; Coppola, V.; Del Conte, F.; Cerulo, M.; Esposito, G.; Farina, A.; Crocetto, F.; Castagnetti, M.; Settimi, A.; Escolino, M. Near-Infrared fluorescence imaging using indocyanine green (ICG): Emerging applications in pediatric urology. *J. Pediatric Urol.* **2020**, *16*, 700–707. [CrossRef] [PubMed]
22. Harke, N.; Schoen, G.; Schiefelbein, F.; Heinrich, E. Selective clamping under the usage of near-infrared fluorescence imaging with indocyanine green in robot-assisted partial nephrectomy: A single-surgeon matched-pair study. *World J. Urol.* **2014**, *32*, 1259–1265. [CrossRef] [PubMed]
23. Lanchon, C.; Arnoux, V.; Fiard, G.; Descotes, J.L.; Rambeaud, J.J.; Lefrancq, J.B.; Poncet, D.; Terrier, N.; Overs, C.; Franquet, Q.; et al. Super-selective robot-assisted partial nephrectomy using near-infrared flurorescence versus early-unclamping of the renal artery: Results of a prospective matched-pair analysis. *Int. Braz. J. Urol.* **2018**, *44*, 53–62. [CrossRef] [PubMed]
24. Sanchez, A.; Feldman, A.S.; Hakimi, A.A. Current Management of Small Renal Masses, Including Patient Selection, Renal Tumor Biopsy, Active Surveillance, and Thermal Ablation. *J. Clin. Oncol.* **2018**, *36*, 3591–3600. [CrossRef] [PubMed]
25. Campbell, S.C.; Uzzo, R.G.; Karam, J.A.; Chang, S.S.; Clark, P.E.; Souter, L. Renal Mass and Localized Renal Cancer: Evaluation, Management, and Follow-up: AUA Guideline: Part II. *J. Urol.* **2021**, *206*, 209–218. [CrossRef]
26. Thompson, R.H.; Lane, B.R.; Lohse, C.M.; Leibovich, B.C.; Fergany, A.; Frank, I.; Gill, I.S.; Blute, M.L.; Campbell, S.C. Every minute counts when the renal hilum is clamped during partial nephrectomy. *Eur. Urol.* **2010**, *58*, 340–345. [CrossRef]
27. Capitanio, U.; Bensalah, K.; Bex, A.; Boorjian, S.A.; Bray, F.; Coleman, J.; Gore, J.L.; Sun, M.; Wood, C.; Russo, P. Epidemiology of Renal Cell Carcinoma. *Eur. Urol.* **2019**, *75*, 74–84. [CrossRef]
28. Chang, Y.H.; Chang, S.W.; Liu, C.Y.; Lin, P.H.; Yu, K.J.; Pang, S.T.; Chuang, C.K.; Kan, H.C.; Shao, I.H. Demographic characteristics and complications of open and minimally invasive surgeries for renal cell carcinoma: A population-based case-control study in Taiwan. *Ther. Clin. Risk Manag.* **2018**, *14*, 1235–1241. [CrossRef]
29. Flum, A.S.; Hamoui, N.; Said, M.A.; Yang, X.J.; Casalino, D.D.; McGuire, B.B.; Perry, K.T.; Nadler, R.B. Update on the Diagnosis and Management of Renal Angiomyolipoma. *J. Urol.* **2016**, *195*, 834–846. [CrossRef]
30. Richard, P.O.; Jewett, M.A.; Bhatt, J.R.; Evans, A.J.; Timilsina, N.; Finelli, A. Active Surveillance for Renal Neoplasms with Oncocytic Features is Safe. *J. Urol.* **2016**, *195*, 581–586. [CrossRef]
31. Flack, C.K.; Calaway, A.C.; Miller, B.L.; Picken, M.M.; Gondim, D.D.; Idrees, M.T.; Abel, E.J.; Gupta, G.N.; Boris, R.S. Comparing oncologic outcomes in patients undergoing surgery for oncocytic neoplasms, conventional oncocytoma, and chromophobe renal cell carcinoma. *Urol. Oncol.* **2019**, *37*, 811.e17–811.e21. [CrossRef] [PubMed]

32. Bjurlin, M.A.; McClintock, T.R.; Stifelman, M.D. Near-infrared fluorescence imaging with intraoperative administration of indocyanine green for robotic partial nephrectomy. *Curr. Urol. Rep.* **2015**, *16*, 20. [CrossRef] [PubMed]
33. Golijanin, D.; Marshall, J.; Cardin, A.; Singer, E.; Wood, R.; Reeder, J.; Wu, G.; Yao, J.; Passamonti, S.; Messing, E. Bilitranslocase (BTL) is immunolocalised in proximal and distal renal tubules and absent in renal cortical tumors accurately corresponding to intraoperative near infrared fluorescence (NIRF) expression of renal cortical tumors using intravenous indocyanine green (ICG). *J. Urol.* **2008**, *179*, 137.
34. Elias, M.M.; Lunazzi, G.C.; Passamonti, S.; Gazzin, B.; Miccio, M.; Stanta, G.; Sottocasa, G.L.; Tiribelli, C. Bilitranslocase localization and function in basolateral plasma membrane of renal proximal tubule in rat. *Am. J. Physiol.* **1990**, *259*, F559–F564. [CrossRef]
35. Mir, M.C.; Campbell, R.A.; Sharma, N.; Remer, E.M.; Simmons, M.N.; Li, J.; Demirjian, S.; Kaouk, J.; Campbell, S.C. Parenchymal volume preservation and ischemia during partial nephrectomy: Functional and volumetric analysis. *Urology* **2013**, *82*, 263–268. [CrossRef]
36. Dagenais, J.; Maurice, M.J.; Mouracade, P.; Kara, O.; Malkoc, E.; Kaouk, J.H. Excisional Precision Matters: Understanding the Influence of Excisional Volume Loss on Renal Function after Partial Nephrectomy. *Eur. Urol.* **2017**, *72*, 168–170. [CrossRef]
37. Chu, W.; Chennamsetty, A.; Toroussian, R.; Lau, C. Anaphylactic Shock after Intravenous Administration of Indocyanine Green during Robotic Partial Nephrectomy. *Urol. Case Rep.* **2017**, *12*, 37–38. [CrossRef]
38. Chapman, D.; Moore, R.; Klarenbach, S.; Braam, B. Residual renal function after partial or radical nephrectomy for renal cell carcinoma. *Can. Urol. Assoc. J.* **2010**, *4*, 337–343. [CrossRef]

cancers

MDPI

Article

The Non-Interventional PAZOREAL Study to Assess the Effectiveness and Safety of Pazopanib in a Real-Life Setting: Reflecting a Changing mRCC Treatment Landscape

Christian Doehn [1,*], Martin Bögemann [2], Viktor Grünwald [3], Manfred Welslau [4], Jens Bedke [5], Martin Schostak [6], Thomas Wolf [7], Rainer Ehneß [8], Elisa Degenkolbe [8], Stefanie Witecy [9] and Peter J. Goebell [10]

1 Urologikum Lübeck, 23566 Lübeck, Germany
2 Department of Urology, University of Münster Medical Center, 48149 Münster, Germany
3 Department of Internal Medicine (Tumor Research) and Department of Urology, West German Cancer Center, University Hospital Essen, 45147 Essen, Germany
4 Klinikum Aschaffenburg, Hämato-Onkologische Schwerpunktpraxis, 63739 Aschaffenburg, Germany
5 Department of Urology, University Hospital Tübingen, 72076 Tübingen, Germany
6 Department of Urology, Urooncology, Robot-Assisted and Focal Therapy, University Hospital Magdeburg, 39120 Magdeburg, Germany
7 Outpatient Centre for Oncology, 01307 Dresden, Germany
8 Novartis Pharma GmbH, 90429 Nürnberg, Germany
9 APOGEPHA Arzneimittel GmbH, 01309 Dresden, Germany
10 Department of Urology, University Hospital Erlangen, 91054 Erlangen, Germany
* Correspondence: doehn@urologikum-luebeck.de

Simple Summary: Clinical trials have demonstrated the effectiveness of pazopanib as a primary treatment for metastatic renal cell carcinoma (mRCC). The approval of tyrosine kinase inhibitors and checkpoint-inhibitors represented further progress in the mRCC treatment landscape. Yet, with the recent changes in treatment options, there are scarce real-world data on these substances, including pazopanib. The PAZOREAL study investigated the effectiveness and safety of pazopanib (first-line), nivolumab (second-line), and everolimus (second- and third-line) in a real-life setting. This study included 376 mRCC patients who received first-line treatment with pazopanib and assessed the treatment's effectiveness, safety, and resultant quality of life. The median time on the drug for the study population was 10.0 months; for primary treatment with pazopanib, it was 6.3 months. The median overall survival (mOS) for the entire study population was 35.9 months. No new safety signals were detected. PAZOREAL provides valuable real-world data for the primary treatment of mRCC with pazopanib.

Abstract: The approval of tyrosine kinase inhibitors and checkpoint inhibitors represented a remarkable progression in the therapeutic landscape for the treatment of metastatic renal cell carcinoma (mRCC). Yet, in the ever-evolving landscape of mRCC treatment, real-world data on these agents, including pazopanib, are scarce. The non-interventional PAZOREAL study investigated the effectiveness and safety of pazopanib (first-line), nivolumab (second-line), and everolimus (second- and third-line) in a real-life setting. The multicentric study included 376 mRCC patients who received first-line treatment with pazopanib and assessed time on the drug (primary endpoint), overall survival, best responses, disease control rates, as well as safety signals and health-related quality of life. The median overall time on the drug was 10.0 months, with first-line pazopanib having a median time on drug of 6.3 months. The median overall survival was 35.9 months. The disease control rate for first-line pazopanib was 56.9%. No new safety signals were detected. PAZOREAL provides valuable real-world data for first-line treatment with pazopanib.

Keywords: pazopanib; renal cell carcinoma; real-world data; non-interventional study; time on drug; nivolumab; everolimus; trial-eligibility

Citation: Doehn, C.; Bögemann, M.; Grünwald, V.; Welslau, M.; Bedke, J.; Schostak, M.; Wolf, T.; Ehneß, R.; Degenkolbe, E.; Witecy, S.; et al. The Non-Interventional PAZOREAL Study to Assess the Effectiveness and Safety of Pazopanib in a Real-Life Setting: Reflecting a Changing mRCC Treatment Landscape. *Cancers* **2022**, *14*, 5486. https://doi.org/10.3390/cancers14225486

Academic Editors: José I. López and Claudia Manini

Received: 18 October 2022
Accepted: 28 October 2022
Published: 8 November 2022

1. Introduction

Renal cell carcinoma (RCC) is considered the most common malignant tumor of the kidney and is diagnosed in about 17,300 patients per year in Germany [1]. At the time of diagnosis, most patients are between 60 and 80 years of age, with considerably more men being afflicted with RCC than women [2]. Clear cell RCC is by far the most common subtype, accounting for about 75% of all RCC cases [3]. When diagnosed, up to 16% of patients already have metastatic RCC (mRCC), and up to 30% recur after curative therapy [4,5].

The introduction of tyrosine kinase inhibitors (TKI), such as pazopanib, was considered a milestone in the first-line treatment of mRCC [6]. Treatment with everolimus, a serine threonine kinase inhibitor (mTOR), commonly followed as a second-line treatment [7,8]. In addition, the approval of nivolumab by the European Medicines Agency (EMA) in April 2016 paved the way for the routine clinical use of checkpoint inhibitors (CPI) for the treatment of mRCC [6,9]. Several CPI-based combination therapies have since been approved for the treatment of mRCC, and are currently considered the standard of care [10]. Despite these other treatment options, pazopanib is still recommended in guidelines as the first-line treatment for all types of RCC when the standard of care is not an option; it is still used in routine clinical practice [11,12].

To achieve a better understanding of the use of pazopanib in routine clinical practice, and the outcomes achieved in this setting, it is important to collect and evaluate real-world data. Additionally, efficacy and safety outcomes should be followed across subsequent treatments to gain information about the entire course of treatment. To date, real-life data for the treatment with pazopanib in first-line and sequential treatment are scarce. Although the safety and efficacy of pazopanib for the first-line treatment of mRCC had been evaluated in pivotal, randomized, double-blind, placebo-controlled, and multinational trials and other clinical studies [13–15], further real-world data are needed to evaluate the effectiveness, safety, tolerability, and quality of life (QoL) in routine clinical practice. Currently, we lack data that reflect the clinical use of pazopanib and associated outcomes within the constantly evolving treatment landscape of mRCC.

The non-interventional study called PAZOREAL, presented here, aims at providing some of the first insights into this knowledge gap.

2. Materials and Methods

2.1. Study Design and Patients

PAZOREAL is a prospective, multi-center, non-interventional observational study that evaluates the effectiveness, tolerability, safety, and QoL in mRCC patients that have been treated with pazopanib as first-line treatment and nivolumab or everolimus as second-line or third-line treatments. A total of 450 patients were planned to be enrolled from about 150 sites in Germany, including patients of oncologists in hospitals and outpatient clinics, as well as in independent oncology practices. PAZOREAL started in December 2015 (first patient enrolled) and lasted until February 2021 (end of observation). Eligible patients were adults with advanced or metastatic RCC and a life expectancy of at least 6 months who started first-line treatment with pazopanib or third-line everolimus no earlier than 8 weeks prior to giving informed consent. The current paper focuses on patients in routine care under first-line treatment with pazopanib at the start of the study. Data on patient demographics, vital signs, concomitant diseases, Karnofsky Performance Status (KPS) or ECOG score (Eastern Cooperative Oncology Group) [16], as well as disease history were assessed at baseline. If available, patient disease history comprised the primary diagnosis of RCC, including the type of histology, tumor staging, and risk assessment, according to the IMDC (International Metastatic RCC Database) Heng score [17,18]. Data concerning nephrectomy and metastasis were also documented at baseline.

2.2. Endpoints and Assessments

The primary endpoint of the study was the time on drug (ToD), defined as the time between the date of first and last administration of study medication in the respective treatment line. Further assessment of effectiveness was based on overall survival (OS). Clinical response rates were determined as SOC every 12 weeks during the course of the study and radiologically or clinically categorized by local investigators according to local clinical practice. Categories included progressive disease (PD), complete response (CR), stable disease (SD, i.e., non-CR, non-PD), and not evaluable.

In addition to the data collected at baseline, further assessments were performed at subsequent visits every 12 weeks throughout the study: ECOG score or KPS [16] and either administered dose of pazopanib (first-line) or nivolumab (second-line). If applicable, any dose modifications, treatment interruptions, or discontinuation of treatment were recorded together with corresponding reasons. Quality of life (QoL) was reported via the EQ-5D-5L questionnaire [19].

To assess for safety, all adverse events (AE), including serious AE (SAE), treatment emergent AE (TEAE), adverse drug reactions (ADR), and serious ADR (SADR), were documented for the study medications from the start of therapy until 30 days after completion of the treatment phase. Any AE that was temporally related to the study medication were considered TEAE. Toxicities were classified and documented according to common terminology criteria for adverse events (CTCAE).

Additional information about adverse events, disease progression, subsequent therapies, and survival status was recorded at follow-up visits every 6 months.

2.3. Statistical Analysis

Descriptive statistical methods were mainly used for the analysis and presentation of data. Time-to-event data, including OS and ToD, were estimated using the Kaplan–Meier method [20], and presented as median and quartiles together with the corresponding 95% confidence intervals (CI) and the frequency and percentage of events and censored cases. For ToD and OS, a sensitivity analysis with trial-eligible patients was performed [14,21,22]. Trial-eligibility was defined as not matching any of the three "trial-ineligibility criteria", i.e., a definition based on Marschner et al. [22] and the exclusion criteria that were frequently applied in mRCC clinical trials during the TKI study period of the pivotal phase III trials [14,21]. These criteria included: a Karnofsky Performance Status < 70%, hemoglobin below the lower normal limit, and non-clear cell carcinoma. On the other hand, patients fulfilling at least one of these criteria were considered "trial-ineligible".

The primary analysis population (i.e., full analysis set, FAS) of this paper included all patients for whom documentation started with first-line treatment with pazopanib (FAS-cohort I) and who received at least one dose of the observed drug. Patients for whom documentation started with third-line treatment with everolimus were not examined in this paper. Where applicable, results focused on the treatment course irrespective of the treatment sequence and in-detail results are presented for first-line treatment with pazopanib; detailed data of the specific treatment sequences are only presented in supplementary manner.

The secondary analysis population (i.e., safety analysis set, SAF) included all patients from FAS-cohort I who received at least one dose of the observed drugs and for whom at least one further post-baseline result (e.g., laboratory tests) was available. Any safety-related analyses, such as those concerning any type of adverse events (AE), were based on the SAF.

3. Results

3.1. Study Population

A total of 398 patients were treated in the study period; 16 of these patients were omitted from analysis (9 due to participation in another clinical trial; 7 due to inspection findings). In total, 382 study subjects were enrolled from 119 sites. Of these, 376 study subjects had documentation beginning with first-line treatment with pazopanib (i.e., FAS-cohort I; see Figure 1). The SAF of FAS-cohort I comprised 375 patients with first-line pazopanib treatment. Of these, 163 patients received nivolumab as second-line treatment, 5 patients received everolimus as second-line treatment, and 9 patients received everolimus as third-line treatment.

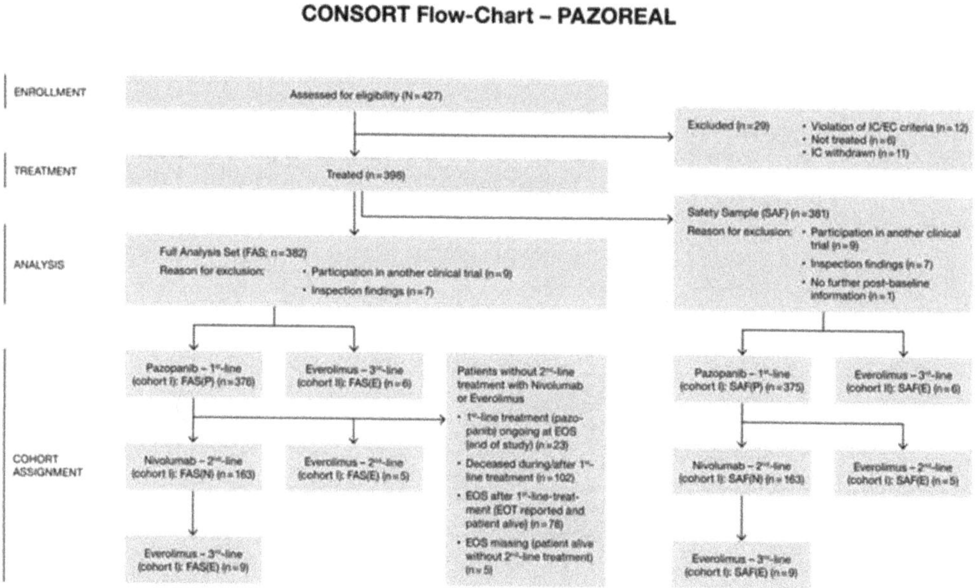

Figure 1. CONSORT flow chart for PAZOREAL with retrospective subset definition.

At baseline, patients in FAS-cohort I were aged between 38 and 89 years, with a median age of 69.7 years. Just over one third of all patients (n = 132, 35.1%) were aged <65 years at the beginning of the study, whereas 244 patients (64.9%) were 65 years or older. Two thirds of the patients were male (n = 257, 68.4%; see Table 1 for all baseline characteristics). The median observation time for FAS-cohort I was 44.6 months (95% CI: 43.2–47.1).

The most common initial doses of pazopanib were 800 mg (n = 248, 66.0%) and 400 mg (n = 100, 26.6%). Other initial doses between 200 and 600 mg were administered in considerably fewer patients (14 patients; less than 3.7% of the total study population). Doses were adjusted in 250 patients (66.5%), increased in 1 patient (0.4%), reduced in 227 patients (90.8%), and interrupted in 114 cases (45.6%).

For first-line treatment with pazopanib, the reasons for discontinuation of treatment were documented for 349 patients (92.8%). The most common reason for the end of treatment was progression of the disease (n = 197, 52.4%), followed by therapy-related toxicity (n = 51, 13.6%), non-therapy related AE (incl. SAE) (n = 22, 5.9%), death (n = 22, 5.9%), and other reasons such as patient's wish or investigator's decision. For 23 patients (6.1%), treatment with pazopanib was ongoing after the end of the observation period, and for 4 patients, there was no documented reason for discontinuation of treatment.

After first-line treatment with pazopanib, 163 patients (43.4%) received second-line treatment with nivolumab. The reasons for a discontinuation of second-line treatment with nivolumab were documented for 143 patients (87.7%); details can be found in Table S1. As the focus of this paper is on the results for pazopanib, the second-line nivolumab treatment is only briefly touched upon in the main text; it can be viewed in more detail in the supplementary section. Due to the small number of patients treated with everolimus in second- (5 patients) or third-line treatments (6 patients), results for sequential treatment with everolimus were omitted.

In FAS-cohort I (n = 376), 146 patients (38.8%) were deemed "trial-eligible" as they did not meet any of the three "trial-ineligibility criteria". A total of 184 patients (48.9%) were classed as trial-ineligible. For 46 patients (12.2%), a response to at least one of the three "trial-ineligibility criteria" was missing, and therefore could not be assigned to one of the groups.

Table 1. Baseline characteristics of FAS-cohort I.

Characteristic	FAS-Cohort I, n = 376
Sex, n (%)	
Women	119 (31.6%)
Men	257 (68.4%)
Median age (range), in years	69.7 (38.5–89.2)
Age group at start of treatment with pazopanib	
<65 years	132 (35.1%)
≥65 years	244 (64.9%)
Median weight at baseline (range), in kg	79.0 (42.0–160.0)
Median BMI at baseline (range), in kg/m^2	26.4 (16.8–58.4)
Number of "trial-eligible" patients, n (%) *	146 (38.8%)
Median time interval from primary diagnosis of RCC to first administration of pazopanib (range), in months	11.0 (0.2–339.3)
ECOG performance status, n (%)	
0	197 (52.4%)
1 (good)	104 (27.7%)
2 (moderate)	40 (10.6%)
3 (poor)	1 (0.3%)
4 (completely disabled)	1 (0.3%)
Not done/Missing	32 (8.5%)/1 (0.3%)
Histology, n (%)	
Clear cell	304 (80.9%)
Non-clear cell	38 (10.1%)
Unknown	34 (9.0%)
Patients with tumor in both kidneys at primary diagnosis, n (%)	16 (4.3%)
Metastatic or non-metastatic disease at enrollment, n (%)	
Metastatic disease	353 (93.9%)
Non-metastatic disease	23 (6.1%)
Number of metastatic sites, n (%)	
0	23 (6.1%)
1–3	322 (85.6%)
4–6	31 (8.2%)
5 most frequent localization of metastases, n (%)	
Lung	218 (58.0%)
Bone	96 (25.5%)
Liver	61 (16.2%)
Lymph nodes, regional	53 (14.1%)
Lymph nodes, distal	45 (12.0%)

* Defined as patients fulfilling none of the three 'trial-ineligibility criteria': (1) Karnofsky Performance Status <70%; (2) hemoglobin < lower limit of normal; (3) non-clear cell carcinoma histology.

3.2. *Effectiveness*

3.2.1. Time on Drug

The median overall time on drug (ToD) for FAS-cohort I was 10.0 months (95% CI: 8.5–11.7), from the start date of the first pazopanib administration until the end date of the last administration of any study medication (i.e., either first-line pazopanib, second-line nivolumab or everolimus, or third-line everolimus). The median ToD of pazopanib was 6.3 months (95% CI: 5.6–7.4), the median ToD of nivolumab was 4.8 months (95% CI: 3.7–6.5), and the median ToD of everolimus was 2.2 months (95% CI: 1.6–NA; second-line) and 3.6 months (95% CI: 0.6–11.2; third-line). The 6-month ToD-rate across all treatment lines was 66.7% (95% CI: 61.6%–71.2%) and the 6-month ToD-rate of pazopanib was 52.2% (95% CI: 47.0–57.1). A third of patients were still receiving pazopanib 12 months after starting first-line treatment with pazopanib (n = 111, 29.5%). Of the remaining patients, 99 patients (26.3%) were deceased, 55 patients (14.6%) ended the study for reasons other than death, and 44 patients (11.7%) received second-line treatment with nivolumab. All other patients were either in-between treatments, in follow-up, or under observation without treatment.

Based on the sensitivity analysis, the median overall ToD was 11.3 months (95% CI: 9.2–14.3) for trial-eligible patients in FAS-cohort I; the median ToD of pazopanib was 7.7 months (95% CI: 6.1–9.0). Further details on nivolumab are listed in Table S2.

3.2.2. Overall Survival

The median overall survival (OS) of FAS-cohort I was 35.9 months (95% CI: 28.2–48.3) (Figure 2). Of the 376 evaluable patients, 174 patients died (46.3%), and the remaining 202 patients (53.7%) were censored at the last date known alive. The sensitivity analysis based on trial-eligible patients revealed a median OS of 53.2 months (95% CI: 38.9-NA). Of the 146 evaluable patients, 59 patients died (40.4%), whereas the remaining patients were censored at the last date known alive (n = 87, 59.6%). However, the 12-month OS rates of FAS-cohort I and trial-eligible patients were comparable, at 71.5 (95% CI: 66.4–76.0%) and 77.9% (95% CI: 69.9–84.0%), respectively. Further sensitivity analyses of OS for FAS-cohort I were performed on sex and age (<65 years, ≥65 years) at start of therapy line (Table 2). The median OS of patients receiving first-line pazopanib and second-line nivolumab, compared with patients with other second-line therapies, is shown in Figure S1.

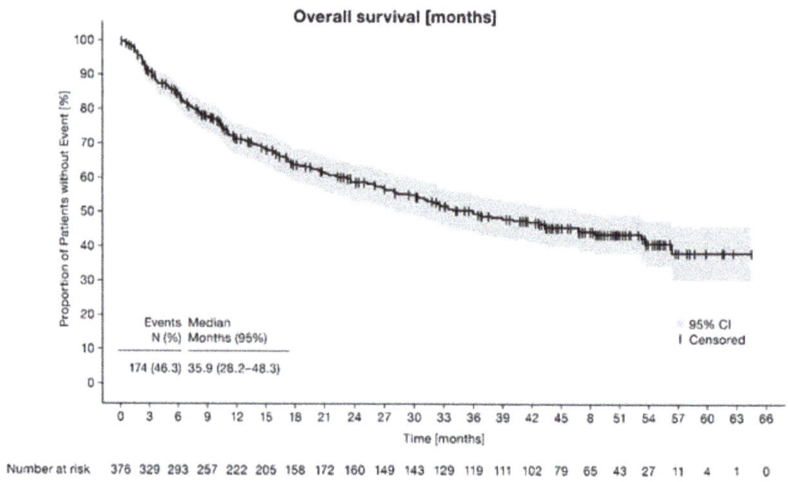

Figure 2. Overall survival of FAS-cohort I.

Table 2. Sensitivity analyses for OS of FAS-cohort I on subgroup sex and age.

	Sex		Age at Start of Therapy Line	
Variable	**Women**	**Men**	**<65 Years**	**≥65 Years**
Patients (n)	119	257	132	244
Events n (%)	64 (53.8%)	110 (42.8%)	69 (52.3%)	105 (43.0%)
Median [95% CI]	31.2 [19.7–35.9]	46.7 [26.9–56.3]	30.4 [23.0–43.4]	43.4 [29.5-NA]
12-month OS rate [95% CI]	67.7% [58.1–75.5]	73.3% [67.1–78.5]	69.6% [60.7–76.9]	72.5% [66.1–77.9]

3.2.3. Best Response and Disease Control Rate

Based on radiological or clinical assessments by local investigators according to local clinical practices, 36 patients (9.6%, 95% CI: 7.0–13.0) receiving first-line treatment with pazopanib achieved a complete response (CR) as their best response, whereas stable disease (SD) was reported as the best response for 178 patients (47.3%, 95% CI: 42.4–52.4). Thus, the disease control rate (DCR), comprising patients with CR and SD, was 56.9% (95% CI: 51.9–61.8). PD was reported in 81 patients (21.5%, 95% CI: 17.7–26.0). The best response was assessed by radiologic assessment (259 patients, 68.9%) or clinical assessment (38 patients, 10.1%). No best response was evaluable for 2 patients, and in 79 patients (21.0%), no assessments were performed. The results of the response analysis for the 163 patients receiving nivolumab as second-line treatment are shown in Table S2.

3.3. *Safety*

Under first-line treatment with pazopanib, 1923 TEAE were documented in 337 patients (89.9%). Of those events, 1038 (54.0%) were judged to be related to pazopanib, which occurred in 270 patients (72.0%). The most common TEAE included diarrhea, nausea, and stomatitis (Table 3). Furthermore, there were 368 grade 3/4 TEAE (19.1%) occurring in 179 patients (47.7%), out of which 151 grade 3/4 TEAE (7.9%) in 95 patients (25.3%) were assessed as being related to pazopanib (Table 4). A total of 129 patients (34.4%) experienced TEAE that led to the discontinuation of treatment, and 66 of these patients (17.6%) were assessed to have TEAE related to pazopanib. In addition, 75 fatal TEAE (3.9%) were reported, of which 3 events (0.2%), occurring in 3 patients (0.8%), were judged as being related to pazopanib by the respective investigators. For each (0.3%) respective patient, the following reasons were reported: death without witnesses, disease progression, and neoplasm progression.

Table 3. Related TEAE of CTCAE severity grade 1/2 under first-line pazopanib with an occurrence of at least 5%.

Primary System Organ Class	Preferred Term	SAF, n = 375
Patients with any event		250 (66.7%)
Gastrointestinal disorders	Patients with any event	167 (44.5%)
	Diarrhea	116 (30.9%)
	Nausea	60 (16.0%)
	Stomatitis	19 (5.1%)
General disorders and administration site conditions	Patients with any event	75 (20.0%)
	Fatigue	47 (12.5%)
Skin and subcutaneous tissue disorders	Patients with any event	75 (20.0%)
	Hair color changes	33 (8.8%)
Nervous system disorders	Patients with any event	71 (18.9%)
	Dysgeusia	39 (10.4%)
Metabolism and nutrition disorders	Patients with any event	40 (10.7%)
	Decreased appetite	35 (9.3%)
Vascular disorders	Patients with any event	34 (9.1%)
	Hypertension	25 (6.7%)

Adverse event terms were encoded according to MedDRA version 20.0.

Table 4. Related TEAE of CTCAE severity grade 3/4 under first-line pazopanib occurring in at least 5 patients.

Primary System organ Class	Preferred Term	SAF, n = 375
Patients with any event		95 (25.3%)
Investigations	Patients with any event	28 (7.5%)
	Gamma-glutamyl transferase increased	8 (2.1%)
	Alanine aminotransferase increased	5 (1.3%)
	Aspartate aminotransferase increased	5 (1.3%)
Vascular disorders	Patients with any event	28 (7.5%)
	Hypertension	16 (4.3%)
	Hypertensive crisis	9 (2.4%)
Gastrointestinal disorders	Patients with any event	19 (5.1%)
	Nausea	7 (1.9%)
	Diarrhea	6 (1.6%)
General disorders and administration site conditions	Patients with any event	11 (2.9%)
	Fatigue	5 (1.3%)

Adverse event terms were encoded according to MedDRA version 20.0.

In 120 out of the 163 patients (73.6%) receiving second-line treatment with nivolumab following first-line treatment with pazopanib, a total of 400 TEAE were reported; further details on nivolumab are listed in Table S3.

3.4. Quality of Life

Among patients receiving first-line pazopanib treatment, 279 patients (74.2%) fulfilled the inclusion criteria for assessment of quality of life (QoL) with the EQ-5D-5L questionnaire—i.e., patients consented to data collection via questionnaire. Out of the 229 questionnaires handed out to patients, 219 (78.5%) questionnaires were available for analysis at baseline. Questionnaires from 58 patients (20.8%) were returned for analysis after 12 months of treatment. The EQ-5D-5L scores generally remained unchanged under treatment. After 12 months, most patients (n = 49, 84.5%) experienced no problems regarding self-care, whereas more patients reported problems ranging from "slight" to "severe problems/extreme discomfort" for the dimensions "Usual activity" (n = 34, 58.6%) and "Pain/Discomfort" (n = 37, 63.8%). About half of the patients reported no problems for the dimensions "Mobility" (n = 28, 48.3%) and "Anxiety/Depression" (n = 31, 53.4%) (Figure 3).

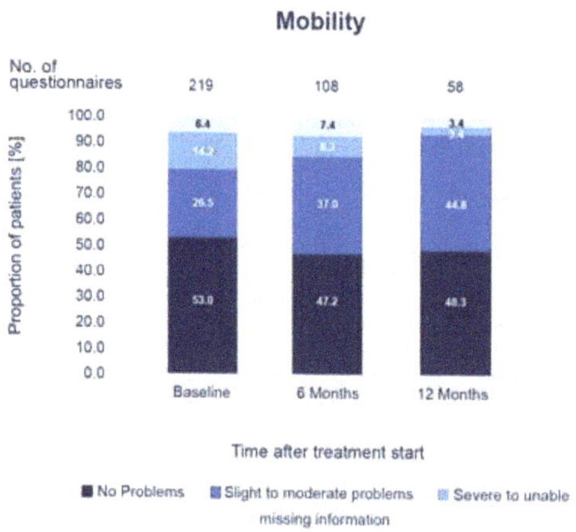

Figure 3. Quality of life under pazopanib for the "mobility" dimension at baseline and after 6 and 12 months of treatment. Percentages refer to the number of returned questionnaires. For other dimensions, see Figures S2–S5.

It should be noted that return rate of questionnaires was low. Therefore, results need to be interpreted with caution and should be understood as a description of the situation, rather than used for generalization. Likewise, for nivolumab as second-line treatment, there was also a low return rate of questionnaires. However, EQ-5D-5L scores generally remained unchanged. Further details on second-line nivolumab are listed in Table S4.

4. Discussion

The results from the non-interventional PAZOREAL study presented here offer first insights into the routine care of mRCC with first-line treatment with pazopanib.

As the primary endpoint, the median ToD for patients receiving first-line treatment with pazopanib was more than 6 months, under routine clinical conditions. Specifically, the median ToD was 6.3 months for first-line pazopanib treatment in FAS-cohort I and median ToD for trial-eligible patients was 7.7 months. This is similar to published results from clinical studies with a ToD of 7.4 months in the pazopanib arm of VEG105192 [13,21], 8.1 months in the overall results for pazopanib of the COMPARZ-study [14,23], and a median ToD of 8.4 months for Asian patients and 7.2 months for non-Asian patients in a subsequent Asian vs. non-Asian subgroup analysis [24]. Overall, this indicates that real-world data are comparable to randomized controlled trial (RCT) settings.

Importantly, although non-interventional studies, including PAZOREAL, cannot derive any conclusions on the superiority of a treatment, they also have certain advantages compared with clinical trials. As data are coming from a less homogeneous patient pool, subpopulations of the patient group can give valuable insights into treatment options for special patient groups, e.g., older patients [25]. One prominent difference between the current real-world data population and those of published RCT studies was the age of the patients. In PAZOREAL, the median age of patients receiving first-line treatment with pazopanib was higher, with a median age of 69.7 years, compared with the median age of 59 (VEG105192) and 65 years (VEG107769) in other clinical trials [13,21,26]. Crucially, despite this difference, results of PAZOREAL are comparable to those of the clinical studies. Still, real-world studies have inherent limitations, such as low internal validity, various biases, and less quality control in data collection, and these factors must be considered when interpreting the results [25].

OS analyses showed a median OS of close to three years; as such, the median OS was longer compared with previous data from the COMPARZ study [23]. Additionally, more than half of the patients achieved either CR or SD. The CR rate under pazopanib was nearly 10%, much higher than expected from previous studies with CR rates of about 0.3 [21] or 0.4% [14]. This could be due to evaluation bias between study physicians as the assessment was not standardized. The CR must therefore be interpreted with caution and should be considered as a clinical complete response rate.

Furthermore, our results suggested that pazopanib was well-tolerated and no new safety signals were detected in PAZOREAL, making the safety profile comparable with findings of clinical studies [27,28]. Additionally, PAZOREAL provided valuable real-world data on trial-ineligible patients, as these patients were not represented in clinical studies in the initial pivotal phase III trials of sunitinib and pazopanib [14,21], despite making up a notable proportion of mRCC patients [29]. It is important to note that criteria on trial ineligibility have changed over time.

Whereas data on ToD is an important effectiveness endpoint of a therapy, it cannot be used to draw conclusions about the well-being of patients on the drug; accordingly, we also examined the quality of life of patients. However, considering the low return rates of the QoL questionnaires, these results must be regarded with caution, as they may be biased and may not be representative of the entire study population. Nevertheless, the results—as a non-generalizable description of the data—are in line with previous findings from pivotal studies that showed no discernible changes in QoL over the course of treatment with pazopanib [21]. Despite the low return rates of questionnaires, PAZOREAL added valuable information to the growing dataset on treatment strategies for mRCC, by assessing health-related QoL in mRCC patients in a real-world setting under treatment with pazopanib. This is particularly important given the lack of data on QoL.

The real-world assessment of routine practice also revealed that Heng scores and MSKCC risk scores were recorded for less than a quarter of all patients. Although the assessment of MSKCC risk scores was still used, further refinement has since been introduced with the IMDC score [11].

Although not the standard of care, pazopanib is approved as the first-line treatment for mRCC when drug combinations including CPI cannot be used and is still recommended in current guidelines [10–12]. The approval of nivolumab for the second-line treatment of mRCC provided a new treatment option for patients with disease progression after first-line treatment with pazopanib [30]. PAZOREAL shows that this sequence is already being used in clinical practice in Germany, as the majority of patients received nivolumab after pazopanib. This is also in line with clinical trials and previous analyses from PAZOREAL [31–33]. However, examination of differences between treatment sequences is still needed.

5. Conclusions

PAZOREAL provides real-world data from a study population encompassing patients receiving first-line pazopanib treatment in routine clinical care. Results from this non-interventional study support findings from clinical studies suggesting that pazopanib is an effective and safe first-line treatment for patients with mRCC who are not eligible for IO-based combinations. Crucially, this was true even though the real-world data showed a less homogenous patient pool than clinical studies; for instance, patients in PAZOREAL were older than in published clinical studies. With the approval of targeted substances for the treatment of mRCC, treatment options have increased and allowed for a more directed approach tailored to patient needs. PAZOREAL provides routine data on the use of pazopanib as the first-line treatment for mRCC.

Supplementary Materials: The following supporting information can be downloaded at: https://www.mdpi.com/article/10.3390/cancers14225486/s1, Figure S1: Kaplan–Meier plot of overall survival under nivolumab, compared with other substances; Figure S2: Quality of life under pazopanib regarding the "self-care" dimension at baseline and after 6 and 12 months of treatment; Figure S3: Quality of life under pazopanib regarding the "usual activity" dimension at baseline and after 6 and 12 months of treatment; Figure S4: Quality of life under pazopanib regarding the "pain/discomfort" dimension at baseline and after 6 and 12 months of treatment; Figure S5: Quality of life under pazopanib regarding the "anxiety/depression" dimension at baseline and after 6 and 12 months of treatment; Table S1: Main reasons for discontinuation under second-line nivolumab; Table S2: Time on drug (ToD), best response, and disease control rate (DCR) under second-line nivolumab; Table S3: Summary of treatment-emergent adverse events (TEAEs) under second-line nivolumab; Table S4: Quality of life under second-line nivolumab at baseline and after 6 and 12 months of treatment.

Author Contributions: Conceptualization, C.D., V.G., M.W., R.E. and P.J.G.; data curation, C.D. and V.G.; investigation, C.D., M.B., V.G., J.B., M.S. and T.W.; methodology, C.D., V.G., M.W., R.E. and P.J.G.; data interpretation, C.D., M.B., V.G., M.W., J.B., M.S., T.W., R.E., E.D., S.W. and P.J.G.; project administration, R.E. and E.D.; supervision, P.J.G.; writing—review and editing, C.D., M.B., V.G., M.W., J.B., M.S., T.W., R.E., E.D., S.W. and P.J.G. All authors have read and agreed to the published version of the manuscript.

Funding: This research was funded by Novartis Pharma GmbH, Nürnberg, Germany. Additional statistical analyses were initiated and funded by APOGEPHA Arzneimittel GmbH. The APC was funded by APOGEPHA Arzneimittel GmbH.

Institutional Review Board Statement: The study was conducted in accordance with the Declaration of Helsinki and approved by the Ethics Committee at the University Hospital Erlangen, Germany (283_15B and date of approval 12 November 2015).

Informed Consent Statement: All patients provided written informed consent before enrollment. The study was conducted in accordance with Good Clinical Practice guidelines.

Data Availability Statement: Data available upon request to the corresponding author or sponsor of the clinical study.

Acknowledgments: We thank the patients who participated in the study and their families. We acknowledge the investigators who contributed to the study. Special thanks to iOMEDICO AG (Germany) for data and project management, including statistical analysis, as well as to co.medical® for original draft preparation.

Conflicts of Interest: E.D. and R.E. are employees of Novartis Pharma GmbH. S.W. is an employee of APOGEPHA Arzneimittel GmbH. Novartis Pharma GmbH had a role in the design of the study, statistical analysis plan, data interpretation, review of the manuscript, and decision to publish. APOGEPHA Arzneimittel GmbH had a role in planning additional statistical analysis, data interpretation, review of the manuscript, and decision to publish. The other authors declare no conflict of interest.

References

1. Padala, S.A.; Barsouk, A.; Thandra, K.C.; Saginala, K.; Mohammed, A.; Vakiti, A.; Rawla, P.; Barsouk, A. Epidemiology of Renal Cell Carcinoma. *World J. Oncol.* **2020**, *11*, 79–87. [CrossRef] [PubMed]
2. Robert Koch Institut. Krebs in Deutschland. Available online: https://www.krebsdaten.de/Krebs/DE/Content/Krebsarten/Nierenkrebs/nierenkrebs_node.html (accessed on 5 November 2021).
3. Linehan, W.M.; Pinto, P.A.; Srinivasan, R.; Merino, M.; Choyke, P.; Choyke, L.; Coleman, J.; Toro, J.; Glenn, G.; Vocke, C.; et al. Identification of the genes for kidney cancer: Opportunity for disease-specific targeted therapeutics. *Clin. Cancer Res.* **2007**, *13*, 671s–679s. [CrossRef] [PubMed]
4. Abdelaziz, A.; Vaishampayan, U. Cabozantinib for Renal Cell Carcinoma: Current and Future Paradigms. *Curr. Treat Options Oncol.* **2017**, *18*, 18. [CrossRef]
5. Capitanio, U.; Bensalah, K.; Bex, A.; Boorjian, S.A.; Bray, F.; Coleman, J.; Gore, J.L.; Sun, M.; Wood, C.; Russo, P. Epidemiology of Renal Cell Carcinoma. *Eur. Urol.* **2019**, *75*, 74–84. [CrossRef]
6. Posadas, E.M.; Limvorasak, S.; Figlin, R.A. Targeted therapies for renal cell carcinoma. *Nat. Rev. Nephrol.* **2017**, *13*, 496–511. [CrossRef] [PubMed]
7. Escudier, B.; Porta, C.; Bono, P.; Powles, T.; Eisen, T.; Sternberg, C.N.; Gschwend, J.E.; De Giorgi, U.; Parikh, O.; Hawkins, R.; et al. Randomized, controlled, double-blind, cross-over trial assessing treatment preference for pazopanib versus sunitinib in patients with metastatic renal cell carcinoma: PISCES study. *J. Clin. Oncol.* **2014**, *32*, 1412–1418. [CrossRef] [PubMed]
8. Ljungberg, B.; Bensalah, K.; Canfield, S.; Dabestani, S.; Hofmann, F.; Hora, M.; Kuczyk, M.A.; Lam, T.; Marconi, L.; Merseburger, A.S.; et al. EAU guidelines on renal cell carcinoma: 2014 update. *Eur. Urol.* **2015**, *67*, 913–924. [CrossRef] [PubMed]
9. Motzer, R.J.; Escudier, B.; McDermott, D.F.; George, S.; Hammers, H.J.; Srinivas, S.; Tykodi, S.S.; Sosman, J.A.; Procopio, G.; Plimack, E.R.; et al. Nivolumab versus Everolimus in Advanced Renal-Cell Carcinoma. *N. Engl. J. Med.* **2015**, *373*, 1803–1813. [CrossRef]
10. Leitlinienprogramm Onkologie, Deutsche Krebsgesellschaft; Deutsche Krebshilfe, A.D., Therapie und Nachsorge des Nierenzellkarzinoms, Kurzversion 2.0, AWMF Registernummer: 043/017OL. Available online: https://www.leitlinienprogramm-onkologie.de/leitlinien/nierenzellkarzinom/ (accessed on 5 November 2021).
11. Escudier, B.; Porta, C.; Schmidinger, M.; Rioux-Leclercq, N.; Bex, A.; Khoo, V.; Grunwald, V.; Gillessen, S.; Horwich, A. Renal cell carcinoma: ESMO Clinical Practice Guidelines for diagnosis, treatment and follow-up. *Ann. Oncol.* **2019**, *30*, 706–720. [CrossRef]
12. Ljungberg, B.; Albiges, L.; Abu-Ghanem, Y.; Bedke, J.; Capitanio, U.; Dabestani, S.; Fernandez-Pello, S.; Giles, R.H.; Hofmann, F.; Hora, M.; et al. European Association of Urology Guidelines on Renal Cell Carcinoma: The 2022 Update. *Eur. Urol.* **2022**, *75*, 799–810. [CrossRef]
13. Sternberg, C.N.; Hawkins, R.E.; Wagstaff, J.; Salman, P.; Mardiak, J.; Barrios, C.H.; Zarba, J.J.; Gladkov, O.A.; Lee, E.; Szczylik, C.; et al. A randomised, double-blind phase III study of pazopanib in patients with advanced and/or metastatic renal cell carcinoma: Final overall survival results and safety update. *Eur. J. Cancer* **2013**, *49*, 1287–1296. [CrossRef] [PubMed]
14. Motzer, R.J.; Hutson, T.E.; Cella, D.; Reeves, J.; Hawkins, R.; Guo, J.; Nathan, P.; Staehler, M.; de Souza, P.; Merchan, J.R.; et al. Pazopanib versus sunitinib in metastatic renal-cell carcinoma. *N. Engl. J. Med.* **2013**, *369*, 722–731. [CrossRef] [PubMed]
15. Welsh, S.J.; Fife, K. Pazopanib for the treatment of renal cell carcinoma. *Future Oncol.* **2015**, *11*, 1169–1179. [CrossRef] [PubMed]
16. Oken, M.M.; Creech, R.H.; Tormey, D.C.; Horton, J.; Davis, T.E.; McFadden, E.T.; Carbone, P.P. Toxicity and response criteria of the Eastern Cooperative Oncology Group. *Am. J. Clin. Oncol.* **1982**, *5*, 649–655. [CrossRef]
17. Motzer, R.J.; Bacik, J.; Murphy, B.A.; Russo, P.; Mazumdar, M. Interferon-alfa as a comparative treatment for clinical trials of new therapies against advanced renal cell carcinoma. *J. Clin. Oncol.* **2002**, *20*, 289–296. [CrossRef]
18. Escudier, B.; Eisen, T.; Porta, C.; Patard, J.J.; Khoo, V.; Algaba, F.; Mulders, P.; Kataja, V.; Group, E.G.W. Renal cell carcinoma: ESMO Clinical Practice Guidelines for diagnosis, treatment and follow-up. *Ann. Oncol.* **2012**, *23* (Suppl. 7), VII65–VII71. [CrossRef]
19. Herdman, M.; Gudex, C.; Lloyd, A.; Janssen, M.; Kind, P.; Parkin, D.; Bonsel, G.; Badia, X. Development and preliminary testing of the new five-level version of EQ-5D (EQ-5D-5L). *Qual. Life Res.* **2011**, *20*, 1727–1736. [CrossRef]
20. Kaplan, E.L.; Meier, P. Nonparametric Estimation from Incomplete Observations. *J. Am. Stat. Assoc.* **1958**, *53*, 457–481. [CrossRef]

21. Sternberg, C.N.; Davis, I.D.; Mardiak, J.; Szczylik, C.; Lee, E.; Wagstaff, J.; Barrios, C.H.; Salman, P.; Gladkov, O.A.; Kavina, A.; et al. Pazopanib in locally advanced or metastatic renal cell carcinoma: Results of a randomized phase III trial. *J. Clin. Oncol.* **2010**, *28*, 1061–1068. [CrossRef]
22. Marschner, N.; Staehler, M.; Muller, L.; Nusch, A.; Harde, J.; Koska, M.; Janicke, M.; Goebell, P.J.; Group, R.C.C.R. Survival of Patients With Advanced or Metastatic Renal Cell Carcinoma in Routine Practice Differs From That in Clinical Trials-Analyses From the German Clinical RCC Registry. *Clin. Genitourin. Cancer* **2017**, *15*, e209–e215. [CrossRef]
23. Motzer, R.J.; Hutson, T.E.; McCann, L.; Deen, K.; Choueiri, T.K. Overall survival in renal-cell carcinoma with pazopanib versus sunitinib. *N. Engl. J. Med.* **2014**, *370*, 1769–1770. [CrossRef] [PubMed]
24. Guo, J.; Jin, J.; Oya, M.; Uemura, H.; Takahashi, S.; Tatsugami, K.; Rha, S.Y.; Lee, J.L.; Chung, J.; Lim, H.Y.; et al. Safety of pazopanib and sunitinib in treatment-naive patients with metastatic renal cell carcinoma: Asian versus non-Asian subgroup analysis of the COMPARZ trial. *J. Hematol. Oncol.* **2018**, *11*, 69. [CrossRef] [PubMed]
25. Di Maio, M.; Perrone, F.; Conte, P. Real-World Evidence in Oncology: Opportunities and Limitations. *Oncologist* **2020**, *25*, e746–e752. [CrossRef] [PubMed]
26. Sternberg, C.N.; Davis, I.D.; Deen, K.C.; Sigal, E.; Hawkins, R.E. An open-label extension study to evaluate safety and efficacy of pazopanib in patients with advanced renal cell carcinoma. *Oncology* **2014**, *87*, 342–350. [CrossRef] [PubMed]
27. McCann, L.; Amit, O.; Pandite, L.; Amado, R.G. An indirect comparison analysis of pazopanib versus other agents in metastatic renal cell carcinoma (mRCC). *J. Clin. Oncol.* **2010**, *28*, e15128. [CrossRef]
28. Gupta, S.; Spiess, P.E. The prospects of pazopanib in advanced renal cell carcinoma. *Ther. Adv. Urol.* **2013**, *5*, 223–232. [CrossRef] [PubMed]
29. Ishihara, H.; Tachibana, H.; Fukuda, H.; Yoshida, K.; Kobayashi, H.; Takagi, T.; Iizuka, J.; Ishida, H.; Kondo, T.; Tanabe, K. Prognostic Impact of Trial-Eligibility Criteria in Patients with Metastatic Renal Cell Carcinoma. *Urol. Int.* **2021**, *106*, 368–375. [CrossRef]
30. Méndez-Vidal, M.J.; Molina, Á.; Anido, U.; Chirivella, I.; Etxaniz, O.; Fernández-Parra, E.; Guix, M.; Hernández, C.; Lambea, J.; Montesa, Á. Pazopanib: Evidence review and clinical practice in the management of advanced renal cell carcinoma. *BMC Pharmacol. Toxicol.* **2018**, *19*, 77.
31. Bedke, J.; Boegemann, M.; Schostak, M.; Hering-Schubert, C.; Welslau, M.; Schleicher, J.; Petzoldt, A.; Doehn, C.; Gruellich, C.; Goebell, P.-J.; et al. Sequential treatment with pazopanib followed by nivolumab in patients with renal cell carcinoma: Updated interim results of the non-interventional study PAZOREAL. *J. Clin. Oncol.* **2020**, *38*, e17075. [CrossRef]
32. Boegemann, M.; Bedke, J.; Schostak, M.; Hering-Schubert, C.; Welslau, M.; Schleicher, J.; Wolf, T.; Petzoldt, A.; Doehn, C.; Grüllich, C.; et al. Sequential treatment with pazopanib (PAZO) followed by nivolumab (NIVO) in patients with advanced or metastatic renal cell carcinoma (mRCC): Third interim results of the non-interventional study PAZOREAL. *J. Clin. Oncol.* **2019**, *37*, 4574. [CrossRef]
33. Boegemann, M.; Bedke, J.; Schostak, M.; Welslau, M.; Hering-Schubert, C.; Wolf, T.; Schleicher, J.; Petzold, A.; Doehn, C.; Grüllich, C.; et al. Effectiveness and safety of pazopanib (PAZO) and everolimus (EVE) in a changing treatment (Tx) landscape: Interim results of the non-interventional study PAZOREAL. *J. Clin. Oncol.* **2018**, *36*, 4584. [CrossRef]

cancers

MDPI

Article

Intravesical Recurrence after Radical Nephroureterectomy in Patients with Upper Tract Urothelial Carcinoma Is Associated with Flexible Diagnostic Ureteroscopy, but Not with Rigid Diagnostic Ureteroscopy

Jee Soo Ha [1], Jinhyung Jeon [1], Jong Cheol Ko [1], Hye Sun Lee [2], Juyeon Yang [2], Daeho Kim [1], June Seok Kim [1], Won Sik Ham [3], Young Deuk Choi [3] and Kang Su Cho [1,4,*]

1 Department of Urology, Prostate Cancer Center, Urological Science Institute, Gangnam Severance Hospital, Yonsei University College of Medicine, Seoul 06273, Republic of Korea
2 Biostatistics Collaboration Unit, Yonsei University College of Medicine, Seoul 03722, Republic of Korea
3 Department of Urology, Urological Science Institute, Severance Hospital, Yonsei University College of Medicine, Seoul 03722, Republic of Korea
4 Center of Evidence Based Medicine, Institute of Convergence Science, Yonsei University, Seoul 03722, Republic of Korea
* Correspondence: kscho99@yuhs.ac; Tel.: +82-2-2019-3471

Simple Summary: Diagnostic ureteroscopy (URS) before radical nephroureterectomy is a risk factor for intravesical recurrence in patients with upper tract urothelial carcinoma. Although flexible URS requires a higher-pressure inflow of irrigation fluid than that of rigid URS, previous studies have not considered mechanical differences in relation to the type of URS. In this manuscript, we assessed the impact of diagnostic URS on intravesical recurrence following radical nephroureterectomy for upper tract urothelial carcinoma according to the type of URS.

Citation: Ha, J.S.; Jeon, J.; Ko, J.C.; Lee, H.S.; Yang, J.; Kim, D.; Kim, J.S.; Ham, W.S.; Choi, Y.D.; Cho, K.S. Intravesical Recurrence after Radical Nephroureterectomy in Patients with Upper Tract Urothelial Carcinoma Is Associated with Flexible Diagnostic Ureteroscopy, but Not with Rigid Diagnostic Ureteroscopy. *Cancers* **2022**, *14*, 5629. https://doi.org/10.3390/cancers14225629

Academic Editors: José I. López and Claudia Manini

Received: 11 October 2022
Accepted: 12 November 2022
Published: 16 November 2022

Abstract: (1) Background: We assessed the impact of diagnostic ureteroscopy (URS) on intravesical recurrence (IVR) following radical nephroureterectomy (RNU) for upper tract urothelial carcinoma according to the type of URS. (2) Methods: Data on 491 consecutive patients who underwent RNU at two institutions between 2016 and 2019 were retrospectively reviewed. The study population was classified according to the type of URS performed before RNU as follows: non-URS, rigid URS, and flexible URS. The study outcome was IVR occurring within 1 year of RNU. Univariable and multivariable Cox proportional hazards models were used to estimate the risk of IVR. (3) Results: Altogether, 396 patients were included for analysis. Rigid and flexible URS were performed in 178 (45%) and 111 (28%) patients, respectively, while 107 (27%) patients did not undergo URS. IVR was identified in 99 (25%) patients. Multivariable Cox regression analysis revealed that the flexible URS group was significantly associated with increased IVR, compared to the non-URS group (HR = 1.807, $p = 0.0416$). No significant difference in IVR was observed between the non-URS and rigid URS groups (HR = 1.301, $p = 0.3388$). (4) Conclusions: In patients with UTUC undergoing RNU, rigid URS may not increase the risk of IVR, whereas flexible URS appears to be associated with a higher risk of IVR.

Keywords: ureteral neoplasms; urinary bladder neoplasms; ureteroscopy

1. Introduction

Upper urinary tract urothelial carcinoma (UTUC) is a rare malignancy, accounting for 5–10% of all urothelial carcinomas [1,2]. According to a recent meta-analysis, computed tomography (CT)-urography shows a 92% sensitivity and 95% accuracy [3] in the diagnoses of UTUC, and diagnostic ureteroscopy (URS) is recommended only when imaging studies and cytology are insufficient for diagnosis [4]. However, although diagnostic URS can reduce the misdiagnosis rate to 2.1% [5], it is reportedly a risk factor for intravesical recurrence (IVR) after radical nephroureterectomy (RNU) [6–8].

The rate of IVR after RNU for UTUC is reportedly 22–47% [9,10]. Downward seeding, as well as field cancerization, is understood to be a mechanism of IVR [11–13]. Shigeta et al. recently revealed that IVR tumors after RNU exhibit molecular characteristics similar to those in UTUC via a comparative study of UTUC origin and primary bladder tumor [14]. Seisen et al. reported several predictors of IVR in their systematic review, including patient-specific (male sex, previous bladder cancer, preoperative chronic kidney disease), tumor-specific (positive preoperative urinary cytology, ureteral location, multifocality, invasive pathological T stage, necrosis), and treatment-specific (laparoscopic RNU, extravesical bladder cuff removal, positive surgical margin) predictors [15]. Other studies suggested that diagnostic URS prior to RNU is a risk factor for higher IVR, although this remains controversial [6,7,16]. Two recent studies, however, demonstrated that URS prior to RNU increases the risk of IVR only when combined with endoscopic biopsy, albeit with no concurrent impact on the other long-term survival outcomes [8,17].

Over the last decade, flexible URS has become a popular modality in the exploration of the entire upper urinary tract, offering improved diagnostic accuracy due to advances in endourologic technologies. Several novel optical diagnostic techniques, such as narrow band imaging, photodynamic diagnosis, and optical coherence tomography, have been introduced in combination with flexible URS to improve UTUC detection [18]. In addition, flexible URS provides clinicians with a means of performing endoscopic treatment for low-risk UTUC, raising new concerns for the selection of eligible patients for kidney-sparing surgery. Apart from the advantages of flexible URS, research suggests that diagnostic URS, including flexible URS, has an adverse effect on cancer control. To date, such research has focused on the impact of diagnostic URS on IVR, regardless of the type of URS. In this retrospective study, we assessed the impact of diagnostic URS on IVR following RNU, according to the type of URS, either flexible or rigid.

2. Materials and Methods

2.1. Ethics

This study was performed in accordance with the tenets of the Declaration of Helsinki. The Institutional Review Board of Gangnam Severance Hospital approved the study protocol (approval number: 3-2020-0199, approval date: 2 July 2020).

2.2. Study Design and Population

This multicenter retrospective study involved two tertiary hospitals in the Republic of Korea. Medical records of 491 consecutive patients who underwent RNU between 2016 and 2019 were reviewed. Patients whose pathological diagnosis was not urothelial carcinoma were excluded. Other exclusion criteria were as follows: patients who underwent radical cystectomy prior to RNU or simultaneously with RNU, those who experienced IVR within 90 days after RNU, and patients whose follow-up period was <90 days. Finally, 396 patients were enrolled for analysis, with a median follow-up of 30 months (interquartile range [IQR]: 19–43).

2.3. Operation Techniques and Postoperative Management

A retrograde pyelography was performed with a 5-Fr ureteral catheter and diluted contrast media. Then, a guide wire was placed under visual and fluoroscopic control. The type of URS was selected at the surgeon's discretion. For rigid URS, gravity-based irrigation was used; the irrigation fluid was placed 60 to 80 cm above the patient. For flexible URS, a ureteral access sheath was introduced along the guide wire below the tumor, and selective ureteral cytology was collected. Irrigation was provided with normal saline and a pressurized pump to ensure clear vision.

RNU was performed via open, laparoscopic, or robotic surgery in compliance with oncological principles. Bladder cuff resection was performed in all patients using an extravesical technique, and the intramural portion of the ureter was completely dissected. Lymph node dissection was performed at the surgeon's discretion. No postoperative dose of intravesical chemotherapy, such as mitomycin, was administered. Adjuvant chemotherapy was administered to patients who had adverse pathological features such as pT3 or higher stage, lymph node involvement, lymphovascular invasion, or aggressive variant histology. All patients who underwent RNU were followed-up with abdomen-pelvis CT and chest CT every 3–6 months and with cystoscopy and urine cytology every 3 months.

2.4. Definitions of Groups, Outcomes, and Covariates

The study cohort was divided into three groups according to the type of diagnostic URS prior to RNU: non-URS, rigid URS, or flexible URS. The study outcome was defined as pathologically proven IVR occurring within 1 year of RNU. Time-to-event was defined as the duration between the date of RNU and the first IVR. The tumor locations for UTUC were divided into the renal pelvis and ureter. Tumor grade and pathological T stages of UTUC were described according to the 2016 World Health Organization/International Society of Urologic Pathology consensus classification and the 2010 American Joint Committee on Cancer and the International Union for Cancer Control tumor–node–metastasis classification [19,20].

2.5. Statistical Analysis

Demographic and clinicopathological factors were described and compared according to the type of diagnostic URS. Differences in categorical variables between the three groups were compared using Pearson's chi-square test or Fisher's exact test. Continuous variables were compared using the Mann–Whitney U test and described in a non-parametric manner. IVR-free survival curves were generated using the Kaplan–Meier method and compared using the log-rank test. A log–log plot and interaction test confirmed that the proportionality assumption was satisfied. Univariable and multivariable Cox proportional hazard models were used to estimate the risk of IVR according to covariates. A multivariable model was generated using the enter method, including factors that were significant ($p < 0.05$) or neared significance ($p < 0.1$) in univariate analysis, as well as factors known to affect IVR. Significance was set at $p < 0.05$, and all statistical tests were two-sided. All study analyses were performed using SAS® System for Windows® (version 9.4; SAS Institute Inc., Cary, NC, USA).

3. Results

In our study cohort, rigid and flexible URS was performed in 178 (45%) and 111 (28%) patients, respectively, and 107 (27%) patients did not undergo diagnostic URS (Table 1). Table 2 summarizes the comparisons of clinicopathological features according to the type of diagnostic URS. Significant differences in sex, urine cytology, tumor location, tumor grade, previous or concurrent bladder cancer, and smoking history were observed among the groups.

IVR was identified in 99 (25%) patients, specifically in 21, 41, and 37 patients in the non-URS, rigid URS, and flexible URS groups, respectively (Table 2). The univariable Cox regression analysis demonstrated that the flexible URS group had a higher IVR rate than the non-URS group (hazard ratio [HR] = 1.866; 95% confidence interval [CI], 1.092–3.188; $p = 0.0224$), with no significant difference between the rigid URS and non-URS groups (HR = 1.161; 95% CI, 0.686–1.966, $p = 0.577$) (Table 3 and Figure 1).

Table 1. Clinicopathological characteristics of 396 patients with UTUC.

Age at RNU *, years	68.0 (60.0–76.0)
Sex	
Female	127 (32.07)
Male	269 (67.93)
HTN	236 (59.60)
DM	87 (21.97)
CAOD	46 (11.62)
COPD, asthma	35 (8.84)
CVA	26 (6.57)
CKD	45 (11.36)
Smoking history	
Never smoked	210 (53.03)
Ex- or current smoker	186 (46.97)
Tumor location	
Renal pelvis	220 (55.56)
Ureter	176 (44.44)
Laterality	
Left	198 (50.00)
Right	198 (50.00)
Previous/concurrent bladder cancer	81 (20.45)
Urine cytology	
Not done	136 (34.34)
Negative	140 (35.35)
Atypical cell	69 (17.43)
Positive	51 (12.88)
Type of URS	
Non-URS	107 (27.02)
Rigid URS	178 (44.95)
Flexible URS	111 (28.03)
Ureteroscopic biopsy	140 (35.35)
Surgical modality	
Open	105 (26.52)
Laparoscope or robot	291 (73.48)
Pathologic T stage	
pTa	35 (8.84)
pT1	126 (31.82)
pT2	80 (20.20)
pT3–4	155 (39.14)
Tumor grade	
Low grade	44 (11.11)
High grade	352 (88.89)
Pathologic N stage	
pN0	47 (11.87)
pNx	332 (83.84)
pN+	17 (4.29)
Tumor size	
<2 cm	66 (16.67)
≥2 cm	330 (83.33)
Lymphovascular invasion	89 (22.59)
Carcinoma in situ	55 (13.89)
Adjuvant chemotherapy	114 (28.79)

UTUC, upper tract urothelial carcinoma; URS, ureteroscopy; RNU, radical nephroureterectomy; HTN, hypertension; DM, diabetes mellitus; CAOD, coronary artery obstructive disease; COPD, chronic obstructive pulmonary disease; CVA, cerebrovascular accident; CKD, chronic kidney disease. Data are presented as *n* (%); * Age at RNU is presented as a median (interquartile range).

Table 2. Comparison of clinicopathological features according to the type of diagnostic URS.

	Non-URS Group	Rigid URS Group	Flexible URS Group	*p*-Value
No. of patients	107 (27%)	178 (45%)	111 (28%)	
Age at RNU *, years	69 (62–76)	68 (60–76)	68 (60–76)	0.633
Sex				<0.0001
Female	45 (42.06)	64 (35.96)	18 (16.22)	
Male	62 (57.94)	114 (64.04)	93 (83.78)	
HTN	58 (54.21)	111 (62.36)	67 (60.36)	0.3901
DM	25 (23.36)	30 (16.85)	32 (28.83)	0.0527
CAOD	6 (5.61)	26 (14.61)	14 (12.61)	0.0665
COPD, asthma	8 (7.48)	18 (10.11)	9 (8.11)	0.7124
CVA	8 (7.48)	13 (7.30)	5 (4.50)	0.5853
CKD	16 (14.95)	14 (7.87)	15 (13.51)	0.1326
Smoking history				0.0343
Never smoked	66 (61.68)	95 (53.37)	49 (44.14)	
Ex- or current smoker	41 (38.32)	83 (46.63)	62 (55.86)	
Tumor location				<0.0001
Renal pelvis	56 (52.34)	69 (38.76)	95 (85.59)	
Ureter	51 (47.66)	109 (61.24)	16 (14.41)	
Laterality				0.4409
Left	57 (53.27)	91 (51.12)	50 (45.05)	
Right	50 (46.73)	87 (48.88)	61 (54.95)	
Previous/concurrent bladder cancer	31 (28.97)	30 (16.85)	20 (18.02)	0.037
Urine cytology				0.0166
Not done	24 (22.43)	69 (38.76)	43 (38.74)	
Negative	41 (38.32)	56 (31.46)	43 (38.74)	
Atypical cell	17 (15.89)	33 (18.54)	19 (17.12)	
Positive	25 (23.36)	20 (11.24)	6 (5.40)	
Ureteroscopic biopsy	0 (0)	85 (47.75)	55 (49.55)	<0.0001
Surgical modality				0.4928
Open	31 (28.97)	42 (23.60)	32 (28.83)	
Laparoscope or robot	76 (71.03)	136 (76.40)	79 (71.17)	
Pathologic T stage				0.2172
pTa	10 (9.35)	13 (7.30)	12 (10.81)	
pT1	31 (28.97)	53 (29.78)	42 (37.84)	
pT2	25 (23.36)	42 (23.60)	13 (11.71)	
pT3–4	41 (38.32)	70 (39.33)	44 (39.64)	
Tumor grade				0.0224
Low grade	10 (9.35)	14 (7.87)	20 (18.02)	
High grade	97 (90.65)	164 (92.13)	91 (81.98)	
Tumor size				0.6782
<2 cm	15 (14.02)	32 (17.98)	19 (17.12)	
≥2 cm	92 (85.98)	146 (82.02)	92 (82.88)	
Lymphovascular invasion	23 (21.50)	41 (23.16)	25 (22.73)	0.9475
Carcinoma in situ	12 (11.21)	30 (16.85)	13 (11.71)	0.303
Adjuvant chemotherapy	26 (24.30)	60 (33.71)	28 (25.23)	0.1466
IVR within 1 year	21 (19.63)	41 (23.03)	37 (33.33)	0.0467

URS, ureteroscopy; RNU, radical nephroureterectomy; HTN, hypertension; DM, diabetes mellitus; CAOD, coronary artery obstructive disease, COPD, chronic obstructive pulmonary disease; CVA, cerebrovascular accident, CKD, chronic kidney disease; IVR, intravesical recurrence. Data are presented as *n* (%); * Age at RNU is presented as a median (interquartile range).

Table 3. Univariable and multivariable Cox regression analysis

	Univariable Model		Multivariable Model	
	HR (95% CI)	***p*-Value**	**HR (95% CI)**	***p*-Value**
Type of URS				
Non-URS	Ref.		Ref.	
Rigid URS	1.161 (0.686–1.966)	0.577	1.301 (0.759–2.230)	0.3388
Flexible URS	1.866 (1.092–3.188)	0.0224	1.807 (1.023–3.192)	0.0416
Age at RNU	0.995 (0.976–1.015)	0.6428		
Sex				
Female	Ref.			
Male	1.150 (0.746–1.772)	0.5279		
HTN	1.428 (0.940–2.170)	0.0945	1.332 (0.869–2.041)	0.1884
DM	1.036 (0.650–1.652)	0.8813		
CAOD	1.536 (0.887–2.661)	0.1258		
COPD, Asthma	1.558 (0.810–2.999)	0.1841		
CVA	0.832 (0.364–1.898)	0.6613		
CKD	1.630 (0.954–2.784)	0.0737	1.795 (1.043–3.092)	0.0348
Smoking history				
Never smoked	Ref.		Ref.	
Ex- or current smoker	1.446 (0.974–2.148)	0.0676	1.390 (0.928–2.080)	0.1099
Tumor location				
Renal pelvis	Ref.			
Ureter	0.742 (0.494–1.115)	0.151		
Laterality				
Left	Ref.			
Right	0.909 (0.613–1.348)	0.6355		
Previous/concurrent bladder cancer	1.341 (0.847–2.124)	0.2105	1.170 (0.724–1.890)	0.5218
Urine cytology				
Negative	Ref.			
Atypical cell	0.959 (0.548–1.676)	0.8823		
Positive	1.378 (0.789–2.410)	0.26		
Ureteroscopic biopsy	1.133 (0.755–1.698)	0.5468		
Surgical modality				
Open	Ref.			
Laparoscope or Robot	0.985 (0.629–1.541)	0.9463		
Pathologic T stage				
pTa	Ref.			
pT1	1.391 (0.614–3.150)	0.4293		
pT2	1.747 (0.755–4.038)	0.1922		
pT3–4	1.277 (0.567–2.875)	0.5547		
Tumor grade				
Low grade	Ref.			
High grade	1.364 (0.688–2.707)	0.3743		
Tumor size				
<2 cm	Ref.			
≥2 cm	1.028 (0.602–1.755)	0.9202		
Lymphovascular invasion	0.731 (0.434–1.234)	0.2412		
Carcinoma in situ	1.582 (0.968–2.584)	0.0669	1.530 (0.926–2.527)	0.0969
Adjuvant chemotherapy	0.630 (0.389–1.021)	0.0605	0.603 (0.368–0.989)	0.0450

URS, ureteroscopy; RNU, radical nephroureterectomy; IVR, intravesical recurrence; HTN, hypertension; DM, diabetes mellitus; CAOD, coronary artery obstructive disease; COPD, chronic obstructive pulmonary disease; CVA, cerebrovascular accident; CKD, chronic kidney disease; DM, diabetes mellitus; CI, confidence interval; HR, hazard ratio; Ref., reference.

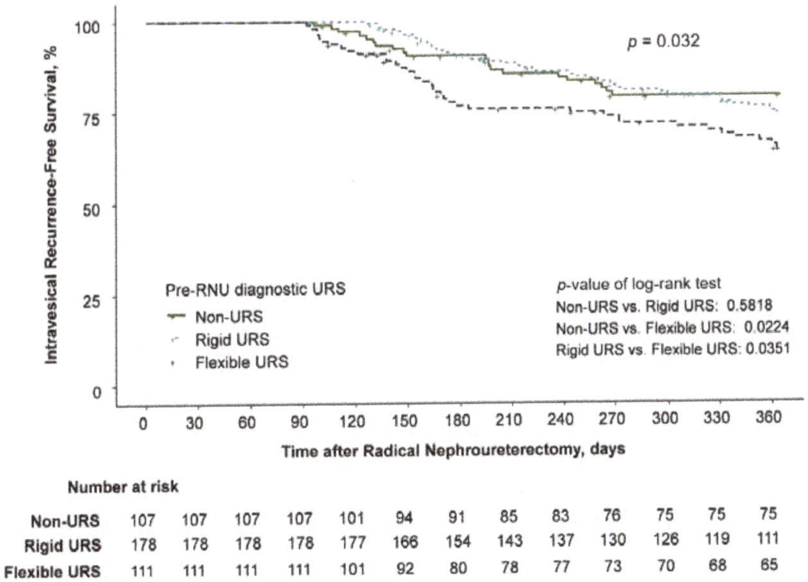

Figure 1. Kaplan–Meier estimates for intravesical recurrence-free survival.

Multivariable Cox regression analysis also revealed that flexible URS was independently associated with an increased risk of IVR (HR = 1.807; 95% CI, 1.023–3.192; $p = 0.0416$), although rigid URS did not increase the risk of IVR (HR = 1.301; 95% CI, 0.759–2.230; $p = 0.3388$). A history of chronic kidney disease was also an independent predictor for an increased risk of IVR (HR = 1.795; 95% CI, 1.043–3.092; $p = 0.0348$). Adjuvant systemic chemotherapy was associated with a decreased risk of IVR (HR = 0.603; 95% CI, 0.368–0.989; $p = 0.045$) (Table 3). Carcinoma in situ, smoking history, and a history of hypertension showed some significance in the univariable analysis, but failed to reach statistical significance in the multivariable analysis.

4. Discussion

The impact of diagnostic URS on IVR remains controversial. Ishikawa et al. first described the impact of diagnostic URS on IVR, and several related studies have been performed [21]. Ishikawa et al. concluded that diagnostic URS has no adverse effects on IVR, and subsequent studies also drew a similar conclusion [16,21]. On the other hand, several studies did report adverse effects for diagnostic URS on IVR [17,22–24]. A recent meta-analysis consistently demonstrated an adverse effect of diagnostic URS on IVR [6–8]. Furthermore, differences according to the timing of diagnostic URS (simultaneously with RNU or before RNU) or whether the accompanied ureteroscopic biopsy was performed were also reported [25,26]. The mechanism through which diagnostic URS accelerates IVR has been explained by intraluminal seeding and cancer cell implantation promoted by either massive irrigation or ureteroscopic biopsy.

With the widespread use of flexible diagnostic URS, flexible URS allows urologists to examine all collecting systems, including the lower calyx, which cannot be visualized using rigid URS. Flexible URS has a smaller working channel than rigid URS, which results in a weaker irrigation capability. To maintain a better view for diagnostic purposes, a higher-pressure inflow of irrigation fluid is required, compared to that of rigid URS. Owing to these mechanical differences, flexible URS and rigid URS might pose different risks of IVR. To the best of our knowledge, this is the first study to investigate differences in the impact of diagnostic URS on IVR according to the type of URS, either rigid or flexible.

Herein, multivariable analysis revealed that flexible URS was significantly associated with an increased risk of IVR, compared to non-URS, however, rigid URS was not. We conducted propensity score matching to adjust for demographic heterogeneity among three groups; it also supported these observations (Tables S1 and S2). Therefore, diagnostic URS should be performed as a guideline only when imaging and cytology are insufficient for the diagnosis of UTUC and/or for risk stratification of a tumor [4]. In particular, surgeons should consider that the risk of IVR might be increased when flexible diagnostic URS is indicated.

The potential mechanisms by which flexible URS increases the risk of IVR can be explained as follows. First, a high-pressure inflow of irrigation fluid is required to secure a clear field of view, owing to their smaller working channel, compared to rigid URS. To maintain a sufficient amount of irrigation fluid, the height of the irrigation fluid should be higher than usual or an irrigation pump is needed. Second, due to the smaller size of forceps used for ureteroscopic biopsy, considerably greater manipulation of cancer cells might occur to obtain a sufficient amount of tissue [27]. Third, an access sheath should be introduced to keep the intrarenal pressure low while performing flexible URS [28,29], resulting in increased irrigation fluid running through the collecting system. Therefore, cancer cells are likely to be exfoliated during flexible URS. From this point of view, the difference between rigid and flexible URS observed in this study may be due to a difference in the irrigation method rather than the difference in the instrument itself. These hypothetical explanations should be proven in clinical and preclinical studies.

Increased intrarenal pressure causes pyelovenous and pyelolymphatic backflow, which is theorized as a potential source of cancer cell transfer [8]. A recent literature review suggested that intrarenal pressure should remain <30 cm H_2O during endoscopic procedures. However, the actual oncological risk promoted by diagnostic endoscopic procedures is uncertain due to the lack of strong evidence in real intrarenal pressure threshold [28]. Results of subsequent studies suggest that URS may not have a negative effect on cancer-specific survival or overall survival [17]. As aforementioned, although mechanical differences exist between flexible URS and rigid URS, previous studies do not distinguish the type of URS as either flexible or rigid. Since this study included relatively recent patients with UTUC to focus on IVR according to the type of URS, the follow-up period was insufficient to estimate either cancer-specific survival or overall survival. A longer follow-up period is required to estimate the effect of flexible URS on oncological outcomes other than IVR.

This study has several advantages in revealing the occurrence of IVR in patients who underwent RNU. Since most previous studies on IVR enrolled patients who underwent RNU for a period close to a decade, several confounding factors, such as surgeon factors and surgical modality, may have inevitably affected the analysis. However, this study was able to reduce heterogeneity because of the relatively short and recent period of study; thus, our study can reflect the current trends in real-world clinical practice. We attempted to minimize the possibility of newly developed primary bladder cancer independent of UTUC by limiting IVR to 1 year within RNU. Therefore, we focused on URS-induced IVR. In the period of this study, the two institutions rarely conducted neoadjuvant chemotherapy even in patients with T3 or higher clinical stage. Additionally, all bladder cuff management procedures were performed using an extravesical approach, and no postoperative intravesical instillation was noted, despite this being a known prevention therapy for IVR [30]. While this hindered us from estimating the effects of bladder cuff management, postoperative intravesical instillation, and neoadjuvant chemotherapy, it did allow us to concentrate on the effect of the type of URS on IVR.

Although we attempted to report IVR according to the type of diagnostic URS, this study has a few limitations. Owing to the retrospective design of this study, the absence of randomization resulted in heterogeneity among the three groups. A randomized clinical trial is needed to obtain definite conclusions on this issue, although it is rarely conducted for practical and ethical reasons. Moreover, we did not have objective data on the duration of the diagnostic URS procedure and the amount of irrigation fluid used during the procedure.

Therefore, further research using more constructed data with a larger patient population is needed to account for differences in such characteristics among the study groups according to the type of URS.

5. Conclusions

In patients with UTUC, rigid URS may not increase the risk of IVR following RNU; however, flexible URS appears to be associated with a higher risk of IVR. When the diagnostic URS is inevitably implemented, physicians should consider that the risk of IVR may increase, especially when flexible URS is required. Our findings should be reproduced in large-population studies. Additionally, clinical and preclinical evidence supporting our results should be accumulated.

Supplementary Materials: The following supporting information can be downloaded at: https://www.mdpi.com/article/10.3390/cancers14225629/s1; Table S1: Patient demographics after PSM in the non-URS and flexible URS groups; Table S2: Patient demographics after PSM in the non-URS and rigid URS groups.

Author Contributions: Conceptualization, J.S.H., J.J., J.C.K. and K.S.C.; methodology, H.S.L., J.Y., and D.K.; software, J.S.H., H.S.L. and J.Y.; validation, W.S.H., Y.D.C. and K.S.C.; formal analysis, J.J., J.C.K., H.S.L. and J.Y.; investigation, J.J., D.K. and J.S.K.; resources, W.S.H., Y.D.C. and K.S.C.; data curation, J.J., J.C.K., D.K. and J.S.K.; writing—original draft preparation, J.S.H., J.C.K. and K.S.C.; writing—review and editing, J.S.H. and K.S.C.; visualization, J.S.H.; supervision, K.S.C.; project administration, J.S.H. and J.J.; funding acquisition, K.S.C. All authors have read and agreed to the published version of the manuscript.

Funding: This study was supported by a faculty research grant from the Yonsei University College of Medicine (6-2020-0061). This organization had no role in the design or conduct of this research.

Institutional Review Board Statement: This study was performed in accordance with the tenets of the Declaration of Helsinki. The Institutional Review Board of Gangnam Severance Hospital approved the study protocol (approval number: 3-2020-0199, approval date: 2 July 2020).

Informed Consent Statement: Patient consent was waived by the ethics committee due to the retrospective analysis.

Data Availability Statement: The datasets generated during and/or analyzed during the current study are available from the corresponding author on reasonable request.

Conflicts of Interest: The authors declare no conflict of interest.

References

1. Redrow, G.P.; Matin, S.F. Upper tract urothelial carcinoma: Epidemiology, high risk populations and detection. *Minerva Urol. Nefrol.* **2016**, *68*, 350–358. [PubMed]
2. Siegel, R.L.; Miller, K.D.; Jemal, A. Cancer statistics, 2020. *CA Cancer J. Clin.* **2020**, *70*, 7–30. [CrossRef] [PubMed]
3. Janisch, F.; Shariat, S.F.; Baltzer, P.; Fajkovic, H.; Kimura, S.; Iwata, T.; Korn, P.; Yang, L.; Glybochko, P.V.; Rink, M.; et al. Diagnostic performance of multidetector computed tomographic (MDCTU) in upper tract urothelial carcinoma (UTUC): A systematic review and meta-analysis. *World J. Urol.* **2020**, *38*, 1165–1175. [CrossRef] [PubMed]
4. Roupret, M.; Babjuk, M.; Burger, M.; Capoun, O.; Cohen, D.; Comperat, E.M.; Cowan, N.C.; Dominguez-Escrig, J.L.; Gontero, P.; Hugh Mostafid, A.; et al. European Association of Urology Guidelines on Upper Urinary Tract Urothelial Carcinoma: 2020 Update. *Eur. Urol.* **2021**, *79*, 62–79. [CrossRef]
5. Soria, F.; Shariat, S.F.; Lerner, S.P.; Fritsche, H.M.; Rink, M.; Kassouf, W.; Spiess, P.E.; Lotan, Y.; Ye, D.; Fernandez, M.I.; et al. Epidemiology, diagnosis, preoperative evaluation and prognostic assessment of upper-tract urothelial carcinoma (UTUC). *World J. Urol.* **2017**, *35*, 379–387. [CrossRef]
6. Marchioni, M.; Primiceri, G.; Cindolo, L.; Hampton, L.J.; Grob, M.B.; Guruli, G.; Schips, L.; Shariat, S.F.; Autorino, R. Impact of diagnostic ureteroscopy on intravesical recurrence in patients undergoing radical nephroureterectomy for upper tract urothelial cancer: A systematic review and meta-analysis. *BJU Int.* **2017**, *120*, 313–319. [CrossRef]
7. Guo, R.Q.; Hong, P.; Xiong, G.Y.; Zhang, L.; Fang, D.; Li, X.S.; Zhang, K.; Zhou, L.Q. Impact of ureteroscopy before radical nephroureterectomy for upper tract urothelial carcinomas on oncological outcomes: A meta-analysis. *BJU Int.* **2018**, *121*, 184–193. [CrossRef]

8. Nowak, L.; Krajewski, W.; Chorbinska, J.; Kielb, P.; Sut, M.; Moschini, M.; Teoh, J.Y.; Mori, K.; Del Giudice, F.; Laukhtina, E.; et al. The Impact of Diagnostic Ureteroscopy Prior to Radical Nephroureterectomy on Oncological Outcomes in Patients with Upper Tract Urothelial Carcinoma: A Comprehensive Systematic Review and Meta-Analysis. *J. Clin. Med.* **2021**, *10*, 4197. [CrossRef]
9. Seisen, T.; Colin, P.; Roupret, M. Risk-adapted strategy for the kidney-sparing management of upper tract tumours. *Nat. Rev. Urol.* **2015**, *12*, 155–166. [CrossRef]
10. Xylinas, E.; Rink, M.; Cha, E.K.; Clozel, T.; Lee, R.K.; Fajkovic, H.; Comploj, E.; Novara, G.; Margulis, V.; Raman, J.D.; et al. Impact of distal ureter management on oncologic outcomes following radical nephroureterectomy for upper tract urothelial carcinoma. *Eur. Urol.* **2014**, *65*, 210–217. [CrossRef]
11. Sidransky, D.; Frost, P.; Von Eschenbach, A.; Oyasu, R.; Preisinger, A.C.; Vogelstein, B. Clonal origin of bladder cancer. *N. Engl. J. Med.* **1992**, *326*, 737–740. [CrossRef] [PubMed]
12. Habuchi, T.; Takahashi, R.; Yamada, H.; Kakehi, Y.; Sugiyama, T.; Yoshida, O. Metachronous multifocal development of urothelial cancers by intraluminal seeding. *Lancet* **1993**, *342*, 1087–1088. [CrossRef]
13. Garcia, S.B.; Park, H.S.; Novelli, M.; Wright, N.A. Field cancerization, clonality, and epithelial stem cells: The spread of mutated clones in epithelial sheets. *J. Pathol.* **1999**, *187*, 61–81. [CrossRef]
14. Shigeta, K.; Matsumoto, K.; Tanaka, N.; Mikami, S.; Kosaka, T.; Yasumizu, Y.; Takeda, T.; Mizuno, R.; Kikuchi, E.; Oya, M. Profiling the Biological Characteristics and Transitions through Upper Tract Tumor Origin, Bladder Recurrence, and Muscle-Invasive Bladder Progression in Upper Tract Urothelial Carcinoma. *Int. J. Mol. Sci.* **2022**, *23*, 5154. [CrossRef]
15. Seisen, T.; Granger, B.; Colin, P.; Leon, P.; Utard, G.; Renard-Penna, R.; Comperat, E.; Mozer, P.; Cussenot, O.; Shariat, S.F.; et al. A Systematic Review and Meta-analysis of Clinicopathologic Factors Linked to Intravesical Recurrence After Radical Nephroureterectomy to Treat Upper Tract Urothelial Carcinoma. *Eur. Urol.* **2015**, *67*, 1122–1133. [CrossRef]
16. Lee, H.Y.; Yeh, H.C.; Wu, W.J.; He, J.S.; Huang, C.N.; Ke, H.L.; Li, W.M.; Li, C.F.; Li, C.C. The diagnostic ureteroscopy before radical nephroureterectomy in upper urinary tract urothelial carcinoma is not associated with higher intravesical recurrence. *World J. Surg. Oncol.* **2018**, *16*, 135. [CrossRef]
17. Veeratterapillay, R.; Geraghty, R.; Pandian, R.; Roy, C.; Stenhouse, G.; Bird, C.; Soomro, N.; Paez, E.; Rogers, A.; Johnson, M.; et al. Ten-year survival outcomes after radical nephroureterectomy with a risk-stratified approach using prior diagnostic ureteroscopy: A single-institution observational retrospective cohort study. *BJU Int.* **2022**, *129*, 744–751. [CrossRef]
18. Bus, M.T.; de Bruin, D.M.; Faber, D.J.; Kamphuis, G.M.; Zondervan, P.J.; Laguna Pes, M.P.; de Reijke, T.M.; Traxer, O.; van Leeuwen, T.G.; de la Rosette, J.J. Optical diagnostics for upper urinary tract urothelial cancer: Technology, thresholds, and clinical applications. *J. Endourol.* **2015**, *29*, 113–123. [CrossRef]
19. Humphrey, P.A.; Moch, H.; Cubilla, A.L.; Ulbright, T.M.; Reuter, V.E. The 2016 WHO Classification of Tumours of the Urinary System and Male Genital Organs-Part B: Prostate and Bladder Tumours. *Eur. Urol.* **2016**, *70*, 106–119. [CrossRef]
20. Edge, S.B.; Compton, C.C. The American Joint Committee on Cancer: The 7th edition of the AJCC cancer staging manual and the future of TNM. *Ann. Surg. Oncol.* **2010**, *17*, 1471–1474. [CrossRef]
21. Ishikawa, S.; Abe, T.; Shinohara, N.; Harabayashi, T.; Sazawa, A.; Maruyama, S.; Kubota, K.; Matsuno, Y.; Osawa, T.; Shinno, Y.; et al. Impact of diagnostic ureteroscopy on intravesical recurrence and survival in patients with urothelial carcinoma of the upper urinary tract. *J. Urol.* **2010**, *184*, 883–887. [CrossRef] [PubMed]
22. Luo, H.L.; Kang, C.H.; Chen, Y.T.; Chuang, Y.C.; Lee, W.C.; Cheng, Y.T.; Chiang, P.H. Diagnostic ureteroscopy independently correlates with intravesical recurrence after nephroureterectomy for upper urinary tract urothelial carcinoma. *Ann. Surg. Oncol.* **2013**, *20*, 3121–3126. [CrossRef] [PubMed]
23. Izol, V.; Deger, M.; Ozden, E.; Bolat, D.; Argun, B.; Baltaci, S.; Celik, O.; Akgul, H.M.; Tinay, I.; Bayazit, Y.; et al. The Effect of Diagnostic Ureterorenoscopy on Intravesical Recurrence in Patients Undergoing Nephroureterectomy for Primary Upper Tract Urinary Carcinoma. *Urol. Int.* **2021**, *105*, 291–297. [CrossRef] [PubMed]
24. Sung, H.H.; Jeon, H.G.; Han, D.H.; Jeong, B.C.; Seo, S.I.; Lee, H.M.; Choi, H.Y.; Jeon, S.S. Diagnostic Ureterorenoscopy Is Associated with Increased Intravesical Recurrence following Radical Nephroureterectomy in Upper Tract Urothelial Carcinoma. *PLoS ONE* **2015**, *10*, e0139976. [CrossRef]
25. Lee, J.K.; Kim, K.B.; Park, Y.H.; Oh, J.J.; Lee, S.; Jeong, C.W.; Jeong, S.J.; Hong, S.K.; Byun, S.S.; Lee, S.E. Correlation Between the Timing of Diagnostic Ureteroscopy and Intravesical Recurrence in Upper Tract Urothelial Cancer. *Clin. Genitourin. Cancer* **2016**, *14*, e37–e41. [CrossRef]
26. Katims, A.B.; Say, R.; Derweesh, I.; Uzzo, R.; Minervini, A.; Wu, Z.; Abdollah, F.; Sundaram, C.; Ferro, M.; Rha, K.; et al. Risk Factors for Intravesical Recurrence after Minimally Invasive Nephroureterectomy for Upper Tract Urothelial Cancer (ROBUUST Collaboration). *J. Urol.* **2021**, *206*, 568–576. [CrossRef]
27. Tavora, F.; Fajardo, D.A.; Lee, T.K.; Lotan, T.; Miller, J.S.; Miyamoto, H.; Epstein, J.I. Small endoscopic biopsies of the ureter and renal pelvis: Pathologic pitfalls. *Am. J. Surg. Pathol.* **2009**, *33*, 1540–1546. [CrossRef]
28. Tokas, T.; Skolarikos, A.; Herrmann, T.R.W.; Nagele, U.; Training and Research in Urological Surgery and Technology (T.R.U.S.T.)-Group. Pressure matters 2: Intrarenal pressure ranges during upper-tract endourological procedures. *World J. Urol.* **2019**, *37*, 133–142. [CrossRef]

29. Doizi, S. Intrarenal Pressure: What Is Acceptable for Flexible Ureteroscopy and Percutaneous Nephrolithotomy? *Eur. Urol. Focus* **2021**, *7*, 31–33. [CrossRef]
30. O'Brien, T.; Ray, E.; Singh, R.; Coker, B.; Beard, R.; British Association of Urological Surgeons Section of Oncology. Prevention of bladder tumours after nephroureterectomy for primary upper urinary tract urothelial carcinoma: A prospective, multicentre, randomised clinical trial of a single postoperative intravesical dose of mitomycin C (the ODMIT-C Trial). *Eur. Urol.* **2011**, *60*, 703–710. [CrossRef]

MDPI
St. Alban-Anlage 66
4052 Basel
Switzerland
Tel. +41 61 683 77 34
Fax +41 61 302 89 18
www.mdpi.com

Cancers Editorial Office
E-mail: cancers@mdpi.com
www.mdpi.com/journal/cancers

www.ingramcontent.com/pod-product-compliance
Lightning Source LLC
Chambersburg PA
CBHW040834250926
51850CB00007BA/1314
9783036569680